D0206129

Kurt Gödel

COLLECTED WORKS

Volume II

Kurt Gödel, 1956

Kurt Gödel

COLLECTED WORKS

Volume II
Publications 1938–1974

EDITED BY

Solomon Feferman
(Editor-in-chief)

John W. Dawson, Jr.
Stephen C. Kleene
Gregory H. Moore
Robert M. Solovay
Jean van Heijenoort

Prepared under the auspices of the Association for Symbolic Logic

New York Oxford
OXFORD UNIVERSITY PRESS
1990

Oxford University Press

Oxford New York Toronto
Delhi Bombay Calcutta Madras Karachi
Petaling Jaya Singapore Hong Kong Tokyo
Nairobi Dar es Salaam Cape Town
Melbourne Auckland

and associated companies in
Berlin Ibadan

Published by Oxford University Press, Inc.,
200 Madison Avenue, New York, New York 10016

Library of Congress Cataloging-in-Publication Data

(Revised for vol. 2)

Gödel, Kurt.
Collected works.

German text, parallel English translation.
Includes bibliographies and indexes.
Contents: v. 1. Publications 1929–1936—
v. 2. Publications, 1938–1974
1. Logic, Symbolic and mathematical.
I. Feferman, Solomon. II. Title.
QA9.G5313 1986 511.3 85–15501
ISBN 0-19-503964-5 (v. 1)
ISBN 0-19-503972-6 (v. 2)

2 4 6 8 9 7 5 3 1

Printed in the United States of America
on acid-free paper

Preface

This second volume of a comprehensive edition of the works of Kurt Gödel contains the remainder of his published work, covering the period 1938–1974. (Volume I included all of his publications from 1929 to 1936; no work of his appeared during 1937.) Succeeding volumes are to contain selections from Gödel's unpublished manuscripts, lectures, lecture notes and correspondence, as well as extracts from his scientific notebooks.

For a detailed explanation of the plan for this edition, the reader should consult the Preface to Volume I of these *Works*. To summarize briefly, each article or closely related group of articles is preceded by an introductory note that elucidates it and places it in historical context. These notes (varying greatly in length) have been written by the members of the editorial board as well as a number of outside experts. Furthermore, the only article originally written in German, *1958*, is accompanied by an English translation on facing pages. As in Volume I, the original articles have been typeset anew in a uniform and more readable format. Finally, the extensive list of references in this volume contains all items referred to either by Gödel or in the introductory notes, and also includes all the items listed in Volume I.

Here again, our aim has been to make the full body of Gödel's work as accessible and useful to as wide an audience as possible, without in any way sacrificing the requirements of historical and scientific accuracy. We expect these volumes to be of interest and value to professionals and students in the areas of logic, mathematics, computer science and even physics, as well as to many non-specialist readers with a broad scientific background. Naturally, even with the assistance of the introductory notes, not all of Gödel's work can be made equally accessible to such a variety of readers; but the general reader should nonetheless be able to gain some appreciation of what Gödel accomplished in each case.

We continue to be indebted to the National Science Foundation and the Sloan Foundation, whose grants have made possible the production of Volumes I and II as well as preparations for succeeding volumes, and to the Association for Symbolic Logic, which has sponsored our project and administered these grants. Our publisher, Oxford University Press, has once more been very accommodating to both our overall plans and our specific wishes. Much of our work was done with the assistance of former Oxford Science Editor Donald Degenhardt; following his return to England, we have received the assistance first of Jeffrey W. House, Vice-President and Executive Editor for Science and Medicine, and more recently of the new Science Editor, Dr. Jacqueline E. Hartt.

For the names of the many other individuals who have helped make these first two volumes of Gödel's *Collected works* possible, the reader should refer to the Preface to Volume I. Our gratitude to all for their assistance is undiminished.

Solomon Feferman

Jean van Heijenoort, our dear friend and co-editor, died on 29 March 1986. His contributions to our work on Kurt Gödel were invaluable at every stage and in every respect. At the outset, his enthusiastic support was instrumental in our decision to embark upon this project. Then, drawing upon his own extensive editorial experience, he helped us to develop our overall plans as well as to make the many detailed choices, and throughout the course of the work he devoted himself unstintingly and with the utmost care to whatever task was at hand.

The present volume was largely completed by the time of van Heijenoort's death; indeed, he had already begun a detailed examination of some of Gödel's unpublished articles for the succeeding volume. His spirit will continue to animate all our work, and we have taken his standards as our own.

Information for the reader

Introductory notes. The purpose of the notes described in the Preface above is (i) to provide a historical context for the items introduced, (ii) to explain their contents to a greater or lesser extent, (iii) to discuss further developments which resulted from them and (iv) in some cases to give a critical analysis. Each note was read in draft form by the editorial board, and then modified by the respective authors in response to criticisms and suggestions, the procedure being repeated as often as necessary in the case of very substantial notes. No attempt, however, has been made to impose uniformity of style, point of view, or even length. While the editorial board actively engaged in a critical and advisory capacity in the preparation of each note and made the final decision as to its acceptability, primary credit and responsibility for the notes rest with the individual authors.

Introductory notes are distinguished typographically by a running vertical line along the left- or right-hand margin and are boxed off at their end.[a] The authorship of each note is given in the Contents and at the end of the note itself.

References. Each volume contains a comprehensive References section which comprises the following three categories of items: (i) a complete bibliography of Gödel's own published work, (ii) all items referred to by Gödel in his publications and (iii) all items referred to in the chapter in Volume I on Gödel's life and work or in the individual introductory notes.

In the list of references, each item is assigned a date with or without a letter suffix, e.g., "1930", "1930a", "1930b", etc.[b] The date is that of publication, where there is a published copy, or of presentation, for unpublished items such as a speech. A suffix is used when there is more than one publication in that year. (The ordering of suffixes does not necessarily correspond to order of publication within any given year.) Date of composition has *not* been used for references, since that is frequently unavailable or only loosely determined.

Within the text of our volumes, all references are supplied by citing author(s) and date in italics, e.g., *Gödel 1930* or *Hilbert and Bernays 1934*. Where no name is specified or determined by the context, the reference is to Gödel's bibliography, as e.g., in "Introductory note to *1929*, *1930* and

[a]A special situation occurs when the note ends in mid-page before facing German and English text. Then the note extends across the top half of the facing pages and is boxed off accordingly.

[b]"198?" is used for articles whose date of publication is to be in 1988 or later, or is not yet known.

1930a". Examples of the use of a name to set the context for a reference are: "Frege's formal system presented in *1879*", "Skolem proved in *1920* that ...", and "Skolem (*1920*) proved that ...".

There are two works by Gödel, *1929* and *1972*, whose dating required special consideration; they appear in Volumes I and II, respectively. The first of these is Gödel's dissertation at the University of Vienna; its date is that of the year in which the dissertation was submitted (as distinguished from the date of its acceptance, which was 1930). The second, a translation and revision of Gödel's paper *1958*, was intended for publication in the journal *Dialectica* but hitherto never actually appeared. It reached the stage of page proofs and was found in that form in Gödel's *Nachlass*. Correspondence surrounding this projected publication shows that Gödel worked on the revision sporadically over a number of years, beginning in 1965; the last date for which we have evidence of his making specific changes is 1972, and that date has therefore been assigned to it in our References. (For more information concerning this work, see the introductory note to *1958* and *1972* in this volume.) Appended to the page proofs of *1972* were three short notes on the incompleteness results; they have been assigned the date *1972a* in the References.

To make the References as useful as possible for historical purposes, authors' names are there supplied with first and/or middle names as well as initials, except when the information could not be determined. Russian names are given both in transliterated form and in their original Cyrillic spelling. In some cases, common variant transliterations of the same author's name, attached to different publications, are also noted.

Editorial annotations and textual notes. Editorial annotations within any of the original texts or their translations or within items quoted from other authors are signaled by double square brackets: ⟦ ⟧. Single square brackets [ ] are used to incorporate corrections supplied by Gödel. In some articles, editorial footnotes are inserted in double square brackets for a further level of annotation. Each volume has, in addition, a separate list of textual notes in which other corrections are supplied. Finally, the following kinds of changes are made uniformly in the original texts: (i) footnote numbers are raised above the line as simple numerals, e.g., ^{2} instead of $^{2)}$; (ii) spacing used for emphasis in the original German is here replaced by italics, e.g., e r f ü l l b a r is replaced by *erfüllbar*; (iii) references are replaced by author(s) and date, as explained above; (iv) initial sub-quotes in German are raised, e.g., „engeren" becomes "engeren".

Translations. The overall aim for the translations, as well as the variety of work required and general responsibility for them has been described in the Preface to Volume I. The only translation in this volume (namely that of *1958*) is the work of Stefan Bauer-Mengelberg and Jean van Heijenoort.

Logical symbols. The logical symbols used in Gödel's original articles are here presented intact, even though these symbols may vary from one article to another. Authors of introductory notes have in some cases followed the notation of the article(s) discussed and in other cases have preferred to make use of other, more current, notation. Finally, logical symbols are sometimes used to abbreviate informal expressions as well as formal operations. No attempt has been made to impose uniformity in this respect. As an aid to the reader, we provide the following glossary of the symbols that are used in one way or another in these volumes, where 'A', 'B' are letters for propositions or formulas and '$A(x)$' is a propositional function of x or a formula with free variable 'x'.

Conjunction ("A and B"): $A \,.\, B$, $A \wedge B$, $A \,\&\, B$

Disjunction ("A or B"): $A \vee B$

Negation ("not A"): $\overline{A}$, $\sim A$, $\neg A$

Conditional, or *Implication* ("if A then B"): $A \supset B$, $A \rightarrow B$

Biconditional ("A if and only if B"): $A \supset\subset B$, $A \equiv B$, $A \sim B$, $A \leftrightarrow B$

Universal quantification ("for all x, $A(x)$"): $(x)A(x)$, $\Pi x A(x)$, $x\Pi(A(x))$, $(\forall x)A(x)$

Existential quantification ("there exists an x such that $A(x)$"): $(Ex)A(x)$, $\Sigma x A(x)$, $(\exists x)A(x)$

Unicity quantification ("there exists a unique x such that $A(x)$"): $(E!x)A(x)$, $\Sigma! x A(x)$, $(\exists! x)A(x)$

Necessity operator ("A is necessary"): $\Box A$, NA

Minimum operator ("the least x such that $A(x)$"): $\epsilon x(A(x))$, $\mu x(A(x))$

Provability relation ("A is provable in the system S"): $S \vdash A$

Note: (i) The "horseshoe" symbol is also used for set-inclusion, i.e., for sets X, Y one writes $X \subset Y$ (or $Y \supset X$) to express that X is a subset of Y. (ii) Dots are sometimes used in lieu of parentheses, e.g., $A \supset . B \supset A$ is written for $A \supset (B \supset A)$.

Typesetting. These volumes have been prepared by the TEX computerized mathematical typesetting system (devised by Donald E. Knuth of Stanford University), as described in the Preface to Volume I. The resulting camera-ready copy was delivered to the publisher for printing. The computerized system was employed because: (i) much material, including the introductory notes and translations, needed to undergo several revisions; (ii) proof-reading was carried on as the project proceeded; (iii) the papers could be prepared in a uniform, very readable form, instead of being photographed from the original articles. Choices of the various typesetting parameters were made by the editors in consultation with the publisher. Primary responsibility for preparing copy for the typesetting system lay with Gregory H. Moore, and the typesetting itself was carried out by Yasuko Kitajima.

For all previously published articles, original pagination is indicated herein by numbers in the margins, with vertical bars in the body of the text used to show the exact page breaks. No page bar or number is used to indicate the initial page of an article.

Footnotes. We use a combination of numbering and lettering, as follows. All footnotes for Gödel's texts and their translations are numbered, with only rare exceptions, as in the original. There is, however, one special case, that of *1972*, in which Gödel provided a second series of footnotes, essentially to preserve the original series from *1958* without change of numbering. The new series is here distinguished by boldface lower-case Roman letters. For all the other material in this volume, footnotes are indicated by lightface lower-case Roman letters.

Gödel's Nachlass. The scientific *Nachlass* of Kurt Gödel was donated to the Institute for Advanced Study in Princeton, N.J., by his widow Adele shortly after his death. The *Nachlass* consists of unpublished manuscripts, lecture notes, course notes, notebooks, memoranda, correspondence and books from Gödel's library. It was catalogued at the Institute for Advanced Study during the years 1982–1984 by John Dawson. Early in 1985 the *Nachlass* with its catalogue was placed on indefinite loan to the Manuscripts Division (located in the Rare Book Room) of the Firestone Library at Princeton University, where the material is available for scholarly examination. All rights for use still reside, however, with the Institute for Advanced Study. Though the *Nachlass* is referred to only here and there in Volumes I and II, it will be the source of almost all the material in subsequent volumes. For further information concerning its general character, see Volume I, pages 26–28.

Photographs. Primary responsibility for securing these lay with John Dawson. Their various individual sources are credited in the Permissions section, which follows directly.

Copyright permissions

The editors are grateful to the following copyright holders, who have granted permission to reprint in this volume the works by Gödel listed below. (See the References for full bibliographic citations.)

The American Mathematical Society, for the following articles:

> "The consistency of the generalized continuum hypothesis" (*1939*), reprinted from *Bulletin of the American Mathematical Society*, vol. 45, p. 93, by permission of the American Mathematical Society.
>
> "Rotating universes in general relativity theory" (*1952*), reprinted from *Proceedings of the International Congress of Mathematicians, Cambridge, Massachusetts, 1950*, pp. 175–181, by permission of the American Mathematical Society.
>
> Postscript to "Provably recursive functionals of analysis: A consistency proof of analysis by an extension of principles formulated in current intuitionistic mathematics", by Clifford Spector (*1962*), reprinted from *Recursive function theory*, Proceedings of symposia in pure mathematics, vol. 5, p. 27, by permission of the American Mathematical Society.

Professors Paul Benacerraf and Hilary Putnam, for the revised version (*1964*) of the article "What is Cantor's continuum problem?", published in their anthology *Philosophy of mathematics: Selected readings*, Copyright © 1964 by Prentice-Hall, Inc., Englewood Cliffs, New Jersey, and Basil Blackwell, Oxford.

Dialectica, for the article "Über eine bisher noch nicht benützte Erweiterung des finiten Standpunktes" (*1958*) and for permission to prepare and publish an English translation of it.

The Institute for Advanced Study, Princeton, literary executors of the estate of Kurt Gödel, for the following articles, in which copyright was retained by Gödel:

> "Remarks before the Princeton bicentennial conference on problems in mathematics" (*1946*)
>
> "On an extension of finitary mathematics which has not yet been used" (*1972*)
>
> "Some remarks on the undecidability results" (*1972a*).

Part of *1972a* appeared earlier in *From mathematics to philosophy*, by Hao Wang, Copyright © 1974 by Routledge and Kegan Paul, London, and Humanities Press, New York. The editors thank Professor Wang for his permission to include that material in this volume.

North-Holland Publishing Company, Amsterdam, for Gödel's untitled remarks *1974*, extracted from the preface to the second edition of *Non-standard analysis*, by Abraham Robinson.

Open Court Publishing Company, La Salle, Illinois, for the following articles:

> "Russell's mathematical logic" (*1944*). Reprinted from *The philosophy of Bertrand Russell*, by Paul A. Schilpp (ed.), by permission of the Open Court Publishing Company, La Salle, Illinois. Copyright 1944, 1951 and © 1970 by The Library of Living Philosophers, Inc.
>
> "A remark about the relationship between relativity theory and idealistic philosophy" (*1949a*). Reprinted from *Albert Einstein, philosopher-scientist*, by Paul A. Schilpp (ed.), by permission of the Open Court Publishing Company, La Salle, Illinois. Copyright 1949, 1951 and © 1970 by The Library of Living Philosophers, Inc.

Princeton University Press, for the monograph *The consistency of the continuum hypothesis*, Annals of Mathematics Studies No. 3. Copyright 1940 by Princeton University Press, © renewed 1968 by Princeton University Press. Reprinted by permission of Princeton University Press.

In addition, the editors thank the following for consenting to the reproduction in this volume of the photographs listed below:

Professor Richard Arens, of the University of California at Los Angeles, for his portrait of Kurt Gödel and Albert Einstein;

The Institute for Advanced Study, Princeton, for the portrait of Kurt Gödel and Alfred Tarski, found in Gödel's *Nachlass*;

The Seeley G. Mudd Manuscript Library, Princeton University, for the group photograph of participants in the Princeton bicentennial conference on problems in mathematics;

Mr. Arnold Newman, New York City, for his portrait of Kurt Gödel;

Dr. Veli Valpola, Espoo, Finland, for the portrait of Kurt Gödel in his office.

Contents

Volume II

List of illustrations

Kurt Gödel

COLLECTED WORKS

Volume II

Introductory note to *1938*, *1939*, *1939a* and *1940*

1. Introduction

The papers discussed in this introductory note deal with Gödel's proof of the relative consistency of the axiom of choice and of the generalized continuum hypothesis with the usual axioms for set theory.[a] The note *1938* announces the results and gives a brief discussion of the ideas underlying their proofs; *1939* is an abstract in which the results are again announced. The paper *1939a* gives considerable technical detail concerning the proofs; essentially complete proofs, along somewhat different lines, are presented in *1940*.

In the next section, I shall describe briefly the historical context of Gödel's work.[b] The third section is devoted to the results themselves, the problems that they solved, and the methods used to obtain them. In the final section, the new problems raised by Gödel's work and the results concerning them obtained by later workers will be discussed.

2. Historical antecedents

Gödel's work bore on two previously considered questions, the axiom of choice and the continuum hypothesis.

2.1 The axiom of choice

Cantor had conjectured the proposition, now called the *well-ordering theorem*, that every set can be well-ordered.[c] In *1904* Zermelo gave a proof of this conjecture, using in an essential way the following mathematical principle: for every set X there is a *choice function*, f, which is defined on the collection of non-empty subsets of X, such that for every set A in its domain we have $f(A) \in A$. Subsequently, in *1908*, Zermelo presented an axiomatic version of set theory in which his proof of the well-ordering theorem could be carried out. One of the axioms

[a]This is here taken to be the Zermelo–Fraenkel (ZF) system of axioms for set theory excluding the axiom of choice, explained in Section 2 below.

[b]I have been greatly helped in preparing Section 2 by conversations and correspondence with Gregory H. Moore.

[c]A *well-ordering* of a set X is a linear ordering of X such that every non-empty subset of X has a least element.

was the principle just stated,[d] which Zermelo referred to as the "Axiom der Auswahl" (the *axiom of choice*, abbreviated AC).

Zermelo's proof was the subject of considerable controversy.[e] The well-ordering theorem is quite remarkable, since, for example, there is no obvious way to *define* a well-ordering of the set of real numbers.[f] Nor is such an explicit well-ordering provided by Zermelo's proof. Thus many people who thought Zermelo's result implausible cast doubt upon the validity of AC. The other set-existence axioms all have the form that some collection of sets, explicitly definable from certain given parameters, is itself a set. The axiom of choice, on the other hand, asserts the existence of a choice function but does not provide an explicit definition of such a choice function. Zermelo was well aware that his axiom had this purely existential character, but many other mathematicians were uncomfortable with existence proofs that did not provide the construction of specific examples of what was asserted to exist.

The work of Gödel dealt with here showed that AC is "safe" in the following sense: If the usual axioms of set theory (including the axiom of foundation but excluding AC) do not lead to a contradiction, then they remain consistent when AC is adjoined as an additional axiom.

2.2 The continuum hypothesis

In his theory of infinite cardinals Cantor proved (making essential but implicit use of AC) that the totality of all infinite cardinal numbers is well-ordered (and in fact is order-isomorphic to the totality of all ordinal numbers). However, an important question was left open by Cantor's work. Let c be the cardinal number of the set of real numbers (or, as this set is sometimes referred to, the *continuum*). Cantor showed that c is not the first infinite cardinal, but he was unable to determine its precise place in the hierarchy of infinite cardinals. He conjectured, however, that c is precisely equal to $\aleph_1$, the second infinite cardinal (*Cantor 1883*). This conjecture became known as the *continuum hypothesis* (*CH*). It is easily shown that $c = 2^{\aleph_0}$, and so *CH* is equivalent to the statement $2^{\aleph_0} = \aleph_1$. A natural generalization, considered later by Hausdorff (*1908*) and called the *generalized continuum hypothesis* (*GCH*), asserts that for every ordinal α, $2^{\aleph_\alpha} = \aleph_{\alpha+1}$.

[d] Zermelo's precise formulation in his Axiom VI was slightly different from the one we have given here, but they are easily proved equivalent in the presence of the other axioms.

[e] *Moore 1982* has an exhaustive discussion of the history of this controversy.

[f] We now know, thanks to the work of Cohen (*1963*, *1964*) and Feferman (*1965*), that it is consistent with all the usual axioms of set theory (including the axioms of choice and foundation) that there is no *definable* well-ordering of the set of real numbers.

Gödel did not succeed in settling whether or not *CH* is true. But he was able to show that the usual axioms of set theory do not disprove *CH*, so that if they settle its truth value at all, it must be a theorem. As it turned out, Cohen was able to show in 1963 that the latter also does not hold. Gödel's method of proof for the consistency of *CH* with the axioms of set theory (including the axiom of choice), to be described below, worked just as well to establish the consistency of *GCH* with those axioms.

2.3 Axiomatizations of set theory

Modern axiomatizations of set theory are all derived from *Zermelo 1908*. However, the systems of Skolem and Fraenkel that have come to replace Zermelo's differ from it in several respects:[g]

(a) Zermelo's original axioms allowed for individuals (or urelements) that are members of sets but are not sets themselves. Thus, Zermelo's version of the axiom of extensionality said that two *sets* with precisely the same members are equal. The modern axiomatization considers only pure sets; thus the variables of the theory range only over sets, and the axiom of extensionality takes the form that whenever x and y have the same members, they are equal.

(b) A key axiom schema of set theory, *separation*, expresses the following: if x is a set and P is a property, then there is a set y whose members are precisely those members of x that have the property P. Clearly some care is needed in the precise formulation of this axiom schema. Zermelo introduced a new undefined notion of "definite property". (He says "A question or assertion ... is said to be *definite* if the fundamental relations of the domain, by means of the axioms and the universally valid laws of logic, determine without arbitrariness whether it holds or not".[h]) He then required that the property P be definite in this sense.

While perhaps adequate for mathematical practice, Zermelo's treatment of this axiom schema was not precise enough for metamathematical investigations. For these purposes, one needs a precise set of axioms in an unambiguously defined formal language. The solution (found by Skolem in his *1923*) is to allow only those properties P that are expressible in the appropriate formal language for set theory.[i] Present-day

[g]Section 4.9 of *Moore 1982* has an excellent discussion of the historical process by which ZF evolved from the axiomatic theory of *Zermelo 1908*.

[h]I am quoting from the translation in *van Heijenoort 1967*, p. 201.

[i]Of course, the idea of a formal first-order language was not as familiar in 1923 as it is today.

versions of axiomatic set theory handle the separation schema in the manner of Skolem.

(c) Even after *Zermelo 1908* is amended as suggested in (a), it still permits the existence of anomalous sets. For example, it is possible to have a model of Zermelo's theory in which there is a set x whose sole member is x itself. This and other related anomalies are ruled out by the *axiom of foundation*, which asserts that every non-empty set x has a member y which has no members in common with x.

Another formulation of the axiom of foundation with a clearer conceptual meaning will be given after we discuss the replacement axiom.

(d) Both Fraenkel and Skolem pointed out that Zermelo's system of axioms could not carry out certain constructions permitted in Cantor's "naive set theory". For example, let Z_0 be the set of non-negative integers and let Z_{i+1} be the power set of Z_i for $i \in \omega$. Then Zermelo's system is unable to prove the existence of the set whose members are the Z_i's.

This defect is handled by adjoining a new axiom schema, that of *replacement*, which asserts roughly the following: let $P(z)$ be a "definite property" in the precise sense above. Suppose that, for every set x, there is precisely one y such that $P(\langle x, y\rangle)$.[j] We can think of P as determining a function F whose domain is the collection of all sets. Then, for every set a, there is a set b whose members are precisely the values of $F(y)$ for $y \in a$.

The modern axiomatization of set theory, ZFC, is obtained by making these four changes in Zermelo's *1908* paper. (The version without the axiom of choice is denoted ZF.) The system obtained from ZFC (respectively, ZF) by dropping the axiom schema for replacement is denoted ZC (respectively, Z).

We can now describe the more conceptual proposition which (in the presence of all the axioms of ZF except the axiom of foundation) is equivalent to the axiom of foundation. (The proof makes essential use of the replacement schema and cannot be carried out in Z or even ZC.)

The levels in the *cumulative hierarchy* are the sets $R(\alpha)$, defined for all ordinals α by transfinite induction on α as follows:

(i) $R(0) = \emptyset$;

(ii) if $\alpha = \beta + 1$, then $R(\alpha) = \mathrm{P}(R(\beta))$ (where $\mathrm{P}(x)$, the *power set* of x, is the collection of all subsets of x);

(iii) if α is a limit ordinal, then $R(\alpha) = \bigcup_{\gamma<\alpha} R(\gamma)$.

The promised equivalent to the axiom of foundation is the proposition that every set is a member of one of the levels $R(\alpha)$ of the cumulative hierarchy.

[j]Here $\langle x, y\rangle$ is the ordered pair of x and y, defined in the usual way due to Kuratowski (*1921*).

Present day research in set theory concentrates on the theory ZFC. However, in *1940* Gödel worked with a different version of axiomatic set theory, due to Bernays and Gödel, which he denoted by Σ but which is now customarily denoted BG.[k] In the system BG there are two different sorts of variables. First of all there are variables which range over *sets*. Intuitively, these *sets* may be identified with the sets which are the range of the variables of ZFC. In addition, there are *class* variables. The intuitive picture is that classes are collections of sets. The axioms of BG ensure that every set is a class, that every member of a class is a set and conversely, and that two classes which have precisely the same members are equal; but in addition to sets, BG provides for the existence of classes, called *proper classes*, which are "too large" to be sets. For example, there is a universal class, V, which has every set as a member.

While every theorem of ZF is a theorem of BG, a striking difference is that BG is finitely axiomatizable while ZF is not. The reason why ZF requires infinitely many axioms is that each of the two axiom schemas (of separation and replacement) has infinitely many instances. Each of these schemas corresponds to a single axiom of BG. (For example, the axiom schema of separation corresponds to the assertion that, if A is a class and x is a set, there is a set y whose members are precisely the sets which are members of both A and x.) BG has only finitely many axioms of class existence, but they suffice to prove that, for any property P of sets which is expressible in the language of ZF (with particular sets allowed as parameters in the definition), BG can prove that there is a class whose members are precisely the sets with property P.[l] (The finitely many axioms of BG that are needed correspond roughly to the finitely many basic predicates and logical connectives of the language of ZF.)

To state the next significant fact about the Bernays–Gödel system, we agree that henceforth BG will refer to the version without the axiom of choice, while BGC will refer to the version with the "global axiom of choice". This principle asserts that there is a single function F (necessarily a proper class) which selects a member from every non-empty set. Mostowski proved that if ϕ is a sentence in the language of ZF, then ϕ is a theorem of ZF if and only if it is a theorem of BG; in other words, BG is a conservative extension of ZF.[m] It turns out that BGC is also

[k]The system BG grew out of earlier work by von Neumann. Cf. *von Neumann 1925, 1928, Bernays 1937, 1941, 1942*, and *Gödel 1940*.

[l]This and more is proved in the metatheorem M1 of Chapter II of *1940*.

[m]Mostowski's proof is sketched in footnote 6 on page 112 of *Mostowski 1950*.

a conservative extension of ZFC. This latter result was proved at about the same time by Kripke, Cohen and the present writer.[n]

Finally, we remark that natural models of the different theories we have discussed can be found at suitable stages of the cumulative hierarchy. (For all the models so considered, we take the ϵ-relation of the model to be the restriction of the usual membership relation to its sets.) In particular, if α is a limit ordinal greater than ω (for example, $\alpha = \omega + \omega$), then $R(\alpha)$ is a model of ZC.

Before stating the corresponding result for natural models of ZFC, we need to recall some notions about infinite cardinals. An infinite cardinal κ is *regular* if κ is not the sum of fewer than κ many cardinals less than κ. An infinite cardinal κ is a *strong limit cardinal* if, whenever λ is a cardinal less than κ, then also $2^\lambda < \kappa$. Finally, an infinite cardinal κ is *strongly inaccessible* if it is regular, is greater than $\aleph_0$, and is a strong limit cardinal. If Ω is a strongly inaccessible cardinal, then $R(\Omega)$ is a model of ZFC. We can obtain a model of BGC by taking the sets of the model to be the members of $R(\Omega)$ and the classes of the model to be the subsets of $R(\Omega)$.[o]

There is another construction of "natural models of set theory" we shall need to refer to later, for which some preliminary definitions are required. A set x is said to be *transitive* if $(\forall y \epsilon x)(y \subseteq x)$. One can prove in ZFC that for every set x there is a smallest transitive set y such that $x \subseteq y$. This y is called the *transitive closure* of x. For λ an infinite cardinal, $H(\lambda)$ is the collection of all sets x whose transitive closure has cardinality less than λ. (One can prove in ZFC that $H(\lambda)$ is a set.) If λ is a regular cardinal greater than $\aleph_0$, then $H(\lambda)$ is a model of all the axioms of ZFC except possibly the power set axiom. If λ is a strong limit cardinal greater than $\aleph_0$, then $H(\lambda)$ is a model of ZC. Finally, if λ is strongly inaccessible, then $H(\lambda)$ is a model of ZFC. (In this last case, one can show that $H(\lambda) = R(\lambda)$.)

This completes our review of the different versions of axiomatic set theory and their simplest models; we now return to our discussion of the historical antecedents of Gödel's work on constructibility.

Gödel's method, which has subsequently become known as the "inner model" method, proceeded as follows: he described a certain collection of sets, called the *constructible* sets, and was able to prove (in axiomatic set theory without the axiom of choice) that each of the axioms of set theory holds in the domain of constructible sets. He also showed that

[n]Subsequently, the proof was rediscovered and published in *Felgner 1971*.

[o]Since the results just stated can be proved in ZFC, it follows by Gödel's second incompleteness theorem that the existence of an inaccessible cardinal cannot be proved in ZFC if ZFC is consistent.

AC and *GCH* hold in this domain. From this it follows easily that if the axioms of set theory became inconsistent after adjoining *AC* and *GCH*, then they must already have been inconsistent without these new axioms. For a proof of $0 = 1$ from the new axioms could be recast as a proof that $0 = 1$ holds in the domain of all constructible sets, an obvious contradiction.

A certain amount of model-theoretic work had been done in set theory prior to Gödel's work. When Zermelo presented his axiomatization of set theory in *1908*, he already raised the question of the system's consistency ("I have not yet even been able to prove rigorously that my axioms are consistent, though this is certainly very essential ..."[p]); he also remarked that his axioms appeared to be independent of each other, but made no attempt at a proof. The first real endeavors to work with models of set theory were by Fraenkel (*1922*, *1922a*) and Skolem (*1923*). Fraenkel attempted[q] to show the independence of a number of Zermelo's axioms, particularly the axioms of choice and separation. However, Fraenkel was definitely not thinking in terms of first-order logic, whereas Skolem was. Naturally, this had profound effects on the sort of models that they considered. Skolem was very interested in countable models of set theory and noted that such a model, even if its natural numbers are standard, will omit some set of natural numbers. He raised the question whether one can add such a set of natural numbers and still have a model of Zermelo set theory. Also, he argued that the continuum hypothesis is probably neither proved nor disproved by Zermelo's axioms.

There matters sat for a while. Then von Neumann in his *1929* gave the first relative consistency proof, that for the axiom of foundation; that is, he showed that if set theory without the axiom of foundation is consistent, then it remains consistent when the axiom is added. Later, Ackermann (*1937*) gave a proof that if number theory is consistent, so is ZFC minus the axiom of infinity. Gödel's discovery of constructible sets and their use in proving the relative consistency of the axiom of choice dates from 1935; the proof of the relative consistency of *CH* (and, in fact, of *GCH*) came later, apparently in 1937.[r]

[p]This passage is quoted from the translation in *van Heijenoort 1967*, pp. 200–201. Of course, in light of Gödel's second incompleteness theorem, it is unreasonable to hope for a proof of the consistency of the Zermelo axioms using means formalizable within Zermelo's system.

[q]Fraenkel's work did not meet modern standards of rigor. Completely adequate versions of Fraenkel's proofs were given in *Lindenbaum and Mostowski 1938*; see *Mostowski 1939*.

[r]For further details on the evolution of Gödel's proof and other references, see p. 158 below as well as these *Works*, Volume I, pp. 9, 21–22; the dating 1937 for the relative consistency of *GCH* comes from an item in Gödel's *Nachlass* (op. cit., p. 36, fn. s). Cf. also *Moore 1982*, pp. 280–283.

As we shall see more clearly in the next section, Gödel's constructible hierarchy (which is used to define the class of constructible sets) can be viewed as a variant of the cumulative rank hierarchy. The rank hierarchy had first been clearly stated by Zermelo in his *1930*, in the context of second-order models of set theory.[s] In a sense, Gödel combined the rank hierarchy of Zermelo with the first-order perspective of Skolem in order to obtain the hierarchy of constructible sets.

Another antecedent of Gödel's constructible hierarchy is the ramified theory of types of Russell and Whitehead. Indeed, Gödel explicitly states that his constructible hierarchy can be viewed as the natural prolongation to transfinite levels of the ramified theory of types (*1944*, page 147). The most striking expression of this connection appears in a letter of Gödel to Hao Wang, dated 7 March 1968 and quoted on page 10 of *Wang 1974*. There Gödel attests to the fruitfulness of his platonistic attitudes for his research in the foundations of mathematics. Referring to his work on the consistency of CH, he says, "However, as far as, in particular, the continuum hypothesis is concerned, there was a special obstacle which *really* made it *practically impossible* for constructivists to discover my consistency proof. It is the fact that the ramified hierarchy, which had been invented *expressly for constructive purposes*, has to be used in an *entirely nonconstructive way*." The essentially nonconstructive element lies in the use of arbitrary ordinals as the levels in Gödel's extension of the ramified theory.

3. Description of the proof

The outline we shall give is substantially that of *Gödel 1939a*. In that brief note, the details of the proofs that the axioms of ZFC hold in L and of the absoluteness arguments needed to establish that $V = L$ holds in L are not given, but all the key notions and ideas are explained.

The notion of *constructible set* is best defined in terms of an auxiliary hierarchy of sets, the L_α's, which are defined, for all ordinals α, by transfinite induction on α, as follows:

(i) $L_0 = \emptyset$;

(ii) if $\alpha = \beta + 1$, then L_α consists of all the subsets of L_β that are definable by a first-order formula of set theory, possibly containing parameters from L_β, when the variables of that formula are interpreted as ranging over L_β;

(iii) if α is a limit ordinal, then $L_\alpha = \bigcup_{\gamma<\alpha} L_\gamma$.

[s] Mirimanoff (*1917*) had introduced the cumulative hierarchy, but was not influential; von Neumann (*1929*) used it, but in a very confusing way.

Finally, a set is *constructible* if and only if it appears in some L_α; while the class L of constructible sets is not in the range of the variables of ZF, the property $L(x)$ of being constructible is definable in ZF, in the form $\exists\alpha(x \in L_\alpha)$.

A number of comments on this definition are in order. First, it should be contrasted with the description of the universe of sets in terms of the cumulative hierarchy. There the form of the definition is exactly the same as the one just given if α is 0 or a limit ordinal, but at the crucial successor case, $R(\alpha+1)$ is the collection of *all* subsets of $R(\alpha)$.

Thus the constructible hierarchy is obtained by modifying the usual definition of the cumulative hierarchy to be far more parsimonious in adding subsets of the collection of sets already defined. Intuitively, at stage $\alpha+1$ one throws in only those subsets of L_α that must appear in any possible model of set theory that contains the set L_α. (However, in contrast to the cumulative hierarchy, some new subsets of L_α may first appear at stages later than $\alpha+1$; indeed, this happens for all infinite stages α.)

We now explain the essentials of Gödel's proof (outlined in *1939a*) that L, the totality of all constructible sets, is a model for all the axioms of ZF together with AC and GCH.[t]

Even though the variables of ZF range only over sets and not over proper classes, it is standard in expositions of ZF set theory to allow limited reference to proper classes when this is done in such a way that the discussion could in principle be expressed solely in terms of sets. Thus, for example, the (true) assertion that every ordinal is constructible might be expressed by $On \subseteq L$. (Here On is the class of all ordinals.) A more detailed discussion of this point can be found in Chapter 1, Section 9, of *Kunen 1980*. Our discussion of absoluteness in the following paragraph should be taken in this spirit.

We first introduce the very important notion of $\phi(x_1,\ldots,x_n)$ being *absolute* from a transitive class M to a transitive subclass N.[u] The formula ϕ is absolute from M to N if and only if, whenever $x_1,\ldots,x_n$

[t]The word "model" has to be taken with a grain of salt. For each particular axiom of ZF, the statement that that axiom holds in the domain of the constructible sets can be formulated in the language of ZF and is in fact a theorem of ZF. Since we cannot, in any obvious way, formulate in a single sentence of the language of ZF the assertion that *all* the axioms of ZF hold in L, we are not in danger of running afoul of Gödel's second incompleteness theorem on the unprovability of consistency. At the same time, this understanding of the model-theoretic approach allows one to establish the consistency of ZFC + GCH relative to ZF.

[u]A class X of sets is called *transitive* if, whenever x is a member of X, then x is a subset of X. For example, L and the class V of all sets are transitive classes.

are members of N, $\phi(x_1, \ldots, x_n)$ holds in the structure N[v] if and only if it holds in the structure M.

It is straightforward to verify that the axioms of ZF hold in L. The axiom of separation is the hardest to check. Here the key idea is that, by the same argument used to prove the reflection principle of set theory,[w] there will be many stages L_α at which some preassigned formula will be absolute from L to L_α. So if $x \in L_\alpha$, a subset of x defined by a formula whose variables are interpreted as ranging over all of L will be defined equally well by the same formula with its variables interpreted as ranging over L_α; but then this subset will appear in $L_{\alpha+1}$.

The remainder of Gödel's argument consists in showing that AC and GCH hold in the constructible universe. This is done in two steps. First, it is shown that these propositions follow from the proposition that every set is constructible, a proposition now customarily referred to as the *axiom of constructibility* and symbolized by the equality $V = L$. Second, it is shown that the proposition $V = L$ holds in the constructible universe.

We take up the second point first. It is natural to think that this is a trivial matter, since from the standpoint of L, the universe of sets consists precisely of those sets lying in L. But there is a subtle difficulty to recognize and deal with. A set x in L might have some property, such as being a cardinal number or being a constructible set, *in* V, but the same property, interpreted in L, might not hold of x. (This possibility definitely can happen for the property "is not a cardinal number".) In fact, one can show that the L_α's, when computed in L, are exactly the same as the L_α's when computed in V. In order to show this, it is necessary to make a detailed study of those operations and notions that are absolute from V to L. It turns out that the operation of forming the set of all first-order definable subsets of a given set is absolute; the operation of forming the full power set is not.

[v]That is, with the ϵ-relation interpreted as usual in N, and with the quantifiers interpreted as ranging over the elements of N.

[w]The reflection principle asserts that for any finite set of formulas Φ (of the language of set theory) there are arbitrarily large ordinals α such that each formula ϕ in Φ is absolute from V to $R(\alpha)$. The proof of this principle runs roughly as follows: We may assume the collection Φ is closed under the taking of subformulas. Assume further that in formalizing first-order logic, we have taken the existential quantifier as basic (and defined the universal quantifier in terms of it). For each formula ϕ of Φ that begins with an existential quantifier, we introduce a corresponding Skolem function, f_ϕ. It is easy to verify the following two facts: (1) If an $R(\alpha)$ is closed under all the functions f_ϕ, then all the formulas in Φ are absolute from V to $R(\alpha)$. (2) There are unboundedly many ordinals α such that $R(\alpha)$ is closed under all the functions f_ϕ. ((2) is an easy consequence of the replacement axiom schema of ZF.)

Gödel 1939a does not discuss the verification of the axioms of ZF in L. However, a very similar use of Skolem functions occurs in the proof of Theorem 2 of *1939a*.

One byproduct of this investigation of absoluteness is the following intrinsic characterization of L: It is the minimal transitive class-model of the axioms of ZF that contains all the ordinals. This characterization shows the fundamental nature of the concept of constructible set.

We now turn to the proof that the axiom of choice and the generalized continuum hypothesis follow from the proposition $V = L$. For AC, this is fairly easy. One can define, by induction on α, a well-ordering W_α of L_α. A little care is needed to make sure that, at a limit stage λ, the union of the W_α's for $\alpha < \lambda$ is a well-ordering for L_λ. But the heart of the argument is to see how to go from a well-ordering of L_α to one for $L_{\alpha+1}$. A definition of a set in $L_{\alpha+1}$ may be viewed as a finite sequence of symbols, each of which is either an integer or a member of L_α. It is easy to well-order the totality of such sequences using the given well-ordering of L_α; one then well-orders $L_{\alpha+1}$ by putting the elements already in L_α first, arranged in the order W_α, and then ordering the new elements in the same order as their minimal definitions.[x]

The proof that $V = L$ implies GCH is more subtle. The key lemma is the following, which appears as Theorem 2 of *1939a*:

> Let λ be an infinite cardinal, and let x be an arbitrary subset of λ; then if $V = L$, x is a member of L_{λ^+}.

Here λ^+ is the least cardinal greater than λ. It is quite easy to show that L_{λ^+} has cardinality equal to λ^+. Thus the lemma will imply that GCH holds in L.

The proof of the lemma is analogous to the proof of the downward Skolem–Löwenheim theorem. Since $V = L$, x will appear in some L_γ, and, by increasing γ if necessary, we may arrange that γ is greater than λ and that L_γ is a model of $V = L$ and of some fixed finite subset T of the axioms of ZF which is sufficiently large to prove the fundamental properties of the L_α's. If M and N are transitive set-models of T with M a subset of N, then the computation of the L_α's will be absolute from N to M. By the downward Skolem–Löwenheim theorem, we can then find an elementary submodel M of L_γ, of cardinality λ, containing x, λ and all the ordinals less than λ. M will in general not be transitive, but it is ϵ-isomorphic to a transitive model, say N (nowadays called the *Mostowski collapse* of M).

Now N is a transitive model of the proposition $V = L$ and of the finite fragment T of ZF. Absoluteness arguments show that, for ordinals

[x]If one just wants to prove the relative consistency of ZFC to ZF, a simpler proof can be given using the notion of *ordinal-definable* set, first introduced by Gödel in *1946*; see the introductory note to *1946* in this volume.

$\delta \in N$, $L_\delta^N = L_\delta$. Since $V = L$ holds in N, N is just the union of the L_δ's for δ an ordinal of N; i.e., $N = L_\theta$, where θ is the least ordinal not in N.

It remains to notice two points. First, the isomorphism between M and N is the identity on the ordinals $\leq \lambda$ and hence carries the set x to itself. (This is fairly easy to verify from the detailed proof that M is isomorphic to a transitive set.) So x lies in $N = L_\theta$. Second, the cardinality of θ is less than or equal to the cardinality of N (or equivalently, the cardinality of M), which is less than or equal to λ. So θ is less than λ^+. This completes the sketch of the proof of the crucial lemma.

Gödel also notes that for $\lambda = \aleph_\omega$, the model L_λ gives a natural model of ZC + GCH. Similarly, if λ is a strongly inaccessible cardinal, then L_λ is a model of ZFC + GCH. These results are closely related to the results about the natural models $H(\lambda)$ discussed at the end of Section 2 of this note. Indeed, it follows by arguments similar to those used to prove GCH in L that if $V = L$, then, for λ an infinite cardinal, $L_\lambda = H(\lambda)$. The result of Gödel just cited follows directly.[y]

The treatment in *Gödel 1940* is significantly different in its details from that outlined above. In the first place, instead of working with the theory ZF as we have done, Gödel works with the Bernays–Gödel set theory BG, discussed on page 5 above.

Second, the definition of L that is used in *1940* is a good deal more ad hoc. A set of "eight fundamental operations" is introduced, and an enumeration of sets, $F\colon On \rightarrow V$, is given which is designed so that (a) the range of F is closed under the eight fundamental operations, and (b) at many limit stages λ, $F(\lambda)$ is the set $\{F(\alpha)\colon \alpha < \lambda\}$. Then L is defined to be the range of this auxiliary function F. (Roughly, the eight fundamental operations are mathematical operations on classes that correspond to basic syntactic operations on formulas. For example, the operation of intersection corresponds to the syntactic operation of taking the conjunction of two formulas.)

Finally, the proof of the fundamental lemma needed to establish GCH in L is presented in a very non-conceptual way that obscures the connection with the Skolem–Löwenheim theorem. While the proofs in *1940* are presented in full detail, very little motivation is given. It is natural to wonder why Gödel presented his results in this way in

[y]We recommend that the reader interested in learning more of the details of Gödel's work on L begin with *1938* and *1939a*. For more detailed proofs, there are good treatments in several modern texts, notably *Kunen 1980, Jech 1978,* and *Devlin 1973.* (The reader of Kunen should note that a knowledge of his Chapter 2 on combinatorics is not needed for an understanding of his treatment of the basic facts about L in a later chapter.)

1940, when it is clear from *1939a* that he was well aware of the more conceptual proof outlined above. My guess is that he wished to avoid a discussion of the technicalities involved in developing the rudiments of model theory within axiomatic set theory. In *1939a* the portions of the argument that would require such a treatment are passed over in silence, while in *1940* an alternative treatment is developed that avoids the necessity for such a formalization.

Besides the results on AC and GCH, Gödel (*1938*) mentions two propositions of descriptive set theory (i.e., the study of definable sets of real numbers) which hold in the model L. In order to state them, we must review some of the standard terminology of descriptive set theory. Let X be one of the spaces $\mathbb{R}^n$, where $\mathbb{R}$ is the set of real numbers.[z] A subset Y of X is *Borel* if it belongs to the smallest family of subsets of X containing the open sets of X and closed under complementation and countable unions. A subset Y of $\mathbb{R}^n$ is $\boldsymbol{\Sigma}^1_1$ (or *analytic*) if it is the projection of a Borel subset of $\mathbb{R}^{n+1}$ (under the map that deletes the last component of an $(n+1)$-tuple). A subset Y of X is $\boldsymbol{\Pi}^1_1$ if it is the complement with respect to X of some $\boldsymbol{\Sigma}^1_1$ subset of X. This hierarchy of subsets is continued as follows: A subset of $\mathbb{R}^n$ is $\boldsymbol{\Sigma}^1_{k+1}$ if it is the projection of a $\boldsymbol{\Pi}^1_k$ subset of $\mathbb{R}^{n+1}$. A subset of $\mathbb{R}^n$ is $\boldsymbol{\Pi}^1_{k+1}$ if it is the complement, relative to $\mathbb{R}^n$, of some $\boldsymbol{\Sigma}^1_{k+1}$ set. A subset Y of X is $\boldsymbol{\Delta}^1_k$ if it is both $\boldsymbol{\Sigma}^1_k$ and $\boldsymbol{\Pi}^1_k$. Finally, Y is *projective* if it is $\boldsymbol{\Sigma}^1_k$ for some integer k.

We can now state the two propositions of descriptive set theory that Gödel showed are valid in L:

(1) There is a $\boldsymbol{\Delta}^1_2$ subset of $\mathbb{R}$ that is not Lebesgue measurable.

(2) There is a $\boldsymbol{\Pi}^1_1$ subset of $\mathbb{R}$ that has cardinality c but contains no perfect subset.

These results should be contrasted with the following theorems of ZFC (cf. *Moschovakis 1980*):

(1′) Every $\boldsymbol{\Sigma}^1_1$ or $\boldsymbol{\Pi}^1_1$ subset of $\mathbb{R}$ is Lebesgue measurable; a fortiori, every Borel subset of $\mathbb{R}$ is Lebesgue measurable.

(2′) Every $\boldsymbol{\Sigma}^1_1$ subset of $\mathbb{R}$ is either countable or contains a perfect subset of cardinality c.

The two results of Gödel just cited are consequences of the fact that, assuming $V = L$, the restriction of the canonical well-ordering of L to the reals gives a *good* Δ^1_2 well-ordering.[aa] Gödel gave no proof of (1) and (2) in *1939a* and a cryptic proof of a few lines, comprehensible only

[z] $\mathbb{R}^n$ is viewed as a topological space in the usual way; with slight modifications, the definitions given here apply to any complete separable metric space X without isolated points. See *Moschovakis 1980*, the standard reference work on modern descriptive set theory.

[aa] The word "good" has a technical meaning here that I shall not stop to explain.

to cognoscenti, in *1940*. (There is a fuller proof in Gödel's handwritten notes for his *1940* lectures.) A detailed proof of (1) and (2) was first published in *Novikov 1951*, and a proof of the result about a good Δ^1_2 well-ordering of $\mathbb{R}$ was first given in *Addison 1959*. For a readable account, see *Jech 1978*, pages 527–530, or *Moschovakis 1980*, pages 274–281.

4. Further work

In this fourth and final section, I shall describe subsequent work done on L and on the questions raised by the work of Gödel. For historical reasons, specific references are given wherever possible; a good pair of general references that cover almost all of the following are *Jech 1978* and *Devlin 1973*.[bb]

After Gödel, the first work on L was done by Kuratowski and Mostowski.[cc] Mostowski reconstructed Gödel's proof that, in L, the reals have a Δ^1_2 well-ordering. Kuratowski showed that, using the projective well-ordering of the reals which Gödel deduced from $V = L$, one could prove that various pathological sets previously constructed using the axiom of choice and the continuum hypothesis would be projective.

In the late 1950s Gödel's "inner model method" was generalized slightly in the work of Hajnal, Levy and Shoenfield.[dd] They introduced a relative version of the notion of constructible set. For example, if x is a set of ordinals then $L[x]$, the class of sets *constructible from* x, can be characterized as the minimal transitive class containing all the ordinals which is a model of ZF and which has x as a member. It may be shown that AC holds in $L[x]$ by the same argument as above, and even that GCH holds in $L[x]$ when $x \subseteq \omega$.[ee] A typical further result is the following theorem of Levy and Shoenfield: If $V = L$ follows from GCH, then $V = L$ is already a theorem of ZF.

It had already become clear in the early 1950s, thanks to the work of Shepherdson (*1951–1953*),[ff] that the inner model method is quite

[bb] *Devlin 1984* is a revised version of *Devlin 1973* that contains a great deal of interesting additional material. For example, it discusses Silver machines and the simplified morasses of Velleman.

[cc] Cf. *Addison 1959*, p. 338. Mostowski's manuscript was destroyed during the Second World War. Kuratowski's work suffered a similar fate, but was later reconstructed and published as *Kuratowski 1948*.

[dd] Cf. *Hajnal 1956, 1961, Levy 1957, 1960b*, and *Shoenfield 1959*.

[ee] By a slightly more difficult argument one can show that GCH continues to hold if $V = L[a]$ and $a \subseteq \aleph_1$.

[ff] Similar results were obtained subsequently but independently by Cohen (*1963a*).

incapable of showing that $V = L$ is *not* a theorem of ZFC. The reason is that one cannot rule out, on the basis of ZFC, the possibility that the universe of sets is *minimal*, i.e., has no non-trivial inner models. The existence of minimal models can be argued as follows. Evidently, a minimal model must satisfy $V = L$. If there are no transitive set-models of $V = L$, then L itself can be shown to be minimal. If there are transitive set-models of ZFC, then by applying the Gödel construction of L within such a model, one can easily conclude that there are transitive models of ZFC of the form L_δ. By taking δ as small as possible, one obtains, once again, a minimal model of ZFC.

Of course, Shepherdson's analysis shows, a fortiori, that the inner model method is incapable of answering the following two natural questions:[gg]

(a) Is the continuum hypothesis a theorem of ZFC?

(b) Can the proposition "The reals have a well-ordering" be proved without the aid of the axiom of choice?[hh]

Thus Gödel's work raised the fundamental new question: Is $V = L$ a theorem of ZFC? But, through the work on relative constructibility, it also provided an important clue to the solution. One could prove that if $V = L$ is not a theorem of ZFC, then there is a model of ZFC of the form $L[x]$ for some set x of ordinals, with x not constructible.[ii] In addition, it seemed highly plausible that x could be taken to be a set of integers. This reformulation was useful because the structure of the model $L[x]$ is quite transparent. In particular, the sets of the model $L[x]$ are naturally parametrized by the ordinals. Thus it was natural to phrase the problem (of showing that $V = L$ is not a theorem of ZFC) as follows: Let M be a countable transitive model of ZFC which has the form L_δ; can we then find a subset x of ω such that $L_\delta[x]$ is again a model of ZFC? Unfortunately, though it is easy to pick x so that $L_\delta[x]$ is *not* a model of the Replacement Axiom of ZFC (simply choose x to encode the ordinal δ), there was no obvious way to ensure that $L_\delta[x]$ *is* a model of ZF (and hence of ZFC).

The questions raised above were all settled by Paul Cohen with his development of the technique of forcing. (Cf. *Cohen 1963, 1964*, and

[gg]More precisely, for each of these questions the expected answer was 'no'. The models that were needed to supply a negative answer could not be constructed by the inner model method. Models constructed by Cohen's method eventually showed the expected answers to be correct.

[hh]As mentioned previously, Mostowski, building on earlier work of Fraenkel, had shown that in the version of set theory that allows individuals as well as pure sets, the axiom of choice could not be proved. But this left open the possiblity that the axiom of choice could be proved for the sets of ordinary mathematical practice, such as the real numbers.

[ii]This follows from the work of Hajnal, Levy and Shoenfield cited in footnote dd.

1966.) Cohen showed (under the assumption that ZFC is consistent) that there are models of ZF in which

(a) AC and GCH hold, but there is a non-constructible set of integers (so $V = L$ is false);

(b) AC holds, but CH is false;

(c) AC fails, and in fact the reals cannot be well-ordered.

Unlike Gödel's inner model method which, by itself, could produce only the single model L, the forcing method has proved to be an extremely flexible and powerful tool for the creation of models in which a wide variety of set-theoretical propositions can be seen to be consistent with the axioms of set theory.

On the other hand, the Cohen forcing method is incapable of showing that a proposition is independent of $V = L$. It does give a systematic method for enlarging a countable transitive model of ZFC, M, to a larger model, N, in which propositions may well hold that do not hold in M. But the notion of "constructible set" is absolute between the two models, and thus the method gives no information about models of ZFC $+ V = L$.

There are several natural propositions of set theory whose status *in* L had been left open by Gödel. The most noteworthy was Suslin's hypothesis (SH), first given in *Suslin 1920*. This asserts that the following four properties characterize the real line as a linearly ordered set:

(1) It is order-dense. That is, if a and b are reals with $a < b$, then for some real c, $a < c < b$.

(2) It is order-complete. That is, every set of reals which is bounded above has a least upper bound.

(3) It has no least element or greatest element.

(4) Every pairwise disjoint collection of open intervals is at most countable.

Condition (4) is an easy consequence of the fact (4′) that $\mathbb{R}$ has a countable order-dense subset. It is quite easy to see, as Cantor showed (*1895*), that conditions (1) through (3) together with (4′) do characterize $\mathbb{R}$ up to order-isomorphism, so SH amounts to the assertion that (4′) can be weakened to (4).

The proposition SH is an extremely natural one that turns up also in the theories of partially ordered sets and Boolean algebras. It was also a natural candidate for a new proposition (other than those considered by Cohen) to be proved independent of the axioms of ZFC. Indeed, shortly after Cohen's technique of forcing was developed, it was shown[jj] that Suslin's hypothesis is both consistent with and independent of the

[jj]Cf. *Jech 1967, Tennenbaum 1968,* and *Solovay and Tennenbaum 1971.*

axioms of ZFC. But that left open the question whether it holds or fails in L.

There is a reformulation of *SH* that makes sense for any regular cardinal κ. Suslin's hypothesis is equivalent to the non-existence of a certain kind of tree (dubbed a Suslin tree) on $\aleph_1$, the first uncountable cardinal. One can then generalize the Suslin problem to the question "For which uncountable regular cardinals κ is there a Suslin tree on κ?"

It was shown by Jensen (*1972*) that Suslin's hypothesis is false in L. Subsequently, with much more effort, Jensen completely determined for which regular cardinals κ there is a κ-Suslin tree in L. It was evident, a priori, that if κ is weakly compact, then no κ-Suslin tree exists. Jensen showed that if $V = L$, there is a κ-Suslin tree for any regular uncountable κ that is not weakly compact.

Jensen's proof that Suslin's hypothesis fails in L proceeds by deducing from $V = L$ a previously unconsidered combinatorial principle, which Jensen dubbed *diamond* ($\diamondsuit$) and which is a considerable strengthening of the continuum hypothesis. In certain circumstances, this principle allows one to meet $\aleph_2$ requirements in the course of a construction of length $\aleph_1$. With its aid, the construction of a Suslin tree is relatively straightforward. This principle and its variants have subsequently had numerous other applications in point-set topology and algebra. One noteworthy example is Shelah's proof that Whitehead's problem is undecidable.[kk] In one direction, Shelah shows that it follows from $\diamondsuit$ that every W-group of size $\aleph_1$ is free. In the other direction, he deduces from Martin's axiom and $2^{\aleph_0} > \aleph_1$ that there is a W-group of size $\aleph_1$ that is not free.

We shall now give a precise statement of $\diamondsuit$. Recall that by definition a subset C of $\aleph_1$ is *club* if and only if C is unbounded in $\aleph_1$ and C contains the least upper bound of each countable subset of C.[ll] A subset S of $\aleph_1$ is *stationary* if the intersection of S with any club subset of $\aleph_1$ is non-empty.[mm] The proposition $\diamondsuit$ asserts the existence of a family $\langle A_\alpha : \alpha < \aleph_1\rangle$ with the following properties:

(a) $A_\alpha \subseteq \alpha$ for all $\alpha < \aleph_1$.

(b) Let $S \subseteq \aleph_1$. Then $\{\alpha : A_\alpha = S \cap \alpha\}$ is stationary.

Thus $\diamondsuit$ gives a "guessing procedure" which, for any subset S of $\aleph_1$, correctly predicts $S \cap \alpha$ a significant portion of the time.

[kk]A W-group is an abelian group G such that $Ext(G, \mathbb{Z}) = 0$, where $\mathbb{Z}$ is the set of integers. Every free abelian group is a W-group, and every countable W-group is free. The version of Whitehead's problem considered by Shelah asks if every W-group of size $\aleph_1$ is free. (Shelah's results appear in his *1974*, but we strongly recommend the exposition of his results in *Eklof 1976*.)

[ll]The word "club" comes from the phrase "closed unbounded".

[mm]The following analogy may be useful. Club subsets of $\aleph_1$ correspond to subsets of the unit interval $[0, 1]$ having Lebesgue measure 1; stationary subsets of $\aleph_1$ correspond to subsets of $[0, 1]$ having positive Lebesgue measure.

While Jensen's solution to the Suslin problem on $\aleph_1$ is classical in spirit (since the proof of $\diamondsuit$ involves the same ideas and techniques used in Gödel's proof that *GCH* holds in L), the situation for the generalization of Suslin's problem to higher regular cardinals is quite different. Again the proof turns on showing that certain remarkable combinatorial principles hold in L. But it required a hitherto unknown detailed level-by-level analysis of the constructible hierarchy. (To facilitate this analysis, Jensen worked with a slight variant of the L_α hierarchy that had better closure properties.) Roughly speaking, by studying the precise place where an ordinal becomes singular, Jensen was able to exploit to good effect the residue of regularity that still remains just before that level. These techniques, which Jensen dubbed "the study of the fine structure of L", have had several other striking applications, notably to the proof of various model-theoretic two-cardinal theorems in L and to the proof of the Jensen covering theorem, which we shall discuss presently.

Subsequent to Jensen's work, Silver developed an alternative approach (the so-called Silver machines) that yields simpler proofs of most of the applications of the fine-structure theory (including all those mentioned here). Silver never published his work, but a presentation can be found in *Devlin 1984*.

Next we consider results having to do with the notion of absoluteness for sentences. A sentence ϕ is said to be ***absolute*** (in ZFC) if we can prove in ZFC that ϕ holds in V if and only if ϕ holds in L.[nn] (Until the work of Cohen referred to above, one could not rule out the possibility that every sentence ϕ in the language of set theory is absolute.) It is easy to see that every arithmetical sentence[oo] is absolute, for L's notion of the integers is identical to that of V. Similarly, by exploiting the connection between Π^1_1 sentences and the concept of well-ordering and the fact that the same ordinals appear in V and in L, one can easily show that Π^1_1 sentences are absolute.

The best possible result in this direction was obtained by Shoenfield (*1961*), who showed that Σ^1_2 formulas are absolute. This easily implies that a Π^1_3 sentence true in V holds also in L.[pp] Shoenfield's theorem is quite useful in descriptive set theory, since it permits one to prove

[nn]This notion is closely related to, but not identical with, the notion of absoluteness introduced in Section 3.

[oo]An arithmetical sentence is one that asserts that some proposition holds in the structure consisting of the non-negative integers equipped with the operations of addition and multiplication.

[pp]The converse need not be true. For example, the assertion that every real is constructible is easily seen to be Π^1_3. In the original forcing models of Cohen, it holds in L but not in V.

the absoluteness of statements that prima facie do not appear to be absolute.

Even the simpler fact of the absoluteness of arithmetical statements is sometimes quite useful. It has the consequence that any arithmetical statement provable using AC and GCH is provable without their aid. For example, in the work (*1965*) of Ax and Kochen on the first-order theory of p-adic fields, a principal tool is the ultraproduct construction of models, and the theory of ultraproducts is much smoother if GCH is assumed. The remark just made ensures that the arithmetical consequences of their investigations (for example, the decidability of the theory of p-adic fields) are outright theorems of ZFC. (This observation is credited to Kreisel.)

We turn next to the implications of large cardinal assumptions for the constructible universe. This part of our subject has had an involved history, starting with Scott's proof in *1961* that if there are measurable cardinals then V is unequal to L, and culminating in the work of Silver, Kunen, and the present author, to be described below. Important intermediate work, which we shall not describe, was done by Gaifman (*1964*, *1974*) and Rowbottom (*1971*).

Before going on, it is worth pausing a moment to note Gödel's own attitude toward large cardinals. In his *1947* he held out the hope that future discoveries in this area might lead to new axioms that would settle the continuum problem. This has not yet happened, and the large cardinal axioms known to date are relatively consistent with both CH and its negation. In conversations with the author, Gödel expressed belief in the existence of measurable cardinals (see pages 167 and 260–261 below) and offered the following heuristic argument in favor of their existence. It is known that every strongly compact cardinal is measurable. But the existence of strongly compact cardinals is equivalent to the statement that a certain property of $\aleph_0$ is also shared by some cardinal greater than $\aleph_0$. Gödel then expressed the belief (which I am unable to present in a coherent way) that reasonable properties possessed by $\aleph_0$ should also be satisfied by some cardinal greater than $\aleph_0$.[qq]

A *measurable cardinal* is, by definition, a cardinal κ such that there is a non-trivial $\{0,1\}$-valued κ-additive measure defined on the collection of all subsets of κ. It has been known since *Tarski 1962* that a measurable cardinal must be very large. In particular, it is a strongly inaccessible cardinal, and if κ is the measurable cardinal in question,

[qq]The restriction to "reasonable properties" is my addition to keep the argument from being blatantly fallacious. (The property of being the least infinite ordinal is satisfied only by $\aleph_0$.) I do not find this particular argument for the existence of measurable cardinals to be convincing.

then κ is the κth strongly inaccessible cardinal. It follows, by Gödel's incompleteness theorem, that the existence of measurable cardinals cannot be proved in ZFC, and that ZFC + "there is a measurable cardinal" cannot be proved consistent from the assumption of the consistency of ZFC.[rr] Nevertheless, many set theorists (including the author) believe that measurable cardinals exist, so that their consequences for the constructible universe are true. Among these are the following:

(a) If λ is an uncountable cardinal (in V), then L_λ is an elementary submodel of L.

(b) Hence (taking the case $\lambda = \aleph_1$ of (a)), if γ is an ordinal definable in L (such as the $\aleph_1$ of L, or the third strongly inaccessible cardinal of L), then γ is countable in V. In particular, there are only countably many constructible sets of integers.

It also follows from (a) that every uncountable cardinal of V is a limit cardinal in L. Hence if γ is an infinite ordinal, then there are precisely card(γ) constructible subsets of γ. Another consequence of (a) is that the satisfaction relation for L is definable in V.[ss]

Further consequences make use of the construction of models generated by a set of indiscernibles. Such models were first considered by Ehrenfeucht and Mostowski (*1956*), and they are in many ways rather special. For example, there are rather few types of elements realized in such models, and they tend to have many elementary monomorphisms into themselves. Silver (*1971*) realized that the techniques of Ehrenfeucht–Mostowski could be applied fruitfully to the study of L, on the assumption that there is a measurable cardinal.

Silver showed that there is a canonical generating class C of indiscernibles for L (now known as the class of Silver indiscernibles). They can be characterized as follows:

(1) C is a closed unbounded class of ordinals.

(2) C generates L. That is, every element of L has a first-order definition in L from a finite number of parameters in C.

(3) The members of C are indiscernible in L. That is, any two increasing n-tuples from C have the same first-order properties in L. In particular, any two members of C look completely alike in L.

One can show that every uncountable cardinal of V is a Silver indiscernible and that the Silver indiscernibles are very large cardinals in L. (They are strongly inaccessible, Mahlo, weakly compact, etc.) Indeed, every large cardinal property that is compatible with $V = L$ holds, in L, of the Silver indiscernibles.

[rr]Provided no arithmetical consequence of ZFC is false.

[ss]This should be contrasted with Tarski's theorem on the undefinability of truth, which implies that the satisfaction relation for L (or V) is not first-order definable in L (or V, respectively).

In view of (3), one can introduce the following set of integers, $0^\#$, which encodes the structure of L using only countably many bits of information. An integer k lies in $0^\#$ if and only if it is the Gödel number of some formula of the language of set theory $\phi(v_1, \ldots, v_n)$ that holds in L when the v_i's are replaced by an increasing n-tuple from C.

The set $0^\#$ is interesting in its own right. Prior to the discovery of $0^\#$, there was no natural example of a definable set of integers that is not constructible.[tt] It was shown in *Solovay 1967* that $0^\#$ is a non-constructible Δ^1_3 set of integers.[uu] It was also proved there that every constructible set of integers is recursive in $0^\#$, and thus is Δ^1_3.

We shall have occasion in the following to refer to the proposition "$0^\#$ exists". One way of expressing this is to say that L has a closed generating class of indiscernibles. This formulation has the drawback, however, that it is not expressible in the usual language of set theory (since it involves bound class variables). One can, however, produce a Π^1_2 formula, $\phi(x)$, that (if a measurable cardinal exists) holds only of $0^\#$. Then "$0^\#$ exists" can be taken to mean $(\exists x)\phi(x)$. One can show that all the consequences of measurable cardinals for the structure of L mentioned above already follow from the proposition "$0^\#$ exists".

Kunen has proved that the proposition "$0^\#$ exists" is equivalent to the existence of a non-trivial elementary monomorphism of L into itself.[vv] This should be compared with the following equivalent of the proposition "a measurable cardinal exists": There is a non-trivial elementary embedding of V into some transitive class M. (In each case, the phrase "non-trivial" means "not the identity map".)

In view of the fact that measurable cardinals are extremely large cardinals whose existence is incompatible with the axiom of constructibility, it is of interest to consider the question, "Which large cardinal axioms are compatible with $V = L$?" Gödel had remarked at the end of his *1938* that

> In this connection, it is important that the consistency proof for A ⟦that is, $V = L$⟧ does not break down if stronger axioms of infinity (e.g., the existence of inaccessible numbers) are adjoined to T. Hence the consistency of A seems to be absolute in some sense, although it is not possible in the present state of affairs to give a precise meaning to this phrase.

[tt]It was known that the existence of a non-constructible ordinal-definable set of integers is consistent. The question was whether a definable non-constructible set of integers could be proved to exist in some reasonable extension of ZFC.

[uu]It follows from the Shoenfield absoluteness theorem mentioned earlier that the definability estimate Δ^1_3 is best possible.

[vv]Kunen never published his proof. A different proof of Kunen's theorem, due to Silver, appears in V.4 of *Devlin 1984*.

We have already indicated that the existence of measurable cardinals contradicts $V = L$ in a strong sense. At the moment, the situation is the following. For large cardinal properties that are not too strong, for example those of being strongly inaccessible, or Mahlo, or weakly compact, the property holds of κ in L if it holds of κ in the universe, and the existence of $0^{\#}$ implies that all the Silver indiscernibles have the property. But stronger properties imply the existence of $0^{\#}$, and so contradict the proposition $V = L$ in a strong way. A description of the precise dividing line would involve an excursion into the subject of partition cardinals; we content ourselves with the remark that there are currently no large cardinal properties for which the status of their compatibility with $V = L$ remains unknown.[ww]

Another remarkable result due to Jensen, his so-called covering theorem, ensures (roughly speaking) that if $0^{\#}$ does not exist, then there are rather tight connections between L and V (*Devlin and Jensen 1975*).

The theorem is as follows: Assume that $0^{\#}$ does not exist. Let X be a set of ordinals. Then there is a set of ordinals Y, lying in L, such that (1) X is a subset of Y and (2) card(Y) is at most the maximum of card(X) and $\aleph_1$. (All cardinals referred to in (2) are computed in V.)

The theorem can be roughly paraphrased as follows. Either $0^{\#}$ exists, whence, by the results cited previously, L is a very sparse subclass of V; or $0^{\#}$ does not exist, in which case every set of ordinals in V can be tightly approximated from above by a constructible set.

The theorem has several striking consequences, of which we mention only the following two, each under the assumption that $0^{\#}$ does not exist:

(1) Let κ be a singular strong limit cardinal. Then $2^{\kappa} = \kappa^{+}$. (That is, *GCH* holds at κ.)

(2) Let κ be a singular cardinal. Then the least cardinal greater than κ, as computed in L, is the same as the least cardinal greater than κ as computed in V.

It is not hard to see that the proposition "$0^{\#}$ does not exist" holds in every forcing extension of L. Thus the Jensen covering theorem can be used to obtain stringent limitations on what one can accomplish merely by forcing, without the use of large cardinals. (For example, without the use of fairly large cardinals one cannot construct a model of ZFC in which the first κ for which $2^{\kappa} \neq \kappa^{+}$ is a singular cardinal.[xx])

[ww]The best results on this dividing line appear in *Baumgartner and Galvin 1978*.

[xx]In *Magidor 1977* a model is constructed by the forcing method in which *GCH* first fails at $\aleph_{\omega}$. Magidor's ground model for this construction contains a "huge" cardinal. (The first huge cardinal is far bigger than the first measurable cardinal.)

Our final topic is the subject of inner models for large cardinals. We have already mentioned that the proposition $V = L$ contradicts the existence of measurable cardinals. However, there is a natural generalization of L, $L[D]$, which has properties closely analogous to L and in which there is a measurable cardinal.

The model $L[D]$ is obtained by a slight generalization of the notion of relative constructibility introduced earlier. Let A be a class. Then one can show that there is a smallest transitive model M of ZF, containing all the ordinals, such that $(\forall x \in M)(A \cap x \in M)$. This M we call $L[A]$. If A is a set of ordinals, then this new notion reduces to that considered on page 14. However, even if A is a set, we need not have $A \in L[A]$.

Suppose now that κ is a measurable cardinal. Then there is a distinguished class of measures on κ, the *normal* measures, defined as follows:

A (two-valued) measure μ is a homomorphism from the Boolean algebra $P(\kappa)$ into the two-element Boolean algebra $\{0, 1\}$. The measure μ is *non-trivial* if $\mu(\kappa) = 1$ and the measure of every one-element subset of κ is 0. The measure μ is *κ-additive* if the union of fewer than κ sets of measure zero itself has measure zero.

A function $f: \kappa \to \kappa$ is *regressive* on a subset $D \subseteq \kappa$ if for every $\alpha \in D$ we have $f(\alpha) < \alpha$. Finally, μ is *normal* if whenever $f: \kappa \to \kappa$ is regressive on a set of measure one, then f is constant on a (possibly smaller) set of measure one.

There is a somewhat more conceptual alternative characterization of normal measures in terms of ultrapowers. If μ is a countably additive measure on κ, then the ultrapower construction gives rise to an elementary embedding $j: V \to M$. Then μ is normal if and only if κ is the least ordinal moved by j and the identity function represents κ in the ultrapower.

Normal measures tend to concentrate on the large cardinals less than κ. For example, one can show that the set of strongly inaccessible cardinals less than κ receives measure one from every normal measure on κ. Moreover, one can show that every measurable cardinal carries a normal measure.

We can now describe the inner model for a measurable cardinal. Let μ be a normal measure on κ and let D be the collection of sets of μ-measure one. It is rather easy to show that, in $L[D]$, κ is a measurable cardinal. It follows from results of Kunen (*1970*) that the model $L[D]$ depends only on κ, that, in $L[D]$, κ is the unique measurable cardinal, and that $D \cap L[D]$ is the collection of sets of measure one with respect to the unique normal measure on κ in $L[D]$.

It turns out that $L[D]$ very closely resembles L. For example, results of Silver show that (a) *GCH* holds in $L[D]$ (*1971a*), and (b) there is a good Δ^1_3 well-ordering of the reals in $L[D]$ (*1971b*). (This should be compared to the result cited on page 13 that there is a good Δ^1_2 well-ordering of the reals in L.) The author showed that there is a "fine structure theory for $L[D]$" quite analogous to the usual fine structure theory of L. (This work was never published, but it is implicit in the subsequent work of Dodd and Jensen (*1981*) on the "core model" K.)

It is certainly an interesting fact in its own right that there is a natural inner model for ZFC + "there is a measurable cardinal" that is quite analogous to the natural inner model L for the theory ZFC. But inner models for large cardinals also have important applications to the problem of establishing lower bounds on the consistency strength of propositions. For example, it is a theorem of Mitchell (*198?*) that if there is a model of ZFC in which GCH first fails at $\aleph_\omega$, then there is a model of ZFC in which there is a measurable cardinal of high order. An essential ingredient in his proof is the construction of inner models for measurable cardinals of high order.

It is therefore an important problem to find L-like models in which there are various large cardinals. Considerable progress has been made on this problem by Mitchell (*1974, 1979*), Dodd and Baldwin. However, recent results of Woodin show that certain large cardinals, if they have inner models at all, only have ones that behave very differently from the inner models discovered to date.

In order to state these results, we shall recall the definitions of some large cardinals.[yy] First, let j be an elementary embedding of V into a transitive class M. If j is not the identity, then one can show that j moves some ordinal. The *critical point of* j is the least ordinal moved by j. The critical point of a non-trivial elementary embedding $j\colon V \to M$ is always a measurable cardinal. The stronger the closure conditions imposed on M, the stronger the corresponding large cardinal property. (We remark that Kunen (*1971*) has shown that there is no non-trivial elementary embedding of V into itself.) Let κ and λ be infinite cardinals. Then κ is λ-*strong* if there is an elementary embedding $j\colon V \to M$ with critical point κ such that $j(\kappa) > \lambda$ and $R(\lambda) \subseteq M$. The cardinal κ is *strong* if it is λ-*strong* for every $\lambda > \kappa$. κ is *superstrong* if there is an elementary embedding $j\colon V \to M$ with critical point κ such that $R(j(\kappa)) \subseteq M$. Finally, κ is λ-*supercompact* if there is an elementary embedding $j\colon V \to M$ with critical point κ such that M is closed under sequences of length λ.

[yy]Two good references on the subject of large cardinals are *Kanamori, Reinhardt and Solovay 1978* and *Kanamori and Magidor 1978*.

As far as consistency strength goes, these concepts are related as follows: If there is a κ which is 2^κ-supercompact, then there is a transitive model of ZFC with a proper class of superstrong cardinals; if there is a superstrong cardinal, then there is a transitive model of ZFC with a proper class of strong cardinals. Building on earlier work of Mitchell (*1979*), Dodd (*198?*) constructed L-like inner models with a proper class of strong cardinals. In these models, the reals have a good Δ^1_3 well-ordering.

It was generally felt that the work of Mitchell and Dodd would eventually lead to inner models for supercompact cardinals. However, in 1984 Woodin proved the following remarkable theorem: If there is a superstrong cardinal, then there is no projective well-ordering of the reals.[zz] But, in all the inner models constructed by Mitchell, Dodd, and Baldwin, there is in fact a Δ^1_3 well-ordering of the reals. Thus inner models for cardinals at least as large as superstrong must in some ways be very different from L.

Woodin's theorem raises many questions. Here are two:

(1) What is the precise dividing line between cardinals which are compatible with a Δ^1_3 well-ordering of the reals and cardinals which are not so compatible?

(2) Suppose that κ is κ^+-supercompact. Is there a transitive class-model of ZFC, containing all the ordinals, in which κ remains κ^+-supercompact and in which *GCH* holds?

We remark that Woodin has constructed models where the first measurable cardinal κ is κ^+-supercompact. However, in a model of ZFC + *GCH* the first κ which is κ^+-supercompact has κ measurable cardinals below it. Thus an affirmative answer to (2) would seem to require some sort of inner model construction for supercompact cardinals of a sort not ruled out by the "anti-inner-model" theorem of Woodin just cited.

To sum up: almost all of the natural questions raised by Gödel's work on L have by now been settled. However, the topic of L-like models for large cardinals is still rife with mystery, though some important progress has been made.

Robert M. Solovay

[zz]Personal communication to the author. Woodin's work relies in an essential way on earlier recent work of Foreman, Magidor and Shelah (*198?*).

The consistency of the axiom of choice and of the generalized continuum hypothesis (*1938*)

Theorem. *Let T be the system of axioms for set theory obtained from von Neumann's system*[1] *S^* by leaving out the axiom of choice* (i.e., replacing Axiom III 3* by Axiom III 3); *then, if T is consistent, it remains so if the following propositions 1–4 are adjoined simultaneously as new axioms*:

1. The axiom of choice (i.e., von Neumann's Axiom III 3*).

2. The generalized continuum hypothesis (i.e., the statement that $2^{\aleph_\alpha} = \aleph_{\alpha+1}$ holds for any ordinal α).

3. The existence of linear non-measurable sets such that both they and their complements are one-to-one projections of two-dimensional complements of analytic sets (and which therefore are B_2-sets in Lusin's terminology[2]).

4. The existence of linear complements of analytic sets, which are of the power of the continuum and contain no perfect subset.

A corresponding theorem holds if T denotes the system of *Principia mathematica*[3] or Fraenkel's system of axioms for set theory,[4] leaving out in both cases the axiom of choice but including the axiom of infinity.

The proof of the above theorems is constructive in the sense that, if a contradiction were obtained in the enlarged system, a contradiction in T could actually be exhibited.

The method of proof consists in constructing on the basis of the axioms[5] of T a model for which the propositions 1–4 are true. This model, roughly speaking, consists of all "mathematically constructible" sets, where the term "constructible" is to be understood in the semi-intuitionistic sense which excludes impredicative procedures. This means "constructible" sets are defined to be those sets which can be obtained by Russell's ramified hierarchy of types, if extended to include transfinite orders. The extension to transfinite orders has the consequence that the model satisfies the impredicative axioms of set theory, because an axiom of reducibility can

[1] Cf. *von Neumann 1929*.

[2] Cf. *Luzin 1930*, p. 270.

[3] Cf. *Tarski 1933*.

[4] Cf. *Fraenkel 1925*.

[5] This means that the model is constructed by essentially transfinite methods and hence gives only a relative proof of consistency, requiring the consistency of T as a hypothesis.

be proved for sufficiently high orders. Furthermore the proposition "Every set is constructible" (which I abbreviate by "A") can be proved to be consistent with the axioms of T, because A turns out to be true for the model consisting of the constructible sets. From A the propositions 1–4 can be deduced. In particular, proposition 2 follows from the fact that all constructible sets of integers are obtained already for orders $< \omega_1$, all constructible sets of sets of integers for orders $< \omega_2$ and so on.

| The proposition A added as a new axiom seems to give a natural completion of the axioms of set theory, in so far as it determines the vague notion of an arbitrary infinite set in a definite way. In this connection it is important that the consistency proof for A does not break down if stronger axioms of infinity (e.g., the existence of inaccessible numbers) are adjoined to T. Hence the consistency of A seems to be absolute in some sense, although it is not possible in the present state of affairs to give a precise meaning to this phrase. 557

The consistency of the generalized continuum hypothesis (*1939*)

We use the following definitions: 1. $M_0 = \Lambda$; 2. $M_{\alpha+1}$ is the set of those subsets of M_α which can be defined by propositional functions containing only the following concepts: $\sim$, $\vee$, the ϵ-relation, elements of M_α, and quantifiers for variables with range M_α; 3. $M_\beta = \Sigma_{\alpha \subset \beta} M_\alpha$ for limit numbers β. Then M_{ω_ω} or M_Ω (Ω being the first inaccessible number) is a model for the system of axioms of set theory (as formulated by A. Fraenkel, J. von Neumann, T. Skolem, P. Bernays) respectively without (or with) the axiom of substitution, the generalized continuum hypothesis ($2^{\aleph_\alpha} = \aleph_{\alpha+1}$) being true in both models. Since the construction of the models can be formalized in the respective systems of set theory themselves, it follows that $2^{\aleph_\alpha} = \aleph_{\alpha+1}$ is consistent with the axioms of set theory, if these axioms are consistent with themselves. The proof is based on the following lemma. Any subset of M_{ω_α} which is an element of some M_β is an element of $M_{\omega_{\alpha+1}}$. This lemma is proved by a generalization of Skolem's method for constructing enumerable models. Since the axiom of choice is not used in the construction of the models, but holds in the models, the consistency of the axiom of choice is obtained as an incidental result.

Consistency proof for the generalized continuum hypothesis[1] (*1939a*)

If M is an arbitrary domain of things in which a binary relation ϵ is defined, call "*propositional function over M*" any expression ϕ containing (besides brackets) only the following symbols: 1. Variables $x, y, \ldots$ whose range is M. 2. Symbols $a_1, \ldots, a_n$ denoting[2] individual elements of M (referred to in the sequel as "*the constants of ϕ*"). 3. ϵ. 4. $\sim$ (not), $\vee$ (or). 5. Quantifiers for the above variables $x, y, \ldots$.[2a] Denote by M' the set of all subsets of M defined by propositional functions $\phi(x)$ over M. Call a function f with s variables a "*function in M*" if for any elements 221 $x_1, \ldots, x_s$ of M | $f(x_1, \ldots, x_s)$ is defined and is an element of M. If $\phi(x)$ is a propositional function over M with the following normal form:

$$(x_1, \ldots, x_n)(\exists y_1, \ldots, y_m)(z_1, \ldots, z_k)(\exists u_1, \ldots, u_e) \ldots \\ L(x, x_1, \ldots, x_n, y_1, \ldots, y_m, z_1, \ldots, z_k, u_1, \ldots, u_e, \ldots)$$

(L containing no more quantifiers) and if $a \epsilon M$, then call "*Skolem functions for ϕ and a*" any functions $f_1, \ldots, f_m, g_1, \ldots, g_e, \ldots$ in M, with respectively $n, \ldots, n, n+k, \ldots, n+k, \ldots$ variables, such that for any elements $x_1, \ldots, x_n$, $z_1, \ldots, z_k, \ldots$ of M the following is true:

$$L(a, x_1, \ldots, x_n, f_1(x_1, \ldots, x_n), \ldots, f_m(x_1, \ldots, x_n), z_1, \ldots, z_k, \\ g_1(x_1, \ldots, x_n, z_1, \ldots, z_k), \ldots, g_e(x_1, \ldots, x_n, z_1, \ldots, z_k), \ldots).$$

The proposition $\phi(a)$ is then equivalent with the existence of Skolem functions for ϕ and a.

Now define: $M_0 = \{\Lambda\}$, $M_{\alpha+1} = M'_\alpha$, $M_\beta = \Sigma_{\alpha<\beta} M_\alpha$ for limit numbers β. Call a set x "*constructible*" if there exists an ordinal α such that $x \epsilon M_\alpha$ and "*constructible of order α*" if $x \epsilon M_{\alpha+1} - M_\alpha$. It follows immediately that $M_\alpha \subset M_\beta$ *and* $M_\alpha \epsilon M_\beta$ *for* $\alpha < \beta$ and that:

[1] This paper gives a sketch of the consistency proof for propositions 1, 2 of *Gödel 1938* if T is Zermelo's system of axioms for set theory (*1908*) with or without axiom of substitution and if Zermelo's notion of "definite Eigenschaft" is identified with "propositional function over the system of all sets". Cf. the first definition of this paper.

[2] It is assumed that for any element of M a symbol denoting it can be introduced.

[2a] Unless explicitly stated otherwise, "propositional function" always means "propositional function with one free variable".

Theorem 1. *$x \epsilon y$ implies that the order of x is smaller than the order of y for any constructible sets x, y.*

It is easy to define a well-ordering of all constructible sets and to associate with each constructible set (of an arbitrary order α) a uniquely determined propositional function $\phi_\alpha(x)$ over M_α as its "*definition*" and furthermore to associate with each pair ϕ_α, a (consisting of a propositional function ϕ_α over M_α and an element a of M_α for which $\phi_\alpha(a)$ is true) uniquely determined "*designated Skolem functions for ϕ_α, a*".[3]

Theorem 2. *Any constructible subset m of M_{ω_μ} has an order $< \omega_{\mu+1}$ (i.e., a constructible set, all of whose elements have orders $< \omega_\mu$ has an order $< \omega_{\mu+1}$).*

Proof: Define a set K of constructible sets, a set O of ordinals and a set F of Skolem functions by the following postulates I–VII:

I. $M_{\omega_\mu} \subseteq K$ and $m \epsilon K$.

II. If $x \epsilon K$, the order of x belongs to O.

III. If $x \epsilon K$, all constants occurring in the definition of x belong to K.

IV. If $\alpha \epsilon O$ and $\phi_\alpha(x)$ is a propositional function over M_α all of whose constants belong to K, then:

1. The subset of M_α defined by ϕ_α belongs to K.
2. For any $y \epsilon K \cdot M_\alpha$ the designated Skolem functions for ϕ_α and y or $\sim\phi_\alpha$ and y (according as $\phi_\alpha(y)$ or $\sim\phi_\alpha(y)$) belong to F.

V. If $f \epsilon F$, $x_1, \ldots, x_n \epsilon K$ and $(x_1, \ldots, x_n)$ belongs to the domain of definition of f, then $f(x_1, \ldots, x_n) \epsilon K$.

VI. If $x, y \epsilon K$ and $x - y \neq \Lambda$ the first[4] element of $x - y$ belongs to K.

VII. No proper subsets of K, O, F satisfy I–VI.

| Theorem 3. *If $x \neq y$ and $x, y \epsilon K \cdot M_{\alpha+1}$, then there exists a $z \epsilon K \cdot M_\alpha$ such that $z \epsilon x - y$ or $z \epsilon y - x$.*[5] 222

(This follows from VI and Theorem 1.)

Theorem 4.[6] $\overline{\overline{K + O + F}} = \aleph_\mu$

since $\overline{\overline{M}}_{\omega_\mu} = \aleph_\mu$ and $K + O + F$ is obtained from $M_{\omega_\mu} + \{m\}$ by forming the closure with respect to the operations expressed by II–VI.

Now denote by η the order type of O and by $\overline{\alpha}$ the ordinal corresponding to α in the similar mapping of O on the set of ordinals $< \eta$. Then we have:

Theorem 5. *There exists a one-to-one mapping x' of K on M_η such that $x \epsilon y \equiv x' \epsilon y'$ for $x, y \epsilon K$ and $x' = x$ for $x \epsilon M_{\omega_\mu}$.*

Proof: The mapping x' (which will carry over the elements of order α of K exactly into all constructible sets of order $\overline{\alpha}$ for any $\alpha \epsilon O$) is

[3] At first, with each ϕ_α an equivalent normal form of the above type has to be associated, which can easily be done.

[4] In the well-ordering of the constructible sets.

[5] Theorems 3, 4, 5, are lemmas for the proof of Theorem 2.

[6] $\overline{\overline{m}}$ means "power of m".

defined by transfinite induction on the order, i.e., we assume that for some $\alpha \epsilon O$ an isomorphic[7] mapping f of $K \cdot M_\alpha$ on $M_{\overline{\alpha}}$[8] has been defined and prove that it can be extended to an isomorphic mapping g of $K \cdot M_{\alpha+1}$ on $M_{\overline{\alpha}+1}$[9] in the following way: At first those propositional functions over M_α whose constants belong to K (hence to $K \cdot M_\alpha$) can be mapped in a one-to-one manner on all propositional functions over $M_{\overline{\alpha}}$ by associating with a propositional function ϕ_α over M_α having the constants $a_1, \ldots, a_n$ the propositional function $\phi_{\overline{\alpha}}$ over $M_{\overline{\alpha}}$ obtained from ϕ_α by replacing a_i by a_i^1 and the quantifiers with the range M_α by quantifiers with the range $M_{\overline{\alpha}}$. Then we have:

Theorem 6. *$\phi_\alpha(x) \equiv \phi_{\overline{\alpha}}(x^1)$ for any $x \epsilon K \cdot M_\alpha$.*

Proof: If $\phi_\alpha(x)$ is true, the designated Skolem functions for ϕ_α and x exist, belong to F (by IV, 2) and are functions in $K \cdot M_\alpha$ (by V). Hence they are carried over by the mapping f into functions in $M_{\overline{\alpha}}$ which are Skolem functions for $\phi_{\overline{\alpha}}, x^1$, because the mapping f is isomorphic with respect to ϵ. Hence $\phi_\alpha(x) \supset \phi_{\overline{\alpha}}(x^1)$.

$\sim\phi_\alpha(x) \supset \sim\phi_{\overline{\alpha}}(x^1)$ is proved in the same way.

Now any ϕ_α over M_α whose constants belong to K defines an element of $K \cdot M_{\alpha+1}$ by IV, 1, and any element b of $K \cdot M_{\alpha+1}$ can be defined by such a ϕ_α (if $b \epsilon M_{\alpha+1} - M_\alpha$, this follows by III; if $b \epsilon M_\alpha$, then "$x \epsilon b$" is such a ϕ_α). Hence the above mapping of the ϕ_α on the $\phi_{\overline{\alpha}}$ gives a mapping g of all elements of $K \cdot M_{\alpha+1}$ on all elements of $M_{\overline{\alpha}+1}$ with the following properties:

A. *g is single-valued*, because, if ϕ_α, ψ_α define the same set, we have $\phi_\alpha(x) \equiv \psi_\alpha(x)$ for $x \epsilon M_\alpha \cdot K$, hence $\phi_{\overline{\alpha}}(x^1) \equiv \psi_{\overline{\alpha}}(x^1)$ by Theorem 6, i.e., $\phi_{\overline{\alpha}}$ and $\psi_{\overline{\alpha}}$ also define the same set.

B. $x \epsilon y \equiv x^1 \epsilon g(y)$ for $x \epsilon K \cdot M_\alpha$, $y \epsilon K \cdot M_{\alpha+1}$ (by Theorem 6).

C. *g is one-to-one*, because if $x, y \epsilon K \cdot M_{\alpha+1}, x \neq y$, then by Theorem 3 there is a $z \epsilon (x-y)+(y-x)$, $z \epsilon K \cdot M_\alpha$, hence $z^1 \epsilon [g(x)-g(y)]+[g(y)-g(x)]$ by B. Hence $g(x) \neq g(y)$.

D. *g is an extension of the mapping f, i.e., $g(x) = x^1$ for $x \epsilon K \cdot M_\alpha$.*

223 | Proof: For any $b \epsilon K \cdot M_\alpha$ a corresponding ϕ_α which defines it is $x \epsilon b$, hence $\phi_{\overline{\alpha}}$ is $x \epsilon b^1$, hence $g(b) = b^1$.

E. *g maps $K \cdot M_\alpha$ exactly on $M_{\overline{\alpha}}$ (by* D*),*[10] *and therefore $K(M_{\alpha+1} - M_\alpha)$ on $M_{\overline{\alpha}+1} - M_{\overline{\alpha}}$ by* C.

F. *g is isomorphic for ϵ, i.e., $g(x) \epsilon g(y) \equiv x \epsilon y$ for any $x, y \epsilon K \cdot M_{\alpha+1}$.*

[7]I.e., $x \epsilon y \equiv f(x) \epsilon f(y)$. In the following proof $f(x)$ is abbreviated by x^1.

[8]I.e., of the elements of order $< \alpha$ of K on the elements of order $< \overline{\alpha}$ of M_η.

[9]I.e., of the elements of order $\leq \alpha$ of K on the elements of order $\leq \overline{\alpha}$ of M_η.

[10]Because f maps $K \cdot M_\alpha$ on $M_{\overline{\alpha}}$ by inductive assumption.

Proof: If $x \in K \cdot M_\alpha$, this follows from B and D; if $x \in K \cdot (M_{\alpha+1} - M_\alpha)$, then $g(x) \in M_{\overline{\alpha}+1} - M_{\overline{\alpha}}$ by E, hence both sides of the equivalence are false by Theorem 1.

By D and F, g is the desired extension of f and hence the existence of an isomorphic mapping x' of K on M_η follows by complete induction. Furthermore, since all ordinals $< \omega_\mu$ belong to O (by I, II) we have $\overline{\beta} = \beta$ for $\beta < \omega_\mu$, from which it follows easily that $x = x'$ for $x \in M_{\omega_\mu}$. This finishes the proof of Theorem 5.

Now, in order to prove Theorem 2, consider the set m' corresponding to m in the isomorphic mapping of K on M_η. Its order is $< \eta < \omega_{\mu+1}$, because $m' \in M_\eta$ and $\overline{\overline{\eta}} = \overline{\overline{O}} \leq \aleph_\mu$ by Theorem 4. Since $x \in m \equiv x' \in m'$ for $x \in K$, we have $x \in m \equiv x \in m'$ for $x \in M_{\omega_\mu}$ by Theorem 5. Since furthermore $m \subseteq M_{\omega_\mu}$, it follows that $m = m' \cdot M_{\omega_\mu}$, i.e., m is an intersection of two sets of order $< \omega_{\mu+1}$, which implies trivially that it has an order $< \omega_{\mu+1}$.

Theorem 7. *M_{ω_ω} considered as a model for set theory satisfies all axioms of Zermelo*[11] *except perhaps the axiom of choice, and M_Ω (Ω being the first inaccessible number) satisfies in addition the axiom of substitution, if in both cases "definite Eigenschaft", respectively "definite Relation", is identified with "propositional function over the class of all sets" (with one, respectively two, free variables).*

Sketch of proof for M_{ω_ω}: Axioms I, II are trivial, Axiom VII is satisfied by $Z = M_\omega$, Axioms III–V have the form $(\exists x)(u)[u \in x \equiv \phi(u)]$, where the ϕ are certain propositional functions over M_{ω_ω}. Hence, by definition of $M_{\alpha+1}$, there exist sets x in $M_{\omega_\omega+1}$ satisfying the axioms. But from Theorem 1 and Theorem 2 it follows easily that the order of x is smaller than ω_ω for the particular ϕ under consideration, so that there exist sets x in the model satisfying the axioms.

For M_Ω Axioms I–V and VII are proved in exactly the same way, and the axiom of substitution is proved by the same method as Axioms III–V. Now denote by "A" the proposition "There exist no non-constructible sets",[12] by "R" the axiom of choice and by "C" the proposition "$2^{\aleph_\alpha} = \aleph_{\alpha+1}$ for any ordinal α". Then we have:

Theorem 8. *$A \supset R$ and $A \supset C$.*

$A \supset R$ follows because for the constructible sets a well-ordering can be defined, and $A \supset C$ holds by Theorem 2, because $\overline{\overline{M_{\omega_\alpha}}} = \aleph_\alpha$.

Now the notion of "constructible set" can be defined and its theory developed in the formal systems of set theory themselves. In particular

[11] Cf. *Zermelo 1908.*

[12] In order to give A an intuitive meaning, one has to understand by "sets" all objects obtained by building up the simplified hierarchy of types on an empty set of individuals (including types of arbitrary transfinite orders).

Theorem 2 and, therefore, Theorem 8 can be proved from the axioms of set theory. Denote the notion of "constructible set" relativized for a model M of set theory (i.e., defined in terms of the ϵ-relation of the model) by *constructible*$_M$; then we have:

224 | Theorem 9. *Any element of* M_{ω_ω} *(respectively,* M_Ω*) is constructible*$_{M_{\omega_\omega}}$ *(respectively, constructible*$_{M_\Omega}$*)*; in other words: *A is true in the models* M_{ω_ω} *and* M_Ω.

The proof is based on the following two facts: 1. The operation M' (defined on p. 220) is absolute in the sense that the operation relativized for the model M_{ω_ω}, applied to an $x \in M_{\omega_\omega}$, gives the same result as the original operation (similarly for M_Ω). 2. The set N_α which has as elements all the M_β (for $\beta < \alpha$) is constructible$_{M_{\omega_\omega}}$ for $\alpha < \omega_\omega$ and constructible$_{M_\Omega}$ for $\alpha < \Omega$, as is easily seen by an induction on α. From Theorem 9 and the provability (from the axioms of set theory) of Theorem 8 there follows:

Theorem 10. *R and C are true for the models* M_{ω_ω} *and* M_Ω.

The construction of M_{ω_ω} and M_Ω and the proof for Theorem 7 and Theorem 9 (therefore also for Theorem 10) can (after certain slight modifications)[13] be accomplished in the respective formal systems of set theory (without the axiom of choice), so that a contradiction derived from C, R, A and the other axioms would lead to a contradiction in set theory without C, R, A.

[13]In particular for the system without the axiom of substitution we have to consider instead of M_{ω_ω} an isomorphic image of it (with some other relation R instead of the ϵ-relation), because M_{ω_ω} contains sets of infinite type, whose existence cannot be proved without the axiom of substitution. The same device is needed for proving the consistency of propositions 3, 4 of the paper quoted in footnote 1.

The consistency of the axiom of choice and of the generalized continuum hypothesis with the axioms of set theory

(*1940*)

Introduction

In these lectures it will be proved that the axiom of choice and Cantor's generalized continuum hypothesis (i.e., the proposition that $2^{\aleph_\alpha} = \aleph_{\alpha+1}$ for any α) are consistent with the other axioms of set theory if these axioms are consistent. The system Σ of axioms for set theory which we adopt includes the axiom of substitution (cf. *Fraenkel 1927*, page 115) and the axiom of "Fundierung" (cf. *Zermelo 1930*, page 31) but of course does not include the axiom of choice. It is essentially due to P. Bernays (cf. *Bernays 1937*) and is equivalent with von Neumann's system S* + VI (cf. *1929*), if the axiom of choice is left out, or, to be more exact, if Axiom III3* is replaced by Axiom III3. What we shall prove is that, if a contradiction from the axiom of choice and the generalized continuum hypothesis were derived in Σ, it could be transformed into a contradiction obtained from the axioms of Σ alone. This result is obtained by constructing within Σ (i.e., using only the primitive terms and axioms of Σ) a model Δ for set theory with the following properties:

1) the propositions which say that the axioms of Σ hold for Δ are theorems demonstrable in Σ,

2) the propositions which say that the axiom of choice and the generalized continuum hypothesis hold in Δ are likewise demonstrable in Σ.

In fact there is a much stronger proposition[1] which can be proved to hold

[1][*Note added in 1951*: In particular, this stronger proposition implies that there exists a *projective* well-ordering of the real numbers (to be more exact, one whose corresponding set of pairs is a PCA-set in the plane). This follows by considering those pairs of relations s, e between integers which, for some $\gamma < \omega_1$ are isomorphic with the pair of relations $<$, $\hat{\alpha}\hat{\beta}$ ($F`\alpha \epsilon F`\beta$) confined to γ. The class M of these pairs s, e can also be defined directly (i.e., without reference to the previously defined F) by requiring that (1) s is to be a well-ordering relation for the integers, and (2) e, with respect to the well-ordering s, satisfies certain recursive postulates, which are the exact analogues of the postulates by which F is defined (cf. Dfn 9.3). The definition of M, in this form, contains quantifiers only for integers and sets of integers (i.e., real numbers) which ensures the projective character of the object defined and makes it possible to determine its projective order by counting the "changes of sign" of the quantifiers for real numbers occurring. In terms of M a projective well-ordering of the real numbers (of the order mentioned) can then be defined. As to consequences of this state of affairs, cf. *Kuratowski 1948*.]

in Δ and which has other interesting consequences besides the axiom of choice and the generalized continuum hypothesis (cf. page 47).

In order to define Δ and to prove the above properties of it from the axioms of Σ, it is necessary first to develop abstract set theory to a certain extent from the axioms of Σ. This is done in Chapters II–IV. Although the definitions and theorems are mostly stated in logistic symbols, the theory developed is not to be considered as a formal system but as an axiomatic theory in which the meaning and the properties of the logical symbols are presupposed to be known. However, to everyone familiar with mathematical logic it will be clear that the proofs could be formalized, using only the rules of Hilbert's "engerer Funktionenkalkul". In several places (in particular for the "general existence theorem" on page 8 and the notions of "relativization" and of "absoluteness" on page 42) we are concerned with metamathematical considerations about the notions and propositions of the system Σ. However, the only purpose of these general metamathematical considerations is to show how the proofs for theorems
2 of a certain kind can be accomplished by | a general method. And, since applications to only a finite number of instances are necessary for proving the properties 1) and 2) of the model Δ, the general metamathematical considerations could be left out entirely, if one took the trouble to carry out the proofs separately for any instance.[2]

In the first introductory part about set theory in general (i.e., in Chapters II–IV) not all proofs are carried out in detail, since many of them can be literally transferred from non-axiomatic set theory and, moreover, an axiomatic treatment on a very similar basis has been given by J. von Neumann (*1928a*).

For the logical notions we use the following symbols: (X), $(\exists X)$, $\sim$, $.$, $\vee$, $\supset$, $\equiv$, $=$, $(E!X)$, which mean respectively: for all X, there is an X, not, and, or, implies, equivalence, identity, there is exactly one X. $X = Y$ means that X and Y are the same object. "For all X" is also expressed by free variables in definitions and theorems.

The system Σ has in addition to the ϵ-relation two primitive notions, namely "class" and "set". Classes are what appear in Zermelo's formulation (*1908*, page 263) as "definite Eigenschaften". However, in the system Σ (unlike Zermelo's) it is stated explicitly by a special group of axioms (group B on page 5) how *definite Eigenschaften* are to be constructed. Classes represent at the same time relations between sets, namely a class A represents the relation which subsists between x and y if the ordered pair $\langle x, y\rangle$ (defined in 1.12) is an element of A. The same ϵ-relation is used

[2]In particular also the complete inductions used in the proofs of Theorems 1.16, M1, M2 are needed only up to a certain definite integer, say 20.

between sets and sets, and sets and classes. The axiom of extensionality (Fraenkel's *Bestimmtheitsaxiom*) is assumed for both sets and classes, and a class for which there exists a set having the same elements is identified with this set, so that every set is a class.[3] On the other hand a class B which is not a set (e.g., the universal class) can never occur as an element owing to Axiom A2, i.e., $B \in X$ is then always false (but meaningful).

| 3

Chapter I
The axioms of abstract set theory

Our primitive notions are: *class*, denoted by $\mathfrak{Cls}$; *set*, denoted by $\mathfrak{M}$; and the diadic *relation* ϵ between class and class, class and set, set and class, or set and set. The primitive notions appear in context as follows:

$$\begin{array}{llll} \mathfrak{Cls}(A), & A \text{ is a class}, & & \\ \mathfrak{M}(A), & A \text{ is a set}, & & \\ X \in Y, & X \in y, & x \in Y, & x \in y, \end{array}$$

where the convention is made that $X, Y, Z, \ldots$ are variables whose range consists of all the classes, and that $x, y, z, \ldots$ are variables whose range is all sets.

The axioms fall into four groups, A, B, C, D.

Group A.

1. $\mathfrak{Cls}(x)$
2. $X \in Y . \supset . \mathfrak{M}(X)$
3. $(u)[u \in X . \equiv . u \in Y] . \supset . X = Y$
4. $(x)(y)(\exists z)(u)[u \in z . \equiv : u = x . \vee . u = y]$

Axiom 1 in the group above states that every set is a class. A class which is not a set is called a *proper class*, i.e.,

1. Dfn $\mathfrak{Pr}(X) = {\sim}\mathfrak{M}(X)$.

Axiom 2 says that every class which is a member of some class is a set. Axiom 3 is the principle of extensionality, that is, two classes are the same

[3]Similarly, *von Neumann 1928a*.

Note. Dots are also used to replace brackets in the well-known manner.

if their elements are the same. Axiom 4 provides for the existence of the set whose members are just x and y, for any sets x and y. Moreover, this set is defined uniquely for given x and y, by Axiom 3. The element z defined by 4 is called the *non-ordered pair* of x and y, denoted by $\{x, y\}$, i.e.,

1.1 Dfn $u \in \{x, y\} \equiv (u = x \vee u = y)$.
1.11 Dfn $\{x\} = \{x, x\}$.

$\{x\}$ is the set whose sole member is x.

4 | 1.12 Dfn $\langle x, y\rangle = \{\{x\}, \{x, y\}\}$.

$\langle x, y\rangle$ is called the *ordered pair* of x and y. We have the following theorem:

1.13 $\langle x, y\rangle = \langle u, v\rangle \,.\supset : x = u \,.\, y = v$,

that is, two ordered pairs are equal if and only if the corresponding elements of each are equal. In this sense, $\langle x, y\rangle$ is an ordered pair. The proof of this theorem is not difficult (cf. *Bernays 1937*, page 69).

The *ordered triple* may now be defined in terms of the ordered pair.

1.14 Dfn $\langle x, y, z\rangle = \langle x, \langle y, z\rangle\rangle$.

The corresponding theorem holds for the ordered triple. The *n-tuple* can be defined by induction as follows:

1.15 Dfn $\langle x_1, x_2, \ldots, x_n\rangle = \langle x_1, \langle x_2, \ldots, x_n\rangle\rangle$.

This gives the theorem

1.16 $\langle x_1, \ldots, x_n, \langle x_{n+1}, \ldots, x_{n+p}\rangle\rangle = \langle x_1, \ldots, x_n, x_{n+1}, \ldots, x_{n+p}\rangle$,

which is proved by induction on n.
In order that $\langle\ \rangle$ be defined for any number of arguments it is convenient to put

1.17 Dfn $\langle x\rangle = x$,

which entails the equation 1.16 also for the case $p = 1$.
We also define *inclusion* $\subseteq$ and *proper inclusion* $\subset$.

1.2 Dfn $X \subseteq Y \,.\equiv.\, (u)[u \in X \,.\supset.\, u \in Y]$;
$X \subset Y \,.\equiv:\, X \subseteq Y \,.\, X \neq Y$.

A class is said to be *empty* if it has no members; "X is empty" is denoted by "$\mathfrak{Em}(X)$", i.e.,

1.22 Dfn $\mathfrak{Em}(X) \equiv (u){\sim}u \in X.$

If X and Y have no members in common, we write "$\mathfrak{Ex}(X, Y)$", that is, "X and Y are *mutually exclusive*", i.e.,

1.23 Dfn $\mathfrak{Ex}(X, Y) \equiv (u){\sim}(u \in X \,.\, u \in Y).$

X is said to be *one-many* (*single-valued*), denoted by "$\mathfrak{Un}(X)$", if for any u there exists at most one v such that $\langle v, u\rangle \in X$, that is:

| 1.3 Dfn $\mathfrak{Un}(X) \equiv (u, v, w)[\langle v, u\rangle \in X \,.\, \langle w, u\rangle \in X :\supset. v = w].$ 5

The axioms of the second group are concerned with the existence of classes:

Group B.

1. $(\exists A)(x, y)[\langle x, y\rangle \in A .\equiv. x \in y]$
2. $(A)(B)(\exists C)(u)[u \in C .\equiv: u \in A \,.\, u \in B]$
3. $(A)(\exists B)(u)[u \in B .\equiv. {\sim}(u \in A)]$
4. $(A)(\exists B)(x)[x \in B .\equiv. (\exists y)(\langle y, x\rangle \in A)]$
5. $(A)(\exists B)(x, y)[\langle y, x\rangle \in B .\equiv. x \in A]$
6. $(A)(\exists B)(x, y)[\langle x, y\rangle \in B .\equiv. \langle y, x\rangle \in A]$
7. $(A)(\exists B)(x, y, z)[\langle x, y, z\rangle \in B .\equiv. \langle y, z, x\rangle \in A]$
8. $(A)(\exists B)(x, y, z)[\langle x, y, z\rangle \in B .\equiv. \langle x, z, y\rangle \in A]$

Axiom B1 is called axiom of the ϵ-relation, B2 axiom of intersection, B3 axiom of the complement, B4 axiom of the domain, B5 axiom of the direct product (because it provides essentially for the existence of $V \times A$, V being the universal class), B6–8 axioms of inversion.[4] Note that the class A in Axiom B1 and the class B in Axioms B5–8 are not uniquely determined, since nothing is said about those sets which are not pairs (triples), whether or not they belong to A (B). On the other hand in Axioms B2–4 the classes C and B are uniquely determined (owing to Axiom A3). These uniquely determined classes in B2–4 are denoted respectively by $A \cdot B$, $-A$, $\mathfrak{D}(A)$ and called *intersection* of A, B, *complement* of A, *domain* of A, respectively. Thus $A \cdot B$, $-A$, $\mathfrak{D}(A)$ are defined by the following properties.

[4]Note that Axioms B7 and B8 have as consequences similar theorems for any permutation of a triple.

1.4 Dfn $x \epsilon A \cdot B \equiv x \epsilon A . x \epsilon B$
1.41 Dfn $x \epsilon -A \equiv \sim x \epsilon A$
1.5 Dfn $x \epsilon \mathfrak{D}(A) \equiv (\exists y)\langle y, x\rangle \epsilon A$

The third group of axioms is concerned with the existence of sets.

Group C.
1. $(\exists a)\{\sim \mathfrak{Em}(a) . (x)[x \epsilon a . \supset . (\exists y)[y \epsilon a . x \subset y]]\}$
2. $(x)(\exists y)(u, v)[u \epsilon v . v \epsilon x :\supset . u \epsilon y]$
3. $(x)(\exists y)(u)[u \subseteq x . \supset . u \epsilon y]$
4. $(x, A)\{\mathfrak{Un}(A) . \supset . (\exists y)(u)[u \epsilon y . \equiv . (\exists v)[v \epsilon x . \langle u, v\rangle \epsilon A]]\}$

Axiom 1 is the so-called *axiom of infinity*. There is a non-void set a such that, given any element x of a, there is another element y of a, of which x is a proper subset. According to Axiom 2, for any set x there is a set y including the sum of all elements of x. Axiom 3 provides for the 6 | existence of a set including the set of all subsets of x. Axiom 4 is the *axiom of substitution*;[5] for any set x and any single-valued A, there is a set y whose elements are just those sets which bear the relation defined by A to members of x. (Instead of C4, Zermelo used the *Aussonderungsaxiom*:

$$(x, A)(\exists y)(u)[u \epsilon y . \equiv : u \epsilon x . u \epsilon A],$$

that is, there is a set whose members are just those elements of x which have the property A.)

The following axiom (proved consistent by J. von Neumann (*1929*)) is not indispensable, but it simplifies considerably the later work:

Axiom D. $\sim \mathfrak{Em}(A) . \supset . (\exists u)[u \epsilon A . \mathfrak{Ex}(u, A)]$,

that is, any non-void class A has some element with no members in common with A.[6] It is a consequence of D that

1.6 $\sim x \epsilon x$,

[5][*Note added in 1951*: The term now in use for Axiom C4 is "axiom of replacement".]

[6]This axiom is equivalent to the non-existence of infinite descending sequences of sets[7] (i.e., such that $x_{i+1} \epsilon x_i$) where however the term "sequence" refers only to sequences representable by sets of the system under consideration. That is (using the definitions 4.65, 7.4, 8.41 below) Axiom D is (owing to the axioms of the groups A, B, C, E) equivalent to the proposition $\sim(\exists y)(i)[y`(i+1) \epsilon y`i]$.

[7][*Note added in 1951*: In this form Axiom D, under the name of "Fundierungsaxiom", was first formulated by E. Zermelo (*1930*).]

for, if there were such an x, x would be a common element of x and $\{x\}$, but, by D, taking $\{x\}$ for A, x can have no element in common with $\{x\}$. Likewise

1.7 $\sim[x \in y \,.\, y \in x]$.

This follows by considering $\{x, y\}$ in an analogous way.

The following axiom is the axiom of choice.[8]

Axiom E. $(\exists A)\{\mathfrak{Un}(A) \,.\, (x)[\sim\mathfrak{Em}(x) \,.\supset.\, (\exists y)[y \in x \,.\, \langle y, x\rangle \in A]]\}$

This is a very strong form of the *axiom of choice*, since it provides for the simultaneous choice, by a single relation, of an element from each set of the universe under consideration. From this form of the axiom, one can prove that the whole universe of sets can be well-ordered. This stronger form of the axiom, if consistent with the other axioms, implies, of course, that a weaker form is also consistent.

| The system of axioms of groups A, B, C, D is called Σ.[9] If a theorem is stated without further specification it means that it follows from Σ. If Axiom E is needed for a theorem or a definition, its number is marked by $*$. 7

| 8

Chapter II
Existence of classes and sets

We now define the metamathematical notion of a *primitive propositional function* (abbreviated ppf). A ppf will be a meaningful formula containing only variables, symbols for special classes $A_1, \ldots, A_k$, ϵ, and logical operators, and such that all bound variables are *set* variables. For example,

$$(u)(u \in X \,.\supset.\, u \in A) \quad \text{and} \quad (u)[u \in x \,.\equiv.\, (v)[v \in u \,.\supset.\, v \in y]]$$

are ppf. A formula is non-primitive if (X) or $(\exists X)$ occurs.

[8][*Note added in 1951*: Using Dfn 4.65, the axiom of choice can be expressed in the following form, equivalent with Axiom E: There exist classes A for which $x \in y \supset A`y \in y$.]

[9]The most important differences between Σ and the system of P. Bernays (*1937*) are:

1. Bernays does not identify sets and classes having the same extension.
2. Bernays assumes a further axiom requiring the existence of the class of all $\{x\}$, which allows B7 and B8 to be replaced by one axiom.

Axiom D is essentially due to von Neumann (*1929*, page 231, Axiom VI 4), whose formulation however is more complicated, because his system has other primitive terms. The concise formulation used in the text is due to P. Bernays.

More precisely, ppf can be defined recursively as follows: Let $\Pi, \Gamma, \ldots,$ denote variables or special classes, then:

(1) $\Pi \in \Gamma$ is a ppf.
(2) If ϕ and ψ are ppf, then so are $\sim\phi$ and $\phi \,.\, \psi$.
(3) If ϕ is a ppf, then $(\exists x)\phi$ is a ppf, and any result of replacing x by another set variable is a ppf.
(4) Only formulas obtained by 1, 2, 3 are ppf.

Logical operators different from $\sim$, ., $\exists$, need not be mentioned since they can be defined in terms of these three.

The following metatheorem says that the extension of any ppf is represented by a class:

M1. *General existence theorem*: If $\phi(x_1, \ldots, x_n)$ is a ppf containing no free variables other than $x_1, \ldots, x_n$ (not necessarily all these), then there exists a class A such that, for any *sets* $x_1, \ldots, x_n$,

$$\langle x_1, \ldots, x_n \rangle \in A \,.\!\equiv\!.\, \phi(x_1, \ldots, x_n).$$

For the proof of this theorem, several preliminary results are needed.

By means of the axioms on intersection and complement, it is possible to prove the existence of a *universal class* V and a *null class* 0. Because of the axiom of extensionality, 0 and V are uniquely determined by the properties

2.1 Dfn $(x)\sim(x \in 0)$,
2.2 Dfn $(x)x \in V$.

As a consequence of Axiom B5, the axiom of the direct product, and B6, the axiom of the inverse relation, we have

9 | 2.3 $(A)(\exists B)(x, y)[\langle x, y \rangle \in B \,.\!\equiv\!.\, x \in A]$.

The following three theorems are also consequeces of B5, B7, and B8.

2.31 $(A)(\exists B)(x, y, z)[\langle z, x, y \rangle \in B \,.\!\equiv\!.\, \langle x, y \rangle \in A]$
2.32 $(A)(\exists B)(x, y, z)[\langle x, z, y \rangle \in B \,.\!\equiv\!.\, \langle x, y \rangle \in A]$
2.33 $(A)(\exists B)(x, y, z)[\langle x, y, z \rangle \in B \,.\!\equiv\!.\, \langle x, y \rangle \in A]$

For example, the first of these theorems is proved by substituting an ordered pair for the second member in the ordered pair appearing in B5, rewriting the variables properly. The other two are obtained by applying to 2.31 the axioms of inversion (B7 and B8).

Substituting $\langle x_1, x_2, \ldots, x_n \rangle$ for x in B5 in a similar way, we get

2.4 $(A)(\exists B)(y, x_1, \ldots, x_n)[\langle y, x_1, x_2, \ldots, x_n\rangle \in B . \equiv . \langle x_1, \ldots, x_n\rangle \in A].$

From this, by iteration,

2.41 $(A)(\exists B)(y_1, \ldots, y_k, x_1, \ldots, x_n)[\langle y_1, \ldots, y_k, x_1, \ldots, x_n\rangle \in B . \equiv . \langle x_1, \ldots, x_n\rangle \in A].$

Similarly,

2.5 $(A)(\exists B)(y_1, \ldots, y_k, x_1, \ldots, x_n)[\langle x_1, y_1, \ldots, y_k, x_2, \ldots, x_n\rangle \in B . \equiv . \langle x_1, \ldots, x_n\rangle \in A].$

This may be obtained by iteration from the case $k = 1$, and this case in turn is a special case of 2.32 obtained by substituting $\langle x_2, \ldots, x_n\rangle$ for y and applying Theorem 1.16.

The following theorems are derived in an analogous fashion, by substituting $\langle y_1, \ldots, y_k\rangle$ for z and y respectively in 2.33, 2.3, and applying 1.16,

2.6 $(A)(\exists B)(x_1, x_2, y_1, \ldots, y_k)[\langle x_1, x_2, y_1, \ldots, y_k\rangle \in B . \equiv . \langle x_1, x_2\rangle \in A],$

2.7 $(A)(\exists B)(x, y_1, \ldots, y_k)[\langle x, y_1, \ldots, y_k\rangle \in B . \equiv . x \in A].$

The next (and for the present, the last) theorem is a generalization of Axiom B4, the axiom of the domain, and is obtained by substituting, in B4, $\langle x_2, \ldots, x_n\rangle$ for x.

2.8 $(A)(\exists B)(x_2, \ldots, x_n)[\langle x_2, \ldots, x_n\rangle \in B . \equiv . (\exists x_1)[\langle x_1, \ldots, x_n\rangle \in A]]$

In particular $B = \mathfrak{D}(A)$ satisfies this equivalence.

In the proof of the general existence theorem, it can be assumed that none of the special classes A_i appears as the first argument of the ϵ-relation, because $A_i \in \Gamma$ can be replaced by $\mid (\exists x)(x = A_i . x \in \Gamma)$ (by Axiom A2) and $x = A_i$ can be replaced by $(u)[u \in x \equiv u \in A_i]$ (by Axiom A3). 10

The proof of M1 is an inductive one, the induction taking place on the number of logical operators in ϕ.

Case 1. ϕ has no logical operators.
In this case ϕ has one of two possible forms, $x_r \in x_s$ and $x_r \in A_k$, where $1 \leq r, s \leq n$. If ϕ is of the form $x_r \in x_s$, we must show that there exists a class A such that $\langle x_1, \ldots, x_n\rangle \in A . \equiv . x_r \in x_s$. If $r = s$, take as A the null class 0, since, by 1.6, $\sim(x_r \in x_r)$. If $r \neq s$, ϕ must be either of the form $x_p \in x_q$ or $x_q \in x_p$, where $p < q$. For $x_p \in x_q$, Axiom B1 provides for

the existence of an F such that $\langle x_p, x_q \rangle \,\epsilon\, F \,.\!\equiv\!.\, x_p \,\epsilon\, x_q$. For $x_q \,\epsilon\, x_p$, B1 followed by B6 provides for the existence of an F such that

$$\langle x_p, x_q \rangle \,\epsilon\, F \,.\!\equiv\!.\, x_q \,\epsilon\, x_p.$$

Therefore, in either case there is an F such that

$$\langle x_p, x_q \rangle \,\epsilon\, F \,.\!\equiv\!.\, \phi(x_1, \ldots, x_n).$$

Now, by 2.6, there is an F_1 with the property:

$$\langle x_p, x_q, x_{q+1}, \ldots, x_n \rangle \,\epsilon\, F_1 \,.\!\equiv\!.\, \langle x_p, x_q \rangle \,\epsilon\, F.$$

Then by 2.5 there exists F_2 such that

$$\langle x_p, \ldots, x_n \rangle \,\epsilon\, F_2 \,.\!\equiv\!.\, \langle x_p, x_q, x_{q+1}, \ldots, x_n \rangle \,\epsilon\, F_1,$$

and finally, by 2.41 there exists a class A such that

$$\langle x_1, \ldots, x_n \rangle \,\epsilon\, A \,.\!\equiv\!.\, \langle x_p, \ldots x_n \rangle \,\epsilon\, F_2.$$

Combining these equivalences the result is:

$$\langle x_1, \ldots, x_n \rangle \,\epsilon\, A \,.\!\equiv\!.\, \phi(x_1, \ldots, x_n).$$

Now suppose ϕ is of the form $x_r \,\epsilon\, A_k$. By 2.3, there is an F such that $\langle x_r, x_{r+1} \rangle \,\epsilon\, F \,.\equiv.\, \phi(x_1, \ldots, x_n)$. (If $r = n$, use Axiom B5 to get $\langle x_{r-1}, x_r \rangle \,\epsilon\, F \,.\equiv.\, \phi(x_1, \ldots, x_n)$.) Now, as above, by means of Theorems 2.6 and 2.41, combining the resulting equivalences establishes the existence of A.

Case 2. ϕ has m logical operators ($m > 0$).
Then ϕ has one of the following three forms:

$$(a)\ {\sim}\psi; \quad (b)\ \psi \,.\, \chi; \quad (c)\ (\exists x)\theta.$$

The hypothesis of the induction is that, for all ppfs $\psi(x_1, \ldots, x_n)$ with $m_1 < m$ logical operators and such that no A_i appears in the context $A_i \,\epsilon\, \Gamma$, there exists an A with the properties required by the theorem. ϕ, χ, and θ are ppfs with fewer than m logical operators. ψ and χ have no other free variables than at most $x_1, \ldots, x_n$, whereas θ has no other free variables than at most $x, x_1, \ldots, x_n$, and A_i cannot appear in the context $A_i \,\epsilon\, \Gamma$ in ψ, χ or θ, because it does not appear in ϕ in this context.

Therefore, by the hypothesis of the induction, there exist classes B, C, D such that

$$\langle x_1, \ldots, x_n \rangle \in B . \equiv . \psi(x_1, \ldots, x_n),$$
$$\langle x_1, \ldots, x_n \rangle \in C . \equiv . \chi(x_1, \ldots, x_n),$$
$$\langle x, x_1, \ldots, x_n \rangle \in D . \equiv . \theta(x, x_1, \ldots, x_n).$$

For (a) take A as $-B$, since, by Axiom B3,

$$\langle x_1, \ldots, x_n \rangle \in -B . \equiv . \sim(\langle x_1, \ldots, x_n \rangle \in B),$$

| so that $\langle x_1, \ldots, x_n \rangle \in -B . \equiv . \sim\psi(x_1, \ldots, x_n)$, that is 11

$$\langle x_1, \ldots, x_n \rangle \in -B . \equiv . \phi(x_1, \ldots, x_n).$$

For (b) take A as $B \cdot C$, since by Axiom B2,

$$\langle x_1, \ldots, x_n \rangle \in B \cdot C . \equiv : \langle x_1, \ldots, x_n \rangle \in B . \langle x_1, \ldots, x_n \rangle \in C,$$

that is,

$$\langle x_1, \ldots, x_n \rangle \in B \cdot C . \equiv : \psi(x_1, \ldots, x_n) . \chi(x_1, \ldots, x_n);$$

therefore

$$\langle x_1, \ldots, x_n \rangle \in B \cdot C . \equiv . \phi(x_1, \ldots, x_n).$$

For (c), take A as the domain $\mathfrak{D}(D)$, since by Theorem 2.8

$$\langle x_1, \ldots, x_n \rangle \in \mathfrak{D}(D) . \equiv . (\exists x)[\langle x, x_1, \ldots, x_n \rangle \in D];$$

therefore

$$\langle x_1, \ldots, x_n \rangle \in \mathfrak{D}(D) . \equiv . (\exists x)\theta(x, x_1, \ldots, x_n),$$

so that

$$\langle x_1, \ldots, x_n \rangle \in \mathfrak{D}(D) . \equiv . \phi(x_1, \ldots, x_n).$$

This completes the proof of the general existence theorem for primitive propositional functions.

The general existence theorem is a *metatheorem*, that is, a theorem about the system, not in the system, and merely indicates once and for all how the formal derivation would proceed in the system for any given ppf.

So far, the existence theorem is proved only for ppfs; but the use of symbols introduced by definition yields a wider class of propositional functions for which it would be desirable to have the existence theorem valid. With this in view, examine the defined symbols introduced thus far. They may be classified into four types, as follows:

1. *Particular classes*: $0, V, \ldots,$

2. *Notions*: $\mathfrak{M}(X)$, $\mathfrak{Pr}(X)$, $\mathfrak{Un}(X)$, $X \subseteq Y, \ldots,$
3. *Operations*: $-X$, $\mathfrak{D}(X)$, $X \cdot Y, \ldots,$
4. *Kinds of variables*: x, $X, \ldots$ (defined by notions).

Henceforth it is to be required that all operations and notions be meaningful, that is, defined, for all classes as arguments. This has been the case hitherto except for the pairs $\{x, y\}$ and $\langle x, y\rangle$, and the n-tuples, which were defined for sets only. The extension for classes as arguments can be accomplished simply by replacing the free set-variables by class-variables in the definitions, i.e.,

3.1 Dfn $(u)\{u \,\epsilon\, \{X, Y\} . \equiv : u = X . \vee . u = Y]$,
3.11 Dfn $\{X\} = \{X, X\}$,
3.12 Dfn $\langle X, Y\rangle = \{\{X\}, \{X, Y\}\}$, etc.

By these definitions, e.g., $\{X, Y\}$ is either $\{X, Y\}$ or $\{X\}$ or $\{Y\}$ or 0 according to whether both or one or none of X, Y are sets.[10] The same procedure of extension is to be applied in Definitions 4.211, 4.65, 6.31, 7.4, where the notions (or operations) under consideration are originally defined only if certain arguments are sets.[11]

12 | The following metamathematical ideas will be useful. A *term* is defined inductively so that (1) any variable is a term, and any symbol denoting a special class is a term; (2) if $\mathfrak{A}$ is an operation with n arguments and $\Gamma_1, \ldots, \Gamma_n$ are terms, then $\mathfrak{A}(\Gamma_1, \ldots, \Gamma_n)$ is a term; (3) there are no terms other than those obtainable from (1) and (2). If $\mathfrak{B}$ is a notion with n arguments and $\Gamma_1, \ldots, \Gamma_n$ are terms, then $\mathfrak{B}(\Gamma_1, \ldots, \Gamma_n)$ is said to be a *minimal*

[10][*Note added in 1951*: One may wish, for aesthetic reasons, that in analogy with Axiom A2 one should have $\langle X, Y\rangle \,\epsilon\, Z . \supset . \mathfrak{M}(X) . \mathfrak{M}(Y)$. This can easily be accomplished by replacing in Dfn 3.1 $u = X$ by $u = X \vee [\mathfrak{Pr}(X) . u \,\epsilon\, X]$, and likewise $u = Y$ by $u = Y \vee [\mathfrak{Pr}(Y) . u \,\epsilon\, Y]$. If this definition is adopted, $\mathfrak{M}(A\text{`}x)$ can be dropped in Dfn 4.65. Otherwise it is indispensable, as was noted by Mr. W. L. Duda, who called my attention to its omission in the first edition. It is not difficult to define $\{X, Y\}$ in such manner that 1.13 also holds for proper classes, but since there is never any occasion of making use of this fact there is no point in doing so.]

[11]Note that in all these definitions it is absolutely unimportant how the notions or operations under consideration are defined for proper classes as arguments.[12] The only purpose of defining them at all for this case is to simplify the metamathematical concepts of "term" and "propositional function" defined on page 12 and the formulation of Theorems M2–M6.

[12][*Note added in 1951*: A similar remark applies to many other concepts which by their usual definition are meaningful only for certain classes, e.g., $\mathfrak{Cnv}$, $\mathfrak{Con}$, etc. only for classes of pairs; $\mathfrak{Max}$, $\mathfrak{Lim}$ only for sets of ordinals (with or without greatest element, respectively), etc. All that is aimed at in the subsequent definitions is that, for those arguments for which, by their usual definitions, the concepts defined are meaningful, the definitions given should agree with the usual ones. For $\mathfrak{Max}$ and $\mathfrak{Lim}$ e.g., this requirement can be satisfied by setting them both equal to $\mathfrak{S}$ (cf. Dfn 7.31).]

propositional function or *minimal formula.* A *propositional function* may be defined recursively as any result of combining minimal propositional functions by means of the logical operators: $\sim$, $\vee$, $.$, $\supset$, $\equiv$ and quantifiers for any kind of variables.

For each of the four types of symbols there is a corresponding kind of definition.

1. A special class A is introduced by a *defining postulate* $\phi(A)$, where ϕ is a propositional function containing only previously defined symbols, and it has to be proved first that there is exactly one class A such that $\phi(A)$.

2. A notion $\mathfrak{B}$ is introduced by the stipulation

$$\mathfrak{B}(X_1, \ldots, X_n) = \phi(X_1, \ldots, X_n),$$

where ϕ is a propositional function containing only previously defined symbols.

3. An operation $\mathfrak{A}$ is introduced by a *defining postulate*

$$(X_1, \ldots, X_n)\phi(\mathfrak{A}(X_1, \ldots, X_n), X_1, \ldots, X_n),$$

where ϕ is a propositional function containing only previously defined symbols, and it has first to be proved that

$$(X_1, \ldots, X_n)(E!Y)\phi(Y, X_1, \ldots, X_n).$$

4. A variable $\mathfrak{x}$ is introduced by a stipulation that for any propositional function ϕ, $(\mathfrak{x})\phi(\mathfrak{x})$ means

$$(X)[\mathfrak{B}(X) \supset \phi(X)],$$

and $(\exists\mathfrak{x})\phi(\mathfrak{x})$ means

$$(\exists X)[\mathfrak{B}(X) \,.\, \phi(X)],$$

where $\mathfrak{B}$ is a previously defined notion, the extension of which is called the *range of the variable* $\mathfrak{x}$.

Special classes, notions and operations are sometimes referred to by the common name "concepts".[13]

All definitions so far introduced are of this type: $\mathfrak{B}$ is called a *normal notion* if there is a ppf ϕ such that

$$\mathfrak{B}(X_1, \ldots, X_n) \,.\!\equiv\!.\, \phi(X_1, \ldots, X_n),$$

[13][*Note added in 1951*: The term "concept" only applies to notions and operations. Special classes should rather be called "objects".]

$\mathfrak{A}$ is called a *normal operation* if there is a ppf ϕ such that

$$Y \epsilon \mathfrak{A}(X_1, \ldots, X_n) . \equiv . \phi(Y, X_1, \ldots, X_n),$$

13 and a *variable is called normal* if its range consists of the | elements of a class. The *propositional function* $\phi(X_1, \ldots, X_n)$ *is called normal* if it contains only normal operations, normal notions, and normal bound variables; and a term is called *normal* if it contains only normal operations.

M2. Any normal propositional function is equivalent to some ppf, and therefore M1 holds also for any normal propositional function $\phi(X_1, \ldots, X_n)$.

Proof: Let $\phi(X_i, \ldots, X_n)$ be the given normal propositional function. Since ϕ contains only normal bound variables, all bound variables not set variables can be replaced by set variables, e.g., $(\exists \mathfrak{x})\chi(\mathfrak{x})$ by $(\exists x)[x \epsilon A . \chi(x)]$, where A defines the range of the variable $\mathfrak{x}$. Next, for any notion $\mathfrak{A}$ occurring in ϕ, since it is normal, the minimal propositional function $\mathfrak{A}(\Gamma_1, \ldots, \Gamma_n)$ can be replaced by the equivalent $\psi(\Gamma_1, \ldots, \Gamma_n)$, where $\psi(X_1, \ldots, X_n)$ is a ppf. Then the only remaining notion is the ϵ-relation. Again all contexts of the form $\Gamma \epsilon \Delta$, where Γ is not a set variable, can be removed by the method explained on page 10 after Theorem 2.8, leaving only minimal formulas of the form $u \epsilon \Gamma$. But Γ, if not a variable or a special class, is of the form $\mathfrak{B}(\Gamma_1, \ldots, \Gamma_n)$, where $\mathfrak{B}$ is a normal operation. But $u \epsilon \mathfrak{B}(\Gamma_1, \ldots, \Gamma_n)$ can be replaced by $\psi(u, \Gamma_1, \ldots, \Gamma_n)$, where the ppf ψ is such that $u \epsilon \mathfrak{B}(\Gamma_1, \ldots, \Gamma_n) . \equiv . \psi(u, \Gamma_1, \ldots, \Gamma_n)$. In this way, ϕ is reduced, getting all operations out. The final result of such reductions can be nothing other than a ppf.

This completes the proof that M1 is valid for normal propositional functions. It remains only to verify that all concepts introduced so far are normal. This will be done by constructing for each of the corresponding expressions $Y \epsilon \mathfrak{A}(X_1, \ldots, X_n)$ and $\mathfrak{B}(X_1, \ldots, X_n)$ equivalent propositional functions containing only notions, operations and bound variables previously shown to be normal. These propositional functions are then equivalent to ppfs by Theorem M2.

$X \epsilon Y$; ϵ is normal, since $X \epsilon Y$ is itself a ppf.
$X = Y . \equiv . (u)[u \epsilon X . \equiv . u \epsilon Y]$
$\mathfrak{M}(X) . \equiv . (\exists u)(u = X)$
$\mathfrak{Pr}(X) . \equiv . \sim \mathfrak{M}(X)$
$Z \epsilon \{X, Y\} . \equiv : (Z = X . \vee . Z = Y) . \mathfrak{M}(Z)$
$Z \epsilon \langle X, Y \rangle . \equiv . Z \epsilon \{\{X\}, \{X, Y\}\}$ and similarly for triples, etc.
$X \subseteq Y . \equiv . (u)(u \epsilon X . \supset . u \epsilon Y)$
$X \subset Y . \equiv . (u)(u \epsilon X . \supset . u \epsilon Y) . \sim(X = Y)$

$\mathfrak{Un}(X) . \equiv . (u, v, w)[\langle u, v\rangle \, \epsilon \, X . \langle w, v\rangle \, \epsilon \, X :\supset . u = w]$
$X \, \epsilon \, (-A) . \equiv : \mathfrak{M}(X) . \sim(X \, \epsilon \, A)$
$X \, \epsilon \, A \cdot B . \equiv : X \, \epsilon \, A . X \, \epsilon \, B$
$X \, \epsilon \, \mathfrak{D}(A) . \equiv : \mathfrak{M}(X) . (\exists y)[\langle y, X\rangle \, \epsilon \, A]$
$\mathfrak{Em}(X) . \equiv . \sim(\exists u)(u \, \epsilon \, X)$
$\mathfrak{Ex}(X, Y) . \equiv . \sim(\exists u)(u \, \epsilon \, X . u \, \epsilon \, Y)$

The general existence theorems M1, M2 (and likewise the later | Theorems M3–M6) are frequently used in these lectures without being quoted explicitly. 14

The particular classes $A_1, \ldots, A_k$ that may appear in the normal propositional function $\phi(x_1, \ldots, x_n)$ are entirely arbitrary, and may therefore be replaced by the general class variables $X_1, \ldots, X_k$, so that the existence theorem takes the form

M3. $$(X_1, \ldots, X_k)(\exists A)(x_1, \ldots, x_n)[\langle x_1, \ldots, x_n\rangle \, \epsilon \, A . \equiv . \phi(x_1, \ldots, x_n, X_1, \ldots, X_k)],$$

if ϕ is normal.

The definitions that follow are mostly based on the existence theorem in this form. In each application of M3 it is apparent upon inspection that ϕ is normal.

The *direct* (*outer*) *product* $A \times B$ is defined by the postulate:

4.1 Dfn $(x)[x \, \epsilon \, A \times B . \equiv . (\exists y, z)[x = \langle y, z\rangle \; : \; y \, \epsilon \, A . z \, \epsilon \, B]]$.

A and B are considered as the constant classes in this application of M3, which assures the existence of $A \times B$ for all A and B. That $A \times B$ is unique is guaranteed by the axiom of extensionality.

4.11 Dfn $A^2 = A \times A$
4.12 Dfn $A^3 = A \times (A^2)$

$A^4, A^5, \ldots$ are defined similarly. Thus V^2 is the class of all ordered pairs, V^3 is the class of all ordered triples, etc. Since every triple is a pair, it follows that

4.13 $V^3 \subseteq V^2$.

Relations are to be defined as classes of ordered pairs, *triadic relations* as classes of ordered triples, etc.

4.2 Dfn $\mathfrak{Rel}(X) . \equiv . X \subseteq V^2$,

4.21 Dfn $\mathfrak{Rel}_3(X) . \equiv . X \subseteq V^3$,

and similarly for all $n \geq 2$. "$\mathfrak{Rel}(X)$" may be written as "$\mathfrak{Rel}_2(X)$".

If A is a relation, then $\langle x, y \rangle \, \epsilon \, A$ is read "x bears the relation A to y", and may be written xAy, i.e.,

4.211 Dfn $xAy . \equiv . \langle x, y \rangle \, \epsilon \, A$.

Relations can be thought of as many-valued functions, so that xAy may be read also as "x is a *value* of A for the argument y" or "x is an *image* of y by A", or "y is an *original* of x, with respect to A". As a corollary of the axiom of extensionality, there is a principle of extensionality for relations:

4.22 $\mathfrak{Rel}(X) . \mathfrak{Rel}(Y) :\supset: (u,v)[\langle u, v \rangle \, \epsilon \, X . \equiv . \langle u, v \rangle \, \epsilon \, Y] . \supset . X = Y$.

15 | The extensionality principle for relations holds also for n-adic relations, in a similar manner. As a result, the existence theorem takes the form:

M4. Given a normal propositional function $\phi(x_1, \ldots, x_n)$, there is exactly one n-adic relation A such that

$$(x_1, \ldots, x_n)[\langle x_1, \ldots, x_n \rangle \, \epsilon \, A . \equiv . \phi(x_1, \ldots, x_n)].$$

The proof is immediate. Take an arbitrary class A' satisfying the condition, and take A as $A' \cdot V^n$. A is an n-adic relation and is unique because of the principle of extensionality, 4.22.

A, as defined by M4, is denoted by $\hat{x}_1, \ldots, \hat{x}_n[\phi(x_1, \ldots, x_n)]$. If $\alpha_1, \ldots, \alpha_n$ are normal variables, $\hat{\alpha}_1, \ldots, \hat{\alpha}_n[\phi(\alpha_1, \ldots, \alpha_n)]$ is by definition the same as $\hat{x}_1, \ldots, \hat{x}_n[\phi(x_1, \ldots, x_n) . x_1 \, \epsilon \, C, \ldots, x_n \, \epsilon \, C]$, where C is the range of the variables α_i. (Note that the symbol ^ belongs to none of the four kinds of symbols introduced on page 11; therefore it must not be used in definitions or in applications of M2–M6.)

The ϵ-relation E and the identity relation I may be defined by means of M4.

4.3 Dfn $\mathfrak{Rel}(E) . (u,v)[\langle u, v \rangle \, \epsilon \, E . \equiv . u \, \epsilon \, v]$

4.31 Dfn $\mathfrak{Rel}(I) . (u,v)[\langle u, v \rangle \, \epsilon \, I . \equiv . u = v]$

I is the class of all pairs $\langle u, u \rangle$.

The following definitions 4.4, 4.41, 4.411 of the *converse relations* correspond to the axioms B6, 7, 8.

4.4 Dfn $\mathfrak{Rel}[\mathfrak{Cnv}(X)] . (u,v)[\langle u, v \rangle \, \epsilon \, \mathfrak{Cnv}(X) . \equiv . \langle v, u \rangle \, \epsilon \, X]$

4.41 Dfn $\mathfrak{Rel}_3[\mathfrak{Cnv}_2(X)] . (u,v,w)[\langle u, v, w \rangle \, \epsilon \, \mathfrak{Cnv}_2(X) . \equiv . \langle v, w, u \rangle \, \epsilon \, X]$

4.411 Dfn $\mathfrak{Rel}_3\,[\mathfrak{Cnv}_3(X)]\,.\,(u,v,w)[\langle u,v,w\rangle \in \mathfrak{Cnv}_3\,(X)\,.\equiv.$
$\langle u,w,v\rangle \in X]$

4.412 Dfn $\mathfrak{Cnv}\,(X)$ is also denoted by $\mathfrak{Cnv}_1(X)$, X^{-1}, and $\breve{X}$.

The binary Boolean operations "+" and "−" are defined in terms of "·" and the complement "−":

4.42 Dfn $X + Y = -[(-X)\cdot(-Y)]$,
4.43 Dfn $X - Y = X \cdot (-Y)$.
4.44 Dfn $\mathfrak{W}\,(X) = \mathfrak{D}(X^{-1})$.

$\mathfrak{W}\,(X)$ is called *domain of values* of X.

The relation "A *confined to* B" is written "$A \restriction B$".

4.5 Dfn $A \restriction B = A \cdot (V \times B)$

$A \restriction B$ consists of all elements of A which are ordered pairs with second member from B. In that sense, "$A \restriction B$" is "A confined to B", since the arguments of A are restricted to lie in B. This gives the theorem:

| 4.51 $\mathfrak{D}(A \restriction B) = B \cdot \mathfrak{D}(A)$. 16
4.512 Dfn $B \upharpoonleft A = A \cdot (B \times V)$
4.52 Dfn $B``X = \mathfrak{W}\,(B \restriction X)$

$B``X$ is the class of all images by B of elements of X.

4.53 Dfn $\langle x,y\rangle \in R|S\,.\equiv.\,(\exists z)(xRz\,.\,zSy)\,.\,\mathfrak{Rel}\,(R|S)$
4.6 Dfn $\mathfrak{Un}_2\,(X)\,.\equiv:\mathfrak{Un}(X)\,.\,\mathfrak{Un}(X^{-1})$

$\mathfrak{Un}_2\,(X)$ means X is *one-to-one*, that is, the relation $X \cdot V^2$ is one-to-one. If X is a relation, and is single-valued, X is said to be a *function.*

4.61 Dfn $\mathfrak{Fnc}(X)\,.\equiv:\mathfrak{Rel}\,(X)\,.\,\mathfrak{Un}(X)$.

A function X whose domain is A is called a *function over* A.

4.63 Dfn $X\,\mathfrak{Fn}\,A\,.\equiv:\mathfrak{Fnc}\,(X)\,.\,\mathfrak{D}\,(X) = A$

$A`x$ (the $A$ of $x$) denotes the $y$ such that $\langle y,x\rangle \in A$, if that $y$ exists and is unique; if $y$ does not exist or is not unique, $A`x = 0$. Hence the defining postulate for $A`x$ reads as follows:

4.65 Dfn $(E!y)[\langle y,x\rangle \in A]\,.\supset.\,\langle A`x,x\rangle \in A$:
$\sim(E!y)[\langle y,x\rangle \in A]\,.\supset.\,A`x = 0\,.:\,\mathfrak{M}\,(A`x)$.

The extensionality principle for relations (4.22) gives the following extensionality principle for functions:

4.67 $X \,\mathfrak{Fn}\, A \,.\, Y \,\mathfrak{Fn}\, A \;:\supset:\; (u)[u \,\epsilon\, A \,.\!\supset:\! X\text{`}u = Y\text{`}u] \,.\!\supset\!.\, X = Y.$

M5. If $\psi(u_1, \ldots, u_n)$ is a normal term, if $B \subseteq V^n$ and if

$$\langle u_1, \ldots, u_n \rangle \,\epsilon\, B \,.\!\supset\!.\, \mathfrak{M}(\psi(u_1, \ldots, u_n)),$$

then there exists exactly one function C over B such that

$$C\text{`}\langle u_1, \ldots, u_n \rangle = \psi(u_1, \ldots, u_n) \text{ for } \langle u_1, \ldots, u_n \rangle \,\epsilon\, B.$$

Proof: Define C by the condition:

$$\langle u, u_1, \ldots, u_n \rangle \,\epsilon\, C \,.\!\equiv:\, u = \psi(u_1, \ldots, u_n) \,.\, \langle u_1, \ldots, u_n \rangle \,\epsilon\, B.$$

Since the right hand side is normal, there is an $(n + 1)$-adic relation C satisfying the condition, by M4. C obviously satisfies the conditions of the theorem.

M5 may be generalized as follows:

M6. If $B_1, \ldots, B_k$ are mutually exclusive, $B_i \subseteq V^n$, and if $\psi_1, \ldots, \psi_k$ are normal terms such that $\mathfrak{M}(\psi_i(u_1, \ldots, u_n))$ for $\langle u_1, \ldots, u_n \rangle \,\epsilon\, B_i$, then there exists exactly one function C over $B_1 + B_2 + \cdots + B_n$ such that
17 $C\text{`}\langle u_1, \ldots, u_n \rangle = \psi_i(u_1, \ldots, u_n)$ for | $\langle u_1, \ldots, u_n \rangle \,\epsilon\, B_i$, $i = 1, 2, \ldots, k$.

We now define five special functions $P_1, \ldots, P_5$ by the following postulates:

4.71 Dfn $P_1\text{`}\langle x, y \rangle = x \,.\, P_1 \,\mathfrak{Fn}\, V^2,$
4.72 Dfn $P_2\text{`}\langle x, y \rangle = y \,.\, P_2 \,\mathfrak{Fn}\, V^2,$
4.73 Dfn $P_3\text{`}\langle x, y \rangle = \langle y, x \rangle \,.\, P_3 \,\mathfrak{Fn}\, V^2,$
4.74 Dfn $P_4\text{`}\langle x, y, z \rangle = \langle z, x, y \rangle \,.\, P_4 \,\mathfrak{Fn}\, V^3,$
4.75 Dfn $P_5\text{`}\langle x, y, z \rangle = \langle x, z, y \rangle \,.\, P_5 \,\mathfrak{Fn}\, V^3.$

Existence and unicity of $P_1, \ldots, P_5$ follow from M5.

4.8 Dfn $u \,\epsilon\, \mathfrak{S}(X) \,.\!\equiv\!.\, (\exists v)[u \,\epsilon\, v \,.\, v \,\epsilon\, X]$

$\mathfrak{S}(X)$ is called the *sum of* X. The following results are immediate:

4.81 $\mathfrak{S}\{x, y\} = x + y,$
4.82 $\mathfrak{S}\{x\} = x,$
4.83 $\mathfrak{S}(X) = E\text{``}X.$

Now define $\mathfrak{P}(X)$, the *power class of* X, the class of subsets of X.

4.84 Dfn $u \in \mathfrak{P}(X) . \equiv . u \subseteq X$

Some of the operations defined have monotonicity properties, e.g.,

4.85 $X \subseteq Y . \supset : \mathfrak{D}(X) \subseteq \mathfrak{D}(Y)$.

It is easily verified that $\mathfrak{W}$, $\mathfrak{S}$, $\mathfrak{P}$, and $\mathfrak{Cnv}_i$ have similar properties. Also

4.86 $A \subseteq B . X \subseteq Y : \supset . A``X \subseteq B``Y$.

1, $\restriction$, $+$, $\cdot$, and $\times$ have similar properties.
We also have some distributivities, such as

4.87 $(A \times B) \cdot (C \times D) = (A \cdot C) \times (B \cdot D)$.

This leads to the special case

4.871 $(A \times V) \cdot (V \times B) = A \times B$.

Likewise

4.88 $\mathfrak{S}(X + Y) = \mathfrak{S}(X) + \mathfrak{S}(Y)$,
4.89 $\mathfrak{S}(X \cdot Y) \subseteq \mathfrak{S}(X) \cdot \mathfrak{S}(Y)$.

| The following theorems result from Definitions 4.71–4.75, and are immediate upon inspection. 18

4.91 $\mathfrak{W}(A) = P_1``A$
4.92 $\mathfrak{D}(A) = P_2``A$
4.93 $\mathfrak{Cnv}(A) = P_3``A$
4.94 $\mathfrak{Cnv}_2(A) = P_4``A$
4.95 $\mathfrak{Cnv}_3(A) = P_5``A$
4.96 $V \times A = \breve{P}_2``A$

The proof for the normality of the notions and operations introduced above and also of those introduced later is contained on page 62.

The results obtained thus far depended on the first two groups of axioms. Theorems on the existence of sets depend, however, on the later axioms. The following theorem depends on Axiom C4, the axiom of substitution.

5.1 $\mathfrak{Un}(A) . \mathfrak{M}(X) : \supset . \mathfrak{M}(A``X)$

Proof: Since $\mathfrak{M}(X)$, there is a set y, by C4, whose elements are just those sets which bear the relation $A \cdot V^2$ to members of X, that is, $(u)[u \in y . \equiv . u \in A``X]$, so that, by the axiom of extensionality, $y$ is identical with $A``X$. Therefore $\mathfrak{M}(A``X)$.

5.11 $\mathfrak{M}(X) . \supset . \mathfrak{M}(X \cdot Y)$

Proof: Substitute $I \upharpoonright Y$ for A in 5.1, obtaining $\mathfrak{M}[(I \upharpoonright Y)``X]$. But $(I \upharpoonright Y)``X = X \cdot Y$.

5.12 $\mathfrak{M}(X) . Y \subseteq X :\supset . \mathfrak{M}(Y)$

Proof: $Y \subseteq X . \supset . Y = X \cdot Y$. Now, by 5.11, the theorem is proved.

5.121 $\mathfrak{M}(X) . \supset . \mathfrak{M}(\mathfrak{P}(X))$

Proof: Axiom C3 provides for the existence of a y such that $\mathfrak{P}(X) \subseteq y$. Therefore by 5.12, $\mathfrak{M}(\mathfrak{P}(X))$.

5.122 $\mathfrak{M}(X) . \supset . \mathfrak{M}(\mathfrak{S}(X))$

Proof: This is proved similarly by using Axiom C2 and 5.12.

5.13 $\mathfrak{M}(X) . \mathfrak{M}(Y) :\supset . \mathfrak{M}(X + Y)$

Proof: If X, Y are sets, we have $X + Y = \mathfrak{S}\{X, Y\}$, and, by Axiom A4, $\{X, Y\}$ is a set. Therefore by 5.122, $\mathfrak{M}(X + Y)$. The next three theorems are proved by 5.1 using 4.91–4.95.

19 | 5.14 $\mathfrak{M}[\mathfrak{D}(x)]$
5.15 $\mathfrak{M}[\mathfrak{Cnv}_i(x)]$ $(i = 1, 2, 3)$
5.16 $\mathfrak{M}[\mathfrak{W}(x)]$

From 5.14 and M5 it follows that there is a function *Do* such that:

5.17 Dfn $Do`x = \mathfrak{D}(x) . Do \, \mathfrak{Fn} \, V$.
5.18 $\mathfrak{M}(x \times y)$

Proof: The members of $x \times y$ are the pairs $\langle u, v \rangle$, where $u \in x$, $v \in y$. In particular, then, u and v are elements of $x + y$, so that $\{u\}$ and $\{u, v\}$ are subsets of $x + y$. Therefore $\{\{u\}, \{u, v\}\}$ is a subset of $\mathfrak{P}(x + y)$, that is, $\langle u, v \rangle \subseteq \mathfrak{P}(x + y)$, so that $\langle u, v \rangle \in \mathfrak{P}[\mathfrak{P}(x + y)]$, i.e., $x \times y \subseteq \mathfrak{P}[\mathfrak{P}(x + y)]$. Therefore $\mathfrak{M}(x \times y)$, by 5.121 and 5.12 and 5.13.

5.19 $F \mathfrak{Fn} x . \supset . \mathfrak{M}(F)$

Proof: $F \mathfrak{Fn} x . \supset . F \subseteq (F``x) \times x$, therefore $\mathfrak{M}(F)$, by 5.1, 5.18, 5.12.

5.2 $\mathfrak{Un}(F) . \supset . \mathfrak{M}(F \upharpoonright x)$

Proof: $F \upharpoonright x$ is a function over $\mathfrak{D}(F \upharpoonright x)$, and $\mathfrak{D}(F \upharpoonright x) \subseteq x$, hence $\mathfrak{D}(F \upharpoonright x)$ is a set. Hence the theorem holds by 5.19.

5.3 $\mathfrak{M}(0)$

Proof: $0 \subseteq x$, therefore $\mathfrak{M}(0)$, by 5.12.

5.31 $\sim\mathfrak{M}(V)$

Proof: $x \in V$; therefore if $\mathfrak{M}(V)$ we would have $V \in V$, but this is impossible, by 1.6.

5.4 $\mathfrak{Pr}(X) . \supset . \mathfrak{Pr}(\mathfrak{S}(X))$

Proof: Suppose $\mathfrak{M}(\mathfrak{S}(X))$; then $\mathfrak{M}(\mathfrak{P}(\mathfrak{S}(X)))$, but $X \subseteq \mathfrak{P}(\mathfrak{S}(X))$, therefore $\mathfrak{M}(X)$, contrary to the hypothesis.

Similarly:

5.41 $\mathfrak{Pr}(X) . \supset . \mathfrak{Pr}(\mathfrak{P}(X))$,
5.42 $\mathfrak{Pr}(X) . \supset . \mathfrak{Pr}(X + Y)$,
5.43 $\mathfrak{Pr}(X) . \sim\mathfrak{Em}(Y) : \supset . \mathfrak{Pr}(X \times Y)$.

Proof: $X \subseteq \mathfrak{S}[\mathfrak{S}(X \times Y)]$, if $Y \neq 0$.

5.44 $\mathfrak{Un}_2(F) . X \subseteq \mathfrak{D}(F) : \supset : \mathfrak{Pr}(X) . \supset . \mathfrak{Pr}(F``X)$,

| that is, a one-to-one image of a proper class is a proper class. The proof follows from the fact that $X \subseteq \breve{F}``(F``X)$, if $X \subseteq \mathfrak{D}(F)$. Therefore, if $F``X$ were a set, X would also be a set by 5.1 and 5.12. 20

5.45 $\mathfrak{Pr}(A) . \supset . \mathfrak{Pr}(A - x)$

This follows from the inclusion $A \subseteq (A - x) + x$, and 5.13.

21 |

Chapter III
Ordinal numbers

Ordinal numbers may now be defined, with the aid of some preliminary definitions.

6.1 Dfn $Y\,\mathfrak{Con}\,X \,.\!\equiv.\, X^2 \subseteq Y + Y^{-1} + I$,

that is, Y *is connex in* X if, for any pair of distinct elements u, v of X, either $\langle u, v\rangle \in Y$ or $\langle v, u\rangle \in Y$.

6.11 Dfn Y is called *transitive* in X if, for *all* elements u, v, w of X,

$$\langle u, v\rangle \in Y \,.\, \langle v, w\rangle \in Y :\supset. \langle u, w\rangle \in Y.$$

6.12 Dfn Y is called *asymmetric* in X if, for *no* elements u, v of X,

$$\langle u, v\rangle \in Y \,.\, \langle v, u\rangle \in Y.$$

6.2 Dfn $X\,\mathfrak{We}\,Y \,.\!\equiv: Y\,\mathfrak{Con}\,X \,.\, (U)[U \neq 0 \,.\, U \subseteq X :\supset$
$.\, (\exists v)[v \in U \,.\, U \cdot Y``\{v\} = 0]]$

that is, X *is well-ordered by* Y if Y is connex in X and any non-void subset U of X has a first element with respect to the ordering Y, since $U \cdot Y``\{v\} = 0$ says that there is no member of U which bears Y to v. Note that the symbol $X\,\mathfrak{We}\,Y$ here introduced is not normal, because of the bound variable U.[14]

6.21 If $X\,\mathfrak{We}\,Y$, then Y is transitive and asymmetric in X.

Proof: Y is asymmetric in X, since if xYy and yYx the class $\{x, y\}$ has no first element. In order to prove the transitivity in X, suppose xYy and yYz;

[14][*Note added in 1951*: The statements made after Dfns 6.2 and 8.1, and on page 62, to the effect that $\mathfrak{We}$ and $\simeq$ are not normal are incorrect, if normality is defined as on page 12. According to *this* definition normality of a concept has nothing to do with the way in which it is defined but only depends on its extension. Therefore all that, prima facie, can be said about $\mathfrak{We}$ and $\simeq$ is that they cannot be proved to be normal by the method applied to the other concepts on page 62. They can however be proved to be normal in a different way, provided the axiom of choice is assumed. For, under this assumption, it can be proved that

$$X \simeq Y \,.\!\equiv.\, X \simeq' Y \,.\vee: \mathfrak{Pr}(X) \,.\, \mathfrak{Pr}(Y)$$

(cf. *von Neumann 1929*). Moreover U can be replaced by u in Dfn 6.2 because the existence of a class without first element implies the existence in it of a descending sequence of type ω. The latter proof requires the singling out of one element in every non-empty class, which however can be accomplished by considering, in every class, the sub*set* of elements of lowest "Stufe" (in the sense of *von Neumann 1929*, page 238).]

then $x \neq z$ because of the asymmetry, hence either xYz or zYx. Consider $U = \{x\} + \{y\} + \{z\}$. If zYx, U will have no first element, therefore xYz.

6.3 Dfn $X \,\mathfrak{Sect}_R\, Y \,.\!\equiv\!:\, X \subseteq Y \,.\, [Y \cdot (R``X) \subseteq X]$,

that is, X *is an* R*-section of* Y if all R-predecessors in Y of members of X also belong to X.

6.30 Dfn X is called a *proper* R-section of Y if it is an R-section of Y and $\neq Y$.

| 6.31 Dfn $\mathfrak{Seg}_R(X, u) = X \cdot R``\{u\}$, 22

that is, if $u \,\epsilon\, X$, *the* R*-segment of* X *generated by* u is the class of elements of X which are R-predecessors of u.

6.32 $\mathfrak{Seg}_R(X, u)$ is an R-section of X, if $u \,\epsilon\, X$ and if R is transitive in X.

Therefore

6.33 If $X \,\mathfrak{We}\, R$, then any R-segment generated by an element of X is an R-section.

Conversely, if $X \,\mathfrak{We}\, R$ and Y is a proper R-section of X, then Y is an R-segment of X, namely the one generated by the first element of $X - Y$.

If R is a one-to-one relation with domain A and converse domain B, then R is called an *isomorphism from* A *to* B *with respect to* S *and* T if for any pair u, v of A such that uSv the corresponding elements of B are in the relation T, and conversely, i.e.,

6.4 Dfn $R \,\mathfrak{Isom}_{S,T}(A, B) \,.\!\equiv\!:\, \mathfrak{Un}_2(R) \,.\, \mathfrak{Rel}\,(R) \,.\, \mathfrak{D}(R) = A \,.$
$\mathfrak{W}\,(R) = B \,.\, (u, v)[u \,\epsilon\, A \,.\, v \,\epsilon\, A \,:\!\supset\!:\, uSv \,.\!\equiv\!.\, (R`u)T(R`v)]$.

If there exists an isomorpism from A to B with respect to S and T, A is called *isomorphic to* B *with respect to* S *and* T. If $S = T$ in 6.4, R is said to be an *isomorphism from* A *to* B *with respect to* S.

6.41 Dfn R is called an *isomorphism with respect to* S if it is an isomorphism from $\mathfrak{D}(R)$ to $\mathfrak{W}(R)$ with respect to S.

"Isomorphism with respect to an n-adic relation S" is defined accordingly.

The method to be used in constructing the ordinals is due essentially to J. von Neumann. The ordinal α will be the class of all ordinals less than α. For instance, $0 =$ the null set, $1 = \{0\}$, $2 = \{0, 1\}$, $\omega =$ the set of all integers, etc. In this way, the class of ordinals will be well-ordered by the ϵ-relation, so that $\alpha \epsilon \beta$ corresponds to $\alpha < \beta$. Any ordinal will itself be well-ordered by the ϵ-relation, since an ordinal is a class of ordinals. Moreover, any element of an ordinal must be identical with the segment generated by itself, since this segment is the set of all smaller ordinals. These considerations lead to the following definition:

Definition: X is an *ordinal* if

1. $X \mathfrak{We} E$,
2. $u \epsilon X :\supset . u = \mathfrak{Seg}_E(X, u)$.

23 | However, as shown by R. M. Robinson (*1937*, page 35; Bernays showed previously that transitivity of E in X and $2'$ are sufficient), conditions 1 and 2 may be replaced, owing to Axiom D, by the weaker conditions:

1′. $E \mathfrak{Con} X$,
2′. $u \epsilon X . \supset . u \subseteq X$.

X is said to be *complete* if it has the property $2'$, i.e., if any element of an element of X is an element of X, that is,

6.5 Dfn $\mathfrak{Comp}(X) . \equiv . (u)[u \epsilon X . \supset . u \subseteq X]$.
6.51 $\mathfrak{Comp}(X) . \equiv . \mathfrak{S}(X) \subseteq X$

The proof is immediate from 6.5 and 4.8.

6.6 Dfn $\mathfrak{Ord}(X) . \equiv : \mathfrak{Comp}(X) . E \mathfrak{Con} X$

This definition combines conditions $1'$ and $2'$. An ordinal which is a set is called an *ordinal number*, denoted by $\mathfrak{O}(X)$.

6.61 Dfn $\mathfrak{O}(X) . \equiv : \mathfrak{Ord}(X) . \mathfrak{M}(X)$

The class of ordinal numbers is denoted by On. (Concerning the normality of $\mathfrak{Ord}$, cf. page 62.)

6.62 Dfn $x \epsilon On . \equiv . \mathfrak{O}(x)$
Dfn The letters $\alpha, \beta, \gamma, \ldots$ will be used to denote variables whose range is the class of ordinal numbers. Evidently these variables are normal.
6.63 Dfn $X < Y . \equiv . X \epsilon Y$
6.64 Dfn $X \leq Y . \equiv : X < Y . \vee . X = Y$
6.65 $\mathfrak{Comp}(X) . \mathfrak{Comp}(Y) :\supset: \mathfrak{Comp}(X + Y) . \mathfrak{Comp}(X \cdot Y)$

Proof: By 4.88, $\mathfrak{S}(X + Y) = \mathfrak{S}(X) + \mathfrak{S}(Y)$. Therefore, by 6.51, we have $\mathfrak{Comp}(X + Y)$. Similarly for $X \cdot Y$ by 4.89.

The next step is to show that the definition 6.6 is equivalent to the stronger definition, i.e.,

6.7 1. $\mathfrak{Ord}(X) \supset X \,\mathfrak{We}\, E$,
2. $\mathfrak{Ord}(X) \,.\, u \in X \supset u = \mathfrak{Seg}_E(X, u)$.

Proof of 1: Given any non-void subset Y of X, there exists u, by Axiom D, such that $u \in Y$ and $Y \cdot u = 0$, that is, $Y \cdot E``\{u\} = 0$, since $u = \mathfrak{S}[\{u\}] = E``\{u\}$ by 4.83, 4.82. Therefore $X \,\mathfrak{We}\, E$, by Definition 6.2, since $E \,\mathfrak{Con}\, X$, by definition of $\mathfrak{Ord}$.

Proof of 2: If $\mathfrak{Ord}(X)$ and $u \in X$, then $\mathfrak{Seg}_E(X, u) = X \cdot E``\{u\} = X \cdot u = u$, by Definition 6.31 and the completeness of X.

| 7.1 $\mathfrak{Ord}(X) \,.\, Y \subset X \,:\supset:\, \mathfrak{Comp}(Y) \,.\supset.\, Y \in X$ 24

Proof: $\mathfrak{S}(Y) \subseteq Y$, so that $E``Y \subseteq Y$ by 4.83. Therefore, by Definition 6.3, Y is a section of X. Hence by 6.33 Y must be a segment of X, generated by some element u of X. But then $Y = u$, by 6.7, hence $Y \in X$.

7.11 $\mathfrak{Ord}(X) \,.\, \mathfrak{Ord}(Y) :\supset: Y \subset X \,.\equiv.\, Y \in X$

Proof: Since Y is an ordinal, it is complete. Therefore 7.1 establishes one half of the equivalence. The other half merely expresses the fact that X is complete, since $Y = X$ is excluded by 1.6.

7.12 If X and Y are ordinals, one and only one of the following relations holds:

$$X \in Y, \quad X = Y, \quad Y \in X.$$

Proof: $X \cdot Y \subseteq X$ and $X \cdot Y \subseteq Y$. Suppose now that $X \cdot Y \subset X$ and $X \cdot Y \subset Y$; then $X \cdot Y \in X$ and $X \cdot Y \in Y$, by 7.1, since the intersection of two complete classes is complete (6.65). But this implies that $X \cdot Y \in X \cdot Y$, which is impossible, by 1.6 and Axiom A2. Therefore either $X \cdot Y = X$ or $X \cdot Y = Y$, i.e., either $Y \subseteq X$ or $X \subseteq Y$, i.e., $X \subset Y \,.\vee.\, X = Y \,.\vee.\, Y \subset X$, hence $X \in Y \,.\vee.\, X = Y \,.\vee.\, Y \in X$ by 7.11. Therefore at least one of the three relations holds. Moreover no two can hold simultaneously, since $X \in X$ or $X \in Y \,.\, Y \in X$ are impossible, by 1.6 and 1.7 and Axiom A2.

7.12 and 6.63 express the fact that any two ordinals are comparable. By 6.1, this implies the statement:

7.13 $E \,\mathfrak{Con}\, On$.
7.14 $\mathfrak{Ord}(A) \,.\supset.\, A \subseteq On$

Proof: Let A be an ordinal and x an element of A. We have to show that $E\,\mathfrak{Con}\,x$ and $\mathfrak{Comp}\,(x)$. Take $z \in y$, $y \in x$; then, since A is complete, $y \in A$; then iterating, $z \in A$. E is a relation of well-ordering for A, therefore transitive in A by 6.21, so that $z \in x$. Therefore x is complete. $E\,\mathfrak{Con}\,A$ and $x \subseteq A$, so that $E\,\mathfrak{Con}\,x$.

7.15 $\mathfrak{Comp}\,(On)$

Proof: By 7.14, $x \in On\,.\supset.\,x \subseteq On$.

7.16 $\mathfrak{Ord}\,(On)$

Proof: 7.13, 7.15, 6.6.

25 7.161 On (and therefore any class of ordinal numbers) is well-|ordered by E.

This follows immediately from 7.16 and 6.7 and allows us to prove properties of ordinal numbers by transfinite induction, if the property under consideration is defined by a normal propositional function, since under this assumption the class of ordinal numbers not having the property exists by M2 and (if not empty) contains a smallest element by 7.161 and Definition 6.2. By an inductive proof is always meant the reductio ad absurdum of the existence of a smallest ordinal not having the property under consideration.

By 7.14, any element of an ordinal number is itself an ordinal number, so that an ordinal number x is identical with the set of ordinals less than x, recalling that the ϵ-relation is the ordering relation for ordinals.

7.17 $\mathfrak{Pr}\,(On)$

Proof: On is an ordinal, so that On would be an ordinal number, if $\mathfrak{M}\,(On)$, hence $On \in On$, which is impossible (1.6).

7.2 $\mathfrak{Ord}\,(X)\,.\supset\colon X \in On\,.\vee.\,X = On$. The only ordinal not an ordinal number is On.

Proof: By 7.14, $X \subseteq On$. If $X \subset On$, by 7.11, $X \in On$.

7.21 Any E-section of an ordinal is an ordinal.

Proof: Any proper E-section of an ordinal X is (by 6.33 and 6.7(2)) an element of X, hence an ordinal by 7.14. A non-proper E-section of X is identical with X.

7.3 $A \subseteq On . \supset . \mathfrak{Ord}\,[\mathfrak{S}(A)]$

Proof: $\mathfrak{S}(A)$ is complete since, if $x \in \mathfrak{S}(A)$, there is an ordinal α such that $x \in \alpha \in A$; then if $y \in x$, $y \in \alpha$, since α is complete, that is $y \in \mathfrak{S}(A)$. Also $E\,\mathfrak{Con}\,\mathfrak{S}(A)$; for take $x \neq y$, elements of $\mathfrak{S}(A)$; then $x \in \alpha \in A$, $y \in \beta \in A$. α and β are comparable so that either $\alpha \subseteq \beta$ or $\beta \subseteq \alpha$. Then both x and y are members of the larger of α and β, so that $x \in y$ or $y \in x$, since $E\,\mathfrak{Con}\,\alpha$ and $E\,\mathfrak{Con}\,\beta$, that is $E\,\mathfrak{Con}\,\mathfrak{S}(A)$. Therefore $\mathfrak{Ord}\,[\mathfrak{S}(A)]$.

$\mathfrak{S}(A)$ is the smallest ordinal greater than or equal to all elements of A, i.e., is either the *maximum* or the *limit* of the ordinals of A according as to whether there is or is not a greatest ordinal in A. Therefore we use "$\mathfrak{Lim}$" and "$\mathfrak{Max}$" to denote the same operation as $\mathfrak{S}$.

7.31 Dfn $\mathfrak{Lim}\,(A) = \mathfrak{S}\,(A)$
$\mathfrak{Max}\,(A) = \mathfrak{S}\,(A)$

| 7.4 Dfn $x \dotplus 1 = x + \{x\}$ 26

This defines the successor relation for ordinal numbers as seen by Theorems 7.41, 7.411.

7.41 $x \dotplus 1 \in On . \equiv . x \in On$

This is easily proved.

7.411 $\sim(\exists\beta)[\alpha < \beta < \alpha \dotplus 1]$

Proof: Suppose $\alpha < \beta < \alpha \dotplus 1$; then $\beta \in \alpha \dotplus 1$, that is $\beta \in \alpha + \{\alpha\}$; that is $\beta \in \alpha$ or $\beta = \alpha$, so that $\beta \leq \alpha$.

Ordinal numbers are to be classified into ordinal numbers of the first kind and ordinal numbers of the second kind, as follows:

7.42 Dfn $x \in K_I . \equiv : (\exists\alpha)[x = \alpha \dotplus 1] . \vee . x = 0.$

x is of the first kind if it is the successor of an ordinal number or 0. Otherwise *x is of the second kind.*

7.43 Dfn $K_{II} = On - K_I$

7.44 Dfn $1 = 0 \dotplus 1$

7.45 Dfn $2 = 1 \dotplus 1$

Likewise $3 = 2 \dotplus 1$, etc. Evidently we have:

7.451 If m is a set of ordinal numbers, the ordinal $\alpha = \mathfrak{S}\,(m) \dotplus 1$ is an ordinal number greater than any element of m.

It will now be shown that it is possible to define functions over On by means of transfinite induction, i.e., determining $F`\alpha$ by means of the behavior of $F$ for ordinal numbers less than $\alpha$. Since $\alpha$ is the class of ordinals less than $\alpha$, $F \upharpoonright \alpha$ is $F$ confined to arguments less than $\alpha$. Therefore the induction should have the form $F`\alpha = G`(F \upharpoonright \alpha)$, where G is a known function. The following theorem, then, is what is needed:

7.5 $\quad (G)(E!F)[F\,\mathfrak{Fn}\,On\,.\,(\alpha)(F`\alpha = G`(F \upharpoonright \alpha))]$.

Proof: Let us construct F. First, by the existence theorem M2, there exists a class K such that:

$$f \,\epsilon\, K \,.\!\equiv.\, (\exists\beta)[f\,\mathfrak{Fn}\,\beta\,.\,(\alpha)[\alpha \,\epsilon\, \beta\,.\!\supset.\, f`\alpha = G`(f \upharpoonright \alpha)]].$$

Now set $F = \mathfrak{S}(K)$. If f, $g \,\epsilon\, K$, where $f\,\mathfrak{Fn}\,\beta\,.\,g\,\mathfrak{Fn}\,\gamma\,.\,\beta \leq \gamma$, it follows that $f = g \upharpoonright \beta$, because for $\alpha \,\epsilon\, \beta$ both f and g satisfy

$$(*) \qquad f`\alpha = G`(f \upharpoonright \alpha);$$

and this equation determines an f over β uniquely, as is seen by an induction on α. This means that any two f, $g \,\epsilon\, K$ coincide within the common part of their domains. Therefore F will be a function and its domain will be the sum | of the domains of all $f \,\epsilon\, K$ (i.e., $\mathfrak{D}(F) = \mathfrak{S}(Do``K)$) and F will coincide with each $f \,\epsilon\, K$ within $\mathfrak{D}(f)$. F will satisfy $(*)$ for each $\alpha \,\epsilon\, \mathfrak{D}(F)$, because $\alpha \,\epsilon\, \mathfrak{D}(F)$ implies $\alpha \,\epsilon\, \mathfrak{D}(f) \,\epsilon\, On$ for some $f \,\epsilon\, K$ where f satisfies $(*)$ in $\mathfrak{D}(f)$ and $f = F \upharpoonright \mathfrak{D}(f)$. Now $\mathfrak{D}(F)$ is an ordinal by 7.3, but cannot be an ordinal number α because otherwise F could be extended to a function H over $\alpha \dotplus 1$, by virtue of $(*)$ and M6. But then $\mathfrak{M}(H)$, by 5.19, hence $H \,\epsilon\, K$, which would imply $\alpha \dotplus 1 \subseteq \alpha$. The unicity of F follows by an induction on α.

27

7.6 Dfn An *ordinal function* is a function G over an ordinal, with ordinal numbers as values, that is $G\,\mathfrak{Fn}\,\alpha$ (for some α) or $G\,\mathfrak{Fn}\,On$, and $\mathfrak{W}(G) \subseteq On$.

7.61 Dfn An ordinal function G is said to be *strictly monotonic* if $\alpha < \beta\,.\!\supset.\, G`\alpha < G`\beta$ for α, $\beta \,\epsilon\, \mathfrak{D}(G)$.

By induction it follows that:

7.611 If G is strictly monotonic, then $G`\alpha \geq \alpha$ for $\alpha \,\epsilon\, \mathfrak{D}(G)$.

From this it follows that no two different ordinals X and Y can be isomorphic with respect to E,

7.62 $\mathfrak{Ord}(X) \,.\, \mathfrak{Ord}(Y) \,.\, H\,\mathfrak{Isom}_{EE}(XY) :\supset: X = Y \cdot H = I \upharpoonright X.$

Proof: By definition of an isomorphism we have: if α, β are elements of X such that $\alpha \,\epsilon\, \beta$, then $H`\alpha \,\epsilon\, H`\beta$, that is, H is strictly monotonic, so that by 7.611 $H`\alpha \geq \alpha$ for $\alpha \,\epsilon\, X$. Likewise, $\breve{H}`(H`\alpha) \geq H`\alpha$, that is, $\alpha \geq H`\alpha$ for $\alpha \,\epsilon\, X$; it follows that $H`\alpha = \alpha$ for $a \,\epsilon\, X$, in other words, $X = Y$, and $H = I \upharpoonright X$.

As a consequence of 7.62, a well-ordered class can be isomorphic to at most one ordinal. Sufficient conditions for a well-ordered class to be isomorphic to an ordinal are given by the following theorem.

7.7 1. If $\mathfrak{Pr}(A)$ and $A\,\mathfrak{We}\,W$, and if any proper W-section of A is a set, then A is isomorphic to On with respect to W and E.
2. If $a\,\mathfrak{We}\,W$, a is isomorphic to an ordinal number with respect to W and E.

Proof of 1: Let $F`\alpha$ be defined by induction as the first element of $A$ which has not yet occurred as a value of $F$, that is $F`\alpha =$ first element of $A - \mathfrak{W}(F \upharpoonright \alpha)$. In order to prove the existence of F by 7.5, this condition must be expressed in the form

$$(*) \qquad F`\alpha = G`(F \upharpoonright \alpha).$$

Define G by the condition:

$$\langle y, x\rangle \,\epsilon\, G \,.\!\equiv\!: y \,\epsilon\, (A - \mathfrak{W}(X)) \,.\, (A - \mathfrak{W}(x)) \,.\, W``\{y\} = 0,$$

and define F by $(*)$ and the condition $F\,\mathfrak{Fn}\,On$. Then $G`x \,\epsilon\, A - \mathfrak{W}(x)$ for any set $x$ because $A - \mathfrak{W}(x)$ is a proper class by 5.45, 5.16, | hence $\neq 0$. Therefore $F`\alpha \,\epsilon\, A - \mathfrak{W}(F \upharpoonright \alpha)$ for any α by $(*)$, hence $\mathfrak{W}(F) \subseteq A$. Moreover F is one-to-one, so that $\mathfrak{W}(F)$, being a one-to-one image of the proper class On, is itself a proper class by 5.44. But $\mathfrak{W}(F)$ is a section of A, hence by the hypothesis cannot be a proper section, i.e., $\mathfrak{W}(F) = A$. In addition, it is easily seen that $\alpha < \beta \,.\!\equiv\!.\, (F`\alpha)W(F`\beta)$. 28

Proof of 2: Construct G and F exactly as in the proof of 1, replacing A by a. Now it can be shown that $a - \mathfrak{W}(F \upharpoonright \alpha) = 0$ for some α. In fact, suppose that $(\alpha)[a - \mathfrak{W}(F \upharpoonright \alpha) \neq 0]$; then we could conclude, as before, that $\mathfrak{W}(F) \subseteq a$. Then $\mathfrak{W}(F)$ would be, as before, a proper class; but this is impossible, since a is a set. Therefore $(\exists\alpha)[a - \mathfrak{W}(F \upharpoonright \alpha) = 0]$. Then, if α is the smallest ordinal of this kind, $F \upharpoonright \alpha$ establishes the isomorphism between a and α. From the axiom of choice it follows:

*7.71 For any set a there exists an ordinal number α and a one-to-one function g over α such that $a = g``\alpha$.

Proof: By Axiom E, the axiom of choice, there is a function C over V such that $x \neq 0 . \supset . C`x \,\epsilon\, x$. Define F by the postulate

$$(\alpha)[F`\alpha = C`(a - \mathfrak{W}(F \restriction \alpha))]$$

and $F \,\mathfrak{Fn}\, On$. Existence and unicity of F follow from 7.5, if first G is defined by $G`x = C`(a - \mathfrak{W}(x))$, $G \,\mathfrak{Fn}\, V$, using M5. As in the second part of 7.7, it is shown that there exists an α such that $(a - \mathfrak{W}(F \restriction \alpha)) = 0$. Then if α is the smallest ordinal of this kind, $F \restriction \alpha$ can be taken as g.

It is desirable to assign a well-ordering for the ordered pairs of ordinal numbers:

7.8 Dfn $\langle \alpha, \beta \rangle Le \langle \gamma, \delta \rangle . \equiv : \beta < \delta . \vee . (\beta = \delta . \alpha < \gamma) .: Le \subseteq (On^2)^2$,

7.81 Dfn $\langle \alpha, \beta \rangle R \langle \gamma, \delta \rangle . \equiv : \mathfrak{Max}\{\alpha, \beta\} < \mathfrak{Max}\{\gamma, \delta\} . \vee$
$. [\mathfrak{Max}\{\alpha, \beta\} = \mathfrak{Max}\{\gamma, \delta\} . \langle \alpha, \beta \rangle Le \langle \gamma, \delta \rangle] .: R \subseteq (On^2)^2$.

The existence of an Le satisfying 7.8 follows from M4 since the relation Le defined by

$$\langle x, y \rangle \,\epsilon\, Le = (\exists \alpha, \beta, \gamma, \delta)[x = \langle \alpha, \beta \rangle . y = \langle \gamma, \delta \rangle \; : \; \beta < \delta . \vee . (\beta = \delta . \alpha < \gamma)]$$

evidently satisfies 7.8. Similarly for R. On^2 is well-ordered by R in such a way that:

7.811 Any proper R-section of On^2 is a set.

Proof: Consider a pair $\langle \mu, \nu \rangle$ such that $\langle \mu, \nu \rangle R \langle \alpha, \beta \rangle$; then

$$\mathfrak{Max}\{\mu, \nu\} \leq \mathfrak{Max}\{\alpha, \beta\} < \mathfrak{Max}\{\alpha, \beta\} \dotplus 1.$$

Therefore μ, $\nu \,\epsilon\, [\mathfrak{Max}\{\alpha, \beta\} \dotplus 1]$, so that $\langle \mu, \nu \rangle \,\epsilon\, a$, where $a = [\mathfrak{Max}\{\alpha, \beta\} \dotplus 1]^2$. a is a set by 5.18. Therefore the class of all pairs $\langle \mu, \nu \rangle$ such that $\langle \mu, \nu \rangle R \langle \alpha, \beta \rangle$ is contained in the set a, hence is itself a set.

29 | Now, applying 7.7 (since On^2 is a proper class by 7.17, 5.43), we have:

7.82 On^2 is isomorphic to On, with respect to R and E. Let the isomorphism from On^2 to On be denoted by P, i.e.,

7.9 Dfn $P \,\mathfrak{Fn}\, On^2 . \mathfrak{W}(P) = On$:
$(\alpha, \beta, \gamma, \delta)[\langle \alpha, \beta \rangle R \langle \gamma, \delta \rangle . \supset . P`\langle \alpha, \beta \rangle < P`\langle \gamma, \delta \rangle]$.

7.91 $P`\langle \alpha, \beta \rangle \geq \mathfrak{Max}\{\alpha, \beta\}$

Proof: Take $\gamma = \mathfrak{Max}\{\alpha, \beta\}$. Then $P`\langle \alpha, \beta \rangle \geq P`\langle \gamma, 0 \rangle$ by 7.9; but, since $P`\langle \gamma, 0 \rangle$ considered as a function of $\gamma$ is strictly monotonic by 7.9, we have $\gamma \leq P`\langle \gamma, 0 \rangle$ by 7.611, i.e., $P`\langle \alpha, \beta \rangle \geq \mathfrak{Max}\{\alpha, \beta\}$.

Chapter IV
Cardinal numbers

30

We can now proceed with the theory of cardinals. Most of the theorems and definitions of this chapter (except those concerning finite cardinals) depend in our development on the axiom of choice, although its use could be avoided in many cases. Two classes X and Y are said to be equivalent if there is a one-to-one correspondence between the elements of each, i.e.,

8.1 Dfn $X \simeq Y . \equiv . (\exists Z)[\mathfrak{Un}_2(Z) . \mathfrak{Rel}(Z) . \mathfrak{D}(Z) = X . \mathfrak{W}(Z) = Y]$.

This notion is not normal;[15] the corresponding normal notion is as follows:

8.12 Dfn $X \simeq' Y . \equiv . (\exists z)[\mathfrak{Un}_2(z) . \mathfrak{Rel}(z) . \mathfrak{D}(z) = X . \mathfrak{W}(z) = Y]$.
8.121 $x \simeq y . \equiv . x \simeq' y$

Proof: A class Z satisfying the right hand side of 8.1 for two sets X, Y is a set by 5.19.

8.13 Dfn $\langle x, y \rangle \,\epsilon\, Aeq . \equiv . x \simeq y \;:\; \mathfrak{Rel}(Aeq)$

The *cardinal of* X, denoted by $\overline{\overline{X}}$, is defined by the postulate:[16]

*8.2 Dfn $x \simeq \overline{\overline{x}} . \overline{\overline{x}} \,\epsilon\, On . (\alpha)[\alpha < \overline{\overline{x}} . \supset . \sim(\alpha \simeq x)] . \mathfrak{Pr}(X) . \supset . \overline{\overline{X}} = On.$

By Theorem 7.71 it is seen that $\overline{\overline{X}}$ exists. The unicity is immediate. $\overline{\overline{X}}$ is a normal operation, since

$$X \,\epsilon\, \overline{\overline{Y}} . \equiv : X \,\epsilon\, On . (\alpha)[\alpha \simeq' Y . \supset . X \,\epsilon\, \alpha].$$

Hence by M5 there exists a function Nc over V such that $Nc`x = \overline{\overline{x}}$ for any set x.

*8.20 Dfn $Nc`x = \overline{\overline{x}} . Nc \,\mathfrak{Fn}\, V$

The cardinal of a set is called a *cardinal number*, i.e., the class N of cardinal numbers is defined by:

[15][*Note added in 1951*: See Note 14 on page 21.]

[16][*Note added in 1951*: Dfn 8.2, for the case that $\mathfrak{Pr}(X)$, is justified by J. von Neumann's result (concerning $\simeq$) quoted in Note 14.]

$*8.21$ Dfn $N = \mathfrak{W}(Nc)$.
$*8.22$ $N \subseteq On$

This follows immediately from 8.2, 8.21.

31 | An ordinal number is a cardinal number if and only if it is equivalent to no smaller ordinal, i.e., if it is an initial number in the usual terminology.[17]

The next five theorems are immediate consequences of the definition of cardinals.

$*8.23$ $x \,\epsilon\, N . \equiv . x = \overline{\overline{x}}$
$*8.24$ $\overline{\overline{x}} \simeq x$
$*8.25$ $x \simeq y . \equiv . \overline{\overline{x}} = \overline{\overline{y}}$
$*8.26$ $\overline{\overline{\alpha}} \leq \alpha$
$*8.27$ $Nc`[Nc`x] = Nc`x$
$*8.28$ $x \subseteq y . \supset . \overline{\overline{x}} \leq \overline{\overline{y}}$

Proof: $y \simeq \overline{\overline{y}}$, therefore there exists a $z \subseteq \overline{\overline{y}}$ such that $x \simeq z$. z is a set of ordinal numbers, hence well-ordered by E, hence is isomorphic to an ordinal number β by 7.7, i.e., there is an h such that $h\,\mathfrak{Isom}_{EE}(\beta, z)$. Hence $\alpha \leq h`\alpha$ for $\alpha \,\epsilon\, \beta$ by 7.611. But $h`\alpha \,\epsilon\, z \subseteq \overline{\overline{y}}$ for $\alpha \,\epsilon\, \beta$. Hence $\alpha \,\epsilon\, \beta . \supset . \alpha \leq h`\alpha \,\epsilon\, \overline{\overline{y}}$, that is $\beta \subseteq \overline{\overline{y}}$. But $\overline{\overline{\beta}} = \overline{\overline{z}}$, therefore $\overline{\overline{z}} = \overline{\overline{\beta}} \leq \beta \leq \overline{\overline{y}}$.

The Schroeder–Bernstein Theorem appears as a consequence. Namely, if $x \simeq t \subseteq y$ and $y \simeq u \subseteq x$, then $\overline{\overline{x}} = \overline{\overline{y}}$, since $\overline{\overline{x}} = \overline{\overline{t}} \leq \overline{\overline{y}}$, and $\overline{\overline{y}} = \overline{\overline{u}} \leq \overline{\overline{x}}$. This proof depends, however, on the axiom of choice.

The proofs of the next three theorems are omitted.

$*8.3$ $\overline{\overline{\alpha \dotplus 1}} \leq \overline{\overline{\alpha^2}}$ for $\alpha > 1$
$*8.31$ $\mathfrak{Un}(A) . \supset . \overline{\overline{A``x}} \leq \overline{\overline{x}}$
$*8.32$ $\overline{\overline{\mathfrak{P}(x)}} > \overline{\overline{x}}$ (Cantor's theorem)
$*8.33$ $\mathfrak{Pr}(N)$

Proof: Take $m \subseteq N$; then by 8.32 $\overline{\overline{\mathfrak{P}(\mathfrak{S}(m))}} > \overline{\overline{\mathfrak{S}(m)}}$. But $\overline{\overline{\mathfrak{S}(m)}} \geq \alpha$, where α is any member of m by 8.28, 8.23. Therefore there is a cardinal number greater than any element of m; hence $m \neq N$, i.e., N can not be a set.

We now define the *class* ω *of integers*:

8.4 Dfn $x \,\epsilon\, \omega . \equiv . x + \{x\} \subseteq K_I$,

[17]This treatment of cardinals is due to von Neumann (cf. *1928a*, p. 731).

i.e., x is an integer if it is an ordinal number of the first kind and if all smaller ordinals are likewise of the first kind. It follows immediately that:

8.41 $\alpha \epsilon \omega . \supset . \alpha \dot{+} 1 \epsilon \omega$ and $\alpha \epsilon \omega . \beta < \alpha : \supset . \beta \epsilon \omega$.
8.42 Dfn i, k are variables whose range is ω.

The principle of induction holds for integers:

8.44 $0 \epsilon A . (k)[k \epsilon A . \supset . k \dot{+} 1 \epsilon A] : \supset . \omega \subseteq A$.

| Proof: If $\omega \subseteq A$ is false, there must be a smallest i such that i is not a member of A. This leads to a contradiction with the hypothesis, since either $i = 0$ or $i = k \dot{+} 1$ by 8.4 and 7.42. 32

Functions over the class of integers may be defined inductively:

8.45 $(a, G)(E!F)[F\, \mathfrak{Fn}\, \omega . F`0 = a . (k)(F`(k \dot{+} 1) = G`(F`k))]$.

This can be proved either by specializing G in 7.5 or by arguments similar to those used in the proof of 7.5.

8.46 $i \neq k . \supset . \sim(i \simeq k)$

This can be proved by induction on integers, since

$$i \dot{+} 1 \simeq k \dot{+} 1 \supset i \simeq k.$$

8.461 $\alpha \neq k . \supset . \sim(\alpha \simeq k)$

Proof by induction on k, since $k \dot{+} 1 \simeq \alpha \geq \omega$ would imply $k \simeq \alpha$.

*8.47 $i \epsilon N$

This follows from 8.46.

A class is called *finite* if it is equivalent to an integer; otherwise *infinite*, i.e.,

8.48 Dfn $\mathfrak{Fin}(x) . \equiv . (\exists \alpha)[\alpha \epsilon \omega . \alpha \simeq' x]$,
8.49 Dfn $\mathfrak{Inf}(x) . \equiv . \sim\mathfrak{Fin}(x)$.
8.491 $\mathfrak{Fin}(x) . z \subseteq x : \supset . \mathfrak{Fin}(z)$
$\mathfrak{Fin}(x) . \mathfrak{Fin}(y) : \supset : \mathfrak{Fin}(x + y) . \mathfrak{Fin}(x \times y)$

This is proved by an induction on the integer i equivalent to x.

8.492 $\mathfrak{Fin}(\alpha) . \equiv . \alpha \epsilon \omega$

This follows from 8.461.

8.5 $\mathfrak{Ord}(\omega)$

Proof: ω is a class of ordinal numbers, hence $E\,\mathfrak{Con}\,\omega$. Moreover, every element of an integer is an integer by Definition 8.4, that is, $\mathfrak{Comp}(\omega)$. Therefore $\mathfrak{Ord}(\omega)$.

8.51 $\mathfrak{M}(\omega)$

Proof: Axiom C1 (the axiom of infinity) provides for the existence of a
33 non-empty set b, such that for every $x \in b$ there is | a $y \in b$ which contains exactly one element more than x; namely take for b the class of all subsets of elements of the set a, whose existence is postulated by Axiom C1 (b is a set because $b \subseteq \mathfrak{P}[\mathfrak{S}(a)]$). Now consider the class c defined by $c = (\omega \upharpoonright Aeq)``b$, i.e., the class of integers equivalent to elements of b. c is a set by 5.1 and 8.46, and $\omega \subseteq c$, as can be shown by induction owing to the above mentioned property of b.

8.52 $\omega \in K_{II}$

Proof: $x \in \omega\,.\supset.\, x \dotplus 1 \in \omega$, by 8.41. If $\omega = \alpha \dotplus 1$, we would have $\alpha \in \omega$ by 7.4, hence $\alpha \dotplus 1 \in \omega$, i.e., $\omega \in \omega$, which is impossible.

8.53 There exists an ordinal number of the second kind.

Proof: By 8.52, ω is such an ordinal.

*8.54 Dfn $N' = N - \omega$
*8.55 $N' \subseteq On$
*8.56 N' is isomorphic to On with respect to E.

Proof: $\mathfrak{Pr}(N')$ by 5.45, since $\mathfrak{M}(\omega)$. Moreover any proper section of N' is generated by an $\alpha \in N'$, hence $\subseteq \alpha$, hence a set. Therefore 7.7 gives the result. The isomorphism from On to N' is denoted by $\aleph$, i.e.,

*8.57 Dfn $\aleph\,\mathfrak{Isom}_{EE}(On, N')$.

It follows:

*8.58 $\aleph`0 = \omega$

since $\omega \in N$ by 8.461. $\aleph_\gamma$ and ω_γ are defined by:

*8.59 Dfn $\aleph_\gamma = \omega_\gamma = \aleph`\gamma$.

$*8.62\ \overline{\overline{\aleph_\alpha^2}} = \aleph_\alpha$

Proof: Assuming γ to be the smallest ordinal number for which $\overline{\overline{\aleph_\gamma^2}} \neq \aleph_\gamma$, we prove $\omega_\gamma^2 \simeq \omega_\gamma$. To this end, owing to the Schroeder–Bernstein Theorem, it is sufficient to show $P``(\omega_\gamma^2) \subseteq \omega_\gamma$, i.e., $P`\langle\alpha,\beta\rangle < \omega_\gamma$ for $\alpha,\beta < \omega_\gamma$, where $P$ is the function defined by 7.9. Since, for every $\delta$, $\delta < \omega_\gamma . \equiv . \overline{\overline{\delta}} < \omega_\gamma$, it is sufficient to show: $\overline{\overline{P`\langle\alpha,\beta\rangle}} < \omega_\gamma$ for $\alpha,\beta < \omega_\gamma$. Now $\overline{\overline{P`\langle\alpha,\beta\rangle}}$ is the power of the set of ordinals $< P`\langle\alpha,\beta\rangle$. This set by definition of P (7.9) is mapped by P on the set m of pairs preceding $\langle\alpha,\beta\rangle$ in the ordering R. Hence $\overline{\overline{P`\langle\alpha,\beta\rangle}} = \overline{\overline{m}}$, but (as seen in the proof of 7.811) $m \subseteq (\mu \dotplus 1)^2$, where $\mu = \mathfrak{Max}\{\alpha,\beta\}$.

| Now we distinguish two cases: 34

1. μ is finite: then $(\mu \dotplus 1)^2 < \omega$ by 8.491. Hence $\overline{\overline{m}} \leq \overline{\overline{(\mu \dotplus 1)^2}} < \omega \leq \omega_\gamma$ in this case.

2. μ is infinite: then, since $\mu < \omega_\gamma$ by assumption, $\overline{\overline{\mu}} = \omega_\delta$ for some $\delta < \gamma$. Hence $\overline{\overline{\mu^2}} = \overline{\overline{\mu}}$ by the inductive assumption. Hence (using $*8.3$) $\overline{\overline{m}} \leq \overline{\overline{(\mu \dotplus 1)^2}} \leq \overline{\overline{(\mu^2)^2}} = \overline{\overline{\mu}} < \omega_\gamma$ also in this case.

It results that:

$*8.621$ For any infinite set x, $\overline{\overline{x^2}} = \overline{\overline{x}}$,

and therefore

$*8.63\quad \mathfrak{Inf}(x) . y \neq 0 :\supset: \overline{\overline{x \times y}} = \overline{\overline{x + y}} = \mathfrak{Max}(\overline{\overline{x}}, \overline{\overline{y}})$.

Furthermore

$*8.64$ If for any $y \in m$, $\overline{\overline{F`y}} \leq \overline{\overline{a}}$, then $\overline{\overline{\mathfrak{S}(F``m)}} \leq \overline{\overline{a \times m}}$.

The proofs of these results on cardinals are not included, since they do not differ from the usual proofs.

8.7 Dfn *A is closed with respect to R* if $R``A \subseteq A$.

8.71 Dfn *A is closed with respect to S as a triadic relation* if $S``(A^2) \subseteq A$.

8.72 Y is called *closure* of X with respect to $R_1,\ldots,R_k$ and with respect to $S_1,\ldots,S_j$ as triadic relations if Y is the smallest class including X which is closed with respect to the R's and closed with respect to the S's as triadic relations.

The existence of this class will be needed only under the following conditions:

$*8.73$ If $\mathfrak{M}(X)$ and if the R's and S's are single-valued, then the closure Y exists and is a set, and if in addition X is infinite then $\overline{\overline{Y}} = \overline{\overline{X}}$.

Proof: Define $G\,\mathfrak{Fn}\,V$ as follows:

$$G\text{`}x = x + R_1\text{``}x + \cdots + R_k\text{``}x + S_1\text{``}(x^2) + \cdots + S_j\text{``}(x^2).$$

The right-hand side is normal and by 5.1, 5.13, 5.18 is a set for any set x; hence G exists by M5. Now define $f\,\mathfrak{Fn}\,\omega$ by 8.45 as follows:

$$f\text{`}0 = x, \quad f\text{`}(k \dotplus 1) = G\text{`}f\text{`}(k).$$

Now consider $\mathfrak{S}(f\text{``}\omega)$; this is a set, and satisfies the requirements of Definition 8.72. Now for any infinite set y we have $\overline{\overline{G\text{`}y}} = \overline{\overline{y}}$ by 8.31, 8.621, 8.63. Therefore, if $x$ is infinite, $\overline{\overline{f\text{`}n}} = \overline{\overline{f\text{`}0}} = \overline{\overline{x}}$, by complete induction on n. Hence

$$\overline{\overline{\mathfrak{S}(f\text{``}\omega)}} \leq \overline{\overline{x \times \omega}} = \mathfrak{Max}\,(\overline{\overline{x}}, \overline{\overline{\omega}}) = \overline{\overline{x}} \text{ by 8.64, 8.63 and}$$
$$\overline{\overline{\mathfrak{S}(f\text{``}\omega)}} \geq \overline{\overline{f\text{`}0}} = \overline{\overline{x}} \text{ by 8.28.}$$

35 |

Chapter V
The model Δ

The classes and sets of the model Δ will form a certain subfamily of the classes and sets of our original system Σ, and the ϵ-relation of the model Δ will be the original ϵ-relation confined to the classes and sets of Δ. We call the classes and sets of Δ *constructible*, and denote the notion of constructible class by $\mathfrak{L}$ and the class of constructible sets by L. Constructible sets are those which can be obtained by iterated application of the operations given by Axioms A4, B1–8, modified so that they yield sets if applied to sets. In addition, at certain stages of this generating process the set of all previously obtained sets will be added as a new constructible set. This permits the generating process to continue into the transfinite. The above mentioned axioms lead to the following eight binary operations $\mathfrak{F}_1, \ldots, \mathfrak{F}_8$ called *fundamental operations*:

9.1 Dfn $\mathfrak{F}_1(X, Y) = \{X, Y\}$,
$\mathfrak{F}_2(X, Y) = E \cdot X$,
$\mathfrak{F}_3(X, Y) = X - Y$,
$\mathfrak{F}_4(X, Y) = X \upharpoonright Y \quad (\text{i.e.}, = X \cdot (V \times Y))$,

$$\begin{aligned}
\mathfrak{F}_5(X,Y) &= X \cdot \mathfrak{D}(Y),\\
\mathfrak{F}_6(X,Y) &= X \cdot Y^{-1},\\
\mathfrak{F}_7(X,Y) &= X \cdot \mathfrak{Cnv}_2\,(Y),\\
\mathfrak{F}_8(X,Y) &= X \cdot \mathfrak{Cnv}_3\,(Y).
\end{aligned}$$

The factor X in $\mathfrak{F}_2, \ldots, \mathfrak{F}_8$ is added for reasons that will appear later (Theorem 9.5). The operation of intersection (given by Axiom B2) is left out because $X \cdot Y = X - (X - Y)$. Owing to 4.92–4.96, $\mathfrak{F}_4, \ldots, \mathfrak{F}_8$ can be expressed differently as follows:

9.11 $$\begin{aligned}
\mathfrak{F}_4(X,Y) &= X \cdot \breve{P}_2\text{``}Y,\\
\mathfrak{F}_5(X,Y) &= X \cdot P_2\text{``}Y,\\
\mathfrak{F}_6(X,Y) &= X \cdot P_3\text{``}Y,\\
\mathfrak{F}_7(X,Y) &= X \cdot P_4\text{``}Y,\\
\mathfrak{F}_8(X,Y) &= X \cdot P_5\text{``}Y.
\end{aligned}$$

In other words,

9.12 $\mathfrak{F}_i(X,Y) = X \cdot Q_i\text{``}Y, \quad i = 4, \ldots, 8,$

where the Q_i are defined by

| 9.14 $Q_4 = \breve{P}_2, \quad Q_5 = P_2, \quad Q_6 = P_3, \quad Q_7 = P_4, \quad Q_8 = P_5.$ 36

By means of Theorem 5.11 it is seen that all the fundamental operations give sets when applied to sets.

Now consider the class $9 \times On^2$ (i.e., the class of triples $\{i, \alpha, \beta\}$, $i < 9$) and define the following well-ordering relation S for it:

9.2 Dfn $\mu, \nu < 9 \,.\supset.:\, \langle \mu, \alpha, \beta \rangle S \langle \nu, \gamma, \delta \rangle \,.\equiv:$
$\langle \alpha, \beta \rangle R \langle \gamma, \delta \rangle \,.\vee.\, (\langle \alpha, \beta \rangle = \langle \gamma, \delta \rangle \,.\, \mu < \nu) :: S \subseteq (9 \times On^2)^2,$

where R is the relation defined by 7.81. Concerning the existence of S, cf. Definition 7.8. Since

$$\langle i, \alpha, \beta \rangle S \langle j, \gamma, \delta \rangle \,.\supset:\, \langle \alpha, \beta \rangle R \langle \gamma, \delta \rangle \,.\vee.\, \langle \alpha, \beta \rangle = \langle \gamma, \delta \rangle,$$

it follows from 7.811 and 5.18 that any proper S-section of $9 \times On^2$ is a set. But $9 \times On^2$ is not a set by 5.43. Hence $9 \times On^2$ is isomorphic to On with respect to S and E by 7.7, i.e., there exists a J satisfying the following defining postulate:

9.21 Dfn $J\,\mathfrak{Fn}\,(9 \times On^2) \,.\, \mathfrak{W}\,(J) = On : \mu, \nu < 9 \,.\supset$
$.\,[\langle \mu, \alpha, \beta \rangle S \langle \nu, \gamma, \delta \rangle \,.\supset.\, J`\langle \mu, \alpha, \beta \rangle < J`\langle \nu, \gamma, \delta \rangle].$

Now we define nine functions $J_0, \ldots, J_8$ over On^2 by:

9.22 Dfn $J_0{}^{\text{`}}\langle\alpha,\beta\rangle = J^{\text{`}}\langle 0,\alpha,\beta\rangle,\quad J_0\mathfrak{Fn}\,On^2,$

$\ldots\ldots\ldots\ldots\ldots\ldots\ldots,$

$J_8{}^{\text{`}}\langle\alpha,\beta\rangle = J^{\text{`}}\langle 8,\alpha,\beta\rangle,\quad J_8\mathfrak{Fn}\,On^2.$

Evidently we have:

9.23 The $\mathfrak{W}(J_i)$, $i = 0,\ldots,8$, are mutually exclusive and their sum is On. (It is easily seen, but not used in the sequel, that the $\mathfrak{W}(J_i)$ are the congruence classes of On mod. 9 and that $J_i{}^{\text{`}}\langle\alpha,\beta\rangle = 9 \dot{\times} P^{\text{`}}\langle\alpha,\beta\rangle \dot{+} i$, where $\dot{+}$ and $\dot{\times}$ denote arithmetic addition and multiplication of ordinals.)

By definition of J there exists for any γ a unique triple $\langle i,\alpha,\beta\rangle$ such that $\gamma = J^{\text{`}}\langle i,\alpha,\beta\rangle$. Hence there are two functions $K_1$, $K_2$ over $On$ such that: $K_1{}^{\text{`}}J_i{}^{\text{`}}\langle\alpha,\beta\rangle = \alpha$, $K_2{}^{\text{`}}J_i{}^{\text{`}}\langle\alpha,\beta\rangle = \beta$, for any $i < 9$. K_1, K_2 are defined by:

9.24 Dfn $\langle\alpha,\gamma\rangle \,\epsilon\, K_1 \equiv (\exists\mu,\beta)[\mu < 9\,.\,\gamma = J^{\text{`}}\langle\mu,\alpha,\beta\rangle]\,.\,K_1 \subseteq On^2,$
$\langle\beta,\gamma\rangle \,\epsilon\, K_2 \equiv (\exists\mu,\alpha)[\mu < 9\,.\,\gamma = J^{\text{`}}\langle\mu,\alpha,\beta\rangle]\,.\,K_2 \subseteq On^2.$

For the J_i and K_i we have the following theorems:

9.25 $J_i{}^{\text{`}}\langle\alpha,\beta\rangle \geq \mathfrak{Max}\{\alpha,\beta\},$
$J_i{}^{\text{`}}\langle\alpha,\beta\rangle > \mathfrak{Max}\{\alpha,\beta\}$ for $i \neq 0,$
$K_1{}^{\text{`}}\alpha \leq \alpha,\quad K_2{}^{\text{`}}\alpha \leq \alpha,$
37 | $K_1{}^{\text{`}}\alpha < \alpha,\quad K_2{}^{\text{`}}\alpha < \alpha$ for $\alpha \,\bar{\epsilon}\, \mathfrak{W}(J_0)$.

Proof: Set $\mathfrak{Max}\{\alpha,\beta\} = \gamma$; then we have $J_0{}^{\text{`}}\langle\alpha,\beta\rangle \geq J_0{}^{\text{`}}\langle\gamma,0\rangle$ by Definition 9.21; $J_0{}^{\text{`}}\langle\gamma,0\rangle \geq \gamma$ by 7.611; $J_i{}^{\text{`}}\langle\alpha,\beta\rangle > J_0{}^{\text{`}}\langle\alpha,\beta\rangle$ for $i \neq 0$ by Definition 9.21. Writing the last three inequalities as one inequality, we obtain (for $i \neq 0$):

$$J_i{}^{\text{`}}\langle\alpha,\beta\rangle > J_0{}^{\text{`}}\langle\alpha,\beta\rangle \geq J_0{}^{\text{`}}\langle\gamma,0\rangle \geq \gamma,$$

which gives the first two statements of 9.25. The last two express the same facts in terms of K_1 and K_2.

*9.26 $\alpha,\beta < \omega_\gamma \supset J_i{}^{\text{`}}\langle\alpha,\beta\rangle < \omega_\gamma$

Proof: By Definition 9.21 J maps the set m of triples preceding $\langle i,\alpha,\beta\rangle$ in the ordering S on the set of ordinals $< J_i{}^{\text{`}}\langle\alpha,\beta\rangle$. Hence $J_i{}^{\text{`}}\langle\alpha,\beta\rangle \simeq m$. But $m \subseteq 9 \times (\mathfrak{Max}\{\alpha,\beta\} \dot{+} 1)^2$ by 9.2 and 7.81. Hence the theorem by 8.491 or 8.63 according as $\gamma = 0$ or $\gamma > 0$ (using 8.492 in the first case). Note that the axiom of choice is not used in the case $\gamma = 0$.

$*9.27 \quad \omega_\alpha \,\epsilon\, \mathfrak{W}(J_0)$

Proof: $\omega_\alpha \leq J`\langle 0, \omega_\alpha, 0\rangle$ by 9.25; but not $\omega_\alpha < J`\langle 0, \omega_\alpha, 0\rangle$, because this would imply $\omega_\alpha = J`\langle i, \gamma, \delta\rangle$ for some triple $\langle i, \gamma, \delta\rangle$ preceding $\langle 0, \omega_\alpha, 0\rangle$ in the ordering S. But $\langle i, \gamma, \delta\rangle S \langle 0, \omega_\alpha, 0\rangle$ implies γ, $\delta < \omega_\alpha$, hence

$$J`\langle i, \gamma, \delta\rangle < \omega_\alpha$$

by 9.26. Hence $\omega_\alpha = J_0` \langle \omega_\alpha, 0\rangle$, i.e., $\omega_\alpha \,\epsilon\, \mathfrak{W}(J_0)$. For $\alpha = 0$ the axiom of choice is not used in this argument.

Now we define by transfinite induction a function F (the letter F is to be used only as a constant from now on. A similar remark applies to R, S, C defined respectively by 7.81, 9.2, 11.81) over On by the following postulates:

$$\begin{array}{lll} 9.3 & \text{Dfn} & \alpha \,\epsilon\, \mathfrak{W}(J_0) \supset F`\alpha = \mathfrak{W}(F \restriction \alpha), \\ & & \alpha \,\epsilon\, \mathfrak{W}(J_1) \supset F`\alpha = \mathfrak{F}_1(F`K_1`\alpha, F`K_2`\alpha), \\ & & \ldots\ldots\ldots\ldots\ldots\ldots\ldots, \\ & & \alpha \,\epsilon\, \mathfrak{W}(J_8) \supset F`\alpha = \mathfrak{F}_8(F`K_1`\alpha, F`K_2`\alpha), \\ & & F \,\mathfrak{Fn}\, On. \end{array}$$

In order to prove the existence of F by 7.5, it is necessary to define first a function G over V by the following postulates: If $\mathfrak{D}(x) \,\epsilon\, \mathfrak{W}(J_0)$, $G`x = \mathfrak{W}(x)$; if $\mathfrak{D}(x) \,\epsilon\, \mathfrak{W}(J_i)$, $i = 1, 2, \ldots, 8$,

$$G`x = \mathfrak{F}_i[x`K_1`\mathfrak{D}(x), x`K_2`\mathfrak{D}(x)];$$

and $G`x = 0$ everywhere else. Since all symbols occurring are normal (cf. page 62), $G$ exists by M6. By 7.5 there exists an $F$ over $On$ satisfying the equation $F`\alpha = G`(F \restriction \alpha)$, which implies that F satisfies 9.3, as is seen by the following proof: Suppose $\alpha \,\epsilon\, \mathfrak{W}(J_i)$, $i \neq 0$. Then, since $\mathfrak{D}(F \restriction \alpha) = \alpha$, $\mathfrak{D}(F \restriction \alpha) \,\epsilon\, \mathfrak{W}(J_i)$. Therefore

$$G`(F \restriction \alpha) = \mathfrak{F}_i[(F \restriction \alpha)`K_1`\alpha, (F \restriction \alpha)`K_2`\alpha]$$

$K_1`\alpha < \alpha$ and $K_2`\alpha < \alpha$, by 9.25, and $(F \restriction \alpha)`\beta = F`\beta$ if $\beta < \alpha$, | there- 38 fore $F`\alpha = G`(F \restriction \alpha) = \mathfrak{F}_i[F`K_1`\alpha, F`K_2`\alpha]$. Similarly, if $\alpha \,\epsilon\, \mathfrak{W}(J_0)$, then $\mathfrak{D}(F \restriction \alpha) \,\epsilon\, \mathfrak{W}(J_0)$, so that

$$F`\alpha = G`(F \restriction \alpha) = \mathfrak{W}(F \restriction \alpha).$$

Hence F exists, and by induction it is seen that F is uniquely determined. The following results are consequences of 9.3 obtained by substituting $J_i`\langle \beta, \gamma\rangle$ for $\alpha$ in the $i^{\text{th}}$ line of 9.3 and applying the equations: $K_1`J_i`\langle \alpha, \beta\rangle = \alpha$, $K_2`J_i`\langle \alpha, \beta\rangle = \beta$, which hold by definition 9.24.

$9.31 \quad F`J_1`\langle \beta, \gamma\rangle = \{F`\beta, F`\gamma\}$

9.32 $F`J_2`\langle\beta,\gamma\rangle = E \cdot F`\beta$
9.33 $F`J_3`\langle\beta,\gamma\rangle = F`\beta - F`\gamma$
9.34 $F`J_i`\langle\beta,\gamma\rangle = F`\beta \cdot Q_i``(F`\gamma), \quad i = 4, 5, \ldots, 8$
9.35 $\alpha \,\epsilon\, \mathfrak{W}(J_0) . \supset . F`\alpha = F``\alpha$

The last set of theorems show how F reflects the nine fundamental operations of 9.1.

A set x is said to be *constructible* if there exists an α such that $x = F`\alpha$. The class of constructible sets is denoted by L, i.e.,

9.4 Dfn $L = \mathfrak{W}(F)$.

A class A is *constructible* if all its elements are constructible sets and if the intersection of A with any constructible set is also a constructible set, i.e.,

9.41 Dfn $\mathfrak{L}(A) . \equiv .: A \subseteq L : x \,\epsilon\, L . \supset . A \cdot x \,\epsilon\, L.$
Dfn $\overline{x}, \ldots, \overline{z}$ will be used as variables for constructible sets and $\overline{X}, \ldots, \overline{Z}$ as variables for constructible classes.
9.42 Dfn The smallest α such that $x = F`\alpha$ is called the *order of* $x$ and is denoted by $Od`x$, i.e.,
9.421 Dfn $\langle y, x\rangle \,\epsilon\, Od \equiv \langle x, y\rangle \,\epsilon\, F . (z)[z \,\epsilon\, y \supset \sim\langle x, z\rangle \,\epsilon\, F] . Od \subseteq V^2.$
9.5 $\mathfrak{Comp}\,(F``\alpha)$

It is sufficient to prove: $F`\alpha \subseteq F``\alpha$, i.e., all elements of a constructible set appear earlier than the set itself.

Proof: Let α be the first ordinal for which $F`\alpha \subseteq F``\alpha$ is false. If $\alpha \,\epsilon\, \mathfrak{W}(J_0)$ then $F`\alpha = F``\alpha$, hence $F`\alpha \subseteq F``\alpha$. If $\alpha \,\epsilon\, \mathfrak{W}(J_i)$, $i \neq 0$, then $\alpha = J_i`\langle\beta,\gamma\rangle$, $i \neq 0$. By Theorems 9.32, 9.33, 9.34, if $i > 1$, $F`\alpha \subseteq F`\beta$. But $\beta < \alpha$, by 9.25, so that the theorem holds for $\beta$, that is, $F`\beta \subseteq F``\beta$. Hence $F`\alpha \subseteq F``\beta$. Again, since $\beta < \alpha$, $F``\beta \subseteq F``\alpha$, therefore $F`\alpha \subseteq F``\alpha$. If $i = 1$, by 9.31 $F`\alpha = \{F`\beta, F`\gamma\}$, where $\alpha = J_i`\langle\beta,\gamma\rangle$. By 9.25, β, $\gamma < \alpha$. Therefore $F`\beta \,\epsilon\, F``\alpha$ and $F`\gamma \,\epsilon\, F``\alpha$, hence $\{F`\beta, F`\gamma\} \subseteq F``\alpha$, i.e., $F`\alpha \subseteq F``\alpha$.

9.51 $\mathfrak{Comp}\,(L)$, i.e., any element of a constructible set is constructible. (For constructible classes the same thing is true by Definition 9.41.)

39 | Proof: Take $x \,\epsilon\, L$ and let $\alpha = Od`x$, so that $F`\alpha = x$. Then by 9.5, $x \subseteq F``\alpha$. Hence $x \subseteq L$, since $F``\alpha \subseteq L$.

The following statement follows from 9.5:

9.52 If $x \in y$, and $x, y \in L$, then $Od`x < Od`y$. In other words $x \in F`\alpha . \supset . Od`x < \alpha$.

$\mathfrak{F}_1, \ldots, \mathfrak{F}_8$ yield constructible sets if applied to constructible sets, i.e.,

9.6 $\mathfrak{F}_i(\overline{x}, \overline{y}) \in L, \quad i = 1, \ldots, 8.$

Proof: There exist β, γ such that $\overline{x} = F`\beta$, $\overline{y} = F`\gamma$; 9.31 to 9.34 give the result.

9.61 $\overline{x} \cdot \overline{y} \in L$

Proof: $\overline{x} \cdot \overline{y} = \overline{x} - (\overline{x} - \overline{y})$; then 9.6 for $i = 3$ gives the theorem.

9.611 $Od`\overline{x} < \omega_\alpha . Od`\overline{y} < \omega_\alpha : \supset . Od`(\overline{x} \cdot \overline{y}) < \omega_\alpha$

Proof by 9.26.

9.62 $x, y \in L . \equiv . \langle x, y \rangle \in L$ and $x, y, z \in L . \equiv . \langle x, y, z \rangle \in L.$

Proof: The implication in one direction results from expressing $\langle x, y \rangle$ as $\{\{x\}, \{x, y\}\}$, then applying 9.6; and the reverse implication is a consequence of 9.51.

9.621 $\langle x, y \rangle \in L . \equiv . \langle y, x \rangle \in L$
$\langle x, y, z \rangle \in L . \equiv . \langle z, x, y \rangle \in L . \equiv . \langle x, z, y \rangle \in L$

(follow immediately from 9.62)

9.623 $Q_i`\overline{x} \in L$ for $i = 5, 6, \ldots, 8$

(follows from 9.62, 9.621)

9.63 $x \subseteq L . \supset . (\exists \overline{y})[x \subseteq \overline{y}]$

Proof: Consider $Od``x$, which is a set of ordinals; by 7.451 there exists an ordinal $\alpha$ greater than every element of $Od``x$, i.e., such that $Od``x \subseteq \alpha$. Moreover, such an $\alpha$ can be found with the additional restriction that $\alpha \in \mathfrak{W}(J_0)$ (e.g., by taking $J_0`\langle 0, \alpha \rangle$ instead of $\alpha$, since $J_0`\langle 0, \alpha \rangle \geq \alpha$ by 9.25), hence $F`\alpha = F``\alpha$ by 9.35; but $x \subseteq F``\alpha$, hence $x \subseteq F`\alpha$, and $F`\alpha$ is a constructible set. It follows that a constructible class which is a set is a constructible set, i.e.:

| 9.64 $\mathfrak{M}(\overline{X}) . \supset . \overline{X} \in L.$ 40

Proof: By 9.41 and 9.63, $\overline{X}$ is contained in some $\overline{y}$. Therefore $\overline{X} \cdot \overline{y} = \overline{X}$, but $\overline{X} \cdot \overline{y}$ is a constructible set by 9.41.

9.65 $\mathfrak{L}(\overline{x})$

Proof: By 9.51, $\overline{x} \subseteq L$; by 9.61, $\overline{x} \cdot \overline{y} \in L$ for any $\overline{y}$.

9.66 $\overline{x} + \overline{y} \in L$

Proof: There is a $\overline{z}$ such that $\overline{x} + \overline{y} \subseteq \overline{z}$, by 9.51 and 9.63. $\overline{x} + \overline{y} = \overline{z} - [(\overline{z} - \overline{x}) - \overline{y}]$. Hence 9.6 gives the theorem.

9.8 $0 \in L$

Proof: $0 = \overline{x} - \overline{x}$, hence constructible, by 9.6.

9.81 $\mathfrak{L}(L)$

Proof: $L \subseteq L$; and because of 9.51, $\overline{x} \cdot L = \overline{x}$, hence $\overline{x} \cdot L \in L$. Therefore $\mathfrak{L}(L)$ by 9.41.

9.82 $\mathfrak{L}(E \cdot L)$

Proof: $E \cdot L \subseteq L$; also $\overline{x} \cdot E \in L$ by 9.6, since $X \cdot E$ is a fundamental operation; but $\overline{x} \cdot E = \overline{x} \cdot E \cdot L$ because $\overline{x} \subseteq L$; hence $\overline{x} \cdot E \cdot L \in L$, and so by 9.41, $\mathfrak{L}(E \cdot L)$.

9.83 $\mathfrak{L}(\overline{A} - \overline{B})$

Proof: $\overline{A} - \overline{B} \subseteq L$; moreover $\overline{x} \cdot \overline{A} - \overline{x} \cdot \overline{B}$ is constructible, by 9.41 and 9.6; but $\overline{x} \cdot \overline{A} - \overline{x} \cdot \overline{B} = \overline{x} \cdot (\overline{A} - \overline{B})$; hence $\overline{x} \cdot (\overline{A} - \overline{B}) \in L$, so that $\mathfrak{L}(\overline{A} - \overline{B})$, by 9.41.

Similarly:

9.84 $\mathfrak{L}(\overline{A} \cdot \overline{B})$,

and

9.85 $\mathfrak{L}(\overline{A} + \overline{B})$.

9.86 $\mathfrak{L}(Q_i``\overline{A})$, $i = 5, \ldots, 8$

and

9.87 $\mathfrak{L}(L \cdot Q_4``\overline{A})$.

The last two theorems are proved as follows: $Q_5, \ldots, Q_8$ take constructible sets into constructible sets, by 9.623; therefore $\mid Q_i``\overline{A} \subseteq L$, $i = 5, \ldots, 8$. 41
In order to prove that $\overline{x} . Q_i``\overline{A} \in L$ for $i = 4, \ldots, 8$, consider an arbitrary $y \in \overline{x} \cdot Q_i``\overline{A}$, $i = 4, \ldots, 8$. y is an image by Q_i of some element of $\overline{A}$; take the element y' of $\overline{A}$ of lowest order of which y is an image. The totality of these y' for all elements y of $\overline{x} \cdot Q_i``\overline{A}$ is a set u of constructible sets and $u \subseteq \overline{A}$. By 9.63 we have $u \subseteq \overline{z}$, for some $\overline{z}$. $\overline{z}$ can be determined so that $\overline{z} \subseteq \overline{A}$, merely by taking $\overline{z} \cdot \overline{A}$. Hence we can assume $u \subseteq \overline{z} \subseteq \overline{A}$. Therefore $\overline{x} \cdot Q_i``\overline{z} \subseteq \overline{x} \cdot Q_i``\overline{A}$ by 4.86; but also $\overline{x} \cdot Q_i``\overline{A} \subseteq \overline{x} \cdot Q_i``\overline{z}$ because any element of $\overline{x} \cdot Q_i``\overline{A}$ has an original in $u$, hence in $\overline{z}$. Hence $\overline{x} \cdot Q_i``\overline{A} = \overline{x} \cdot Q_i``\overline{z}$, but $(\overline{x} \cdot Q_i``\overline{z}) \in L$ by 9.6.

By means of Theorems 4.92 to 4.96, Theorems 9.86 and 9.87 take the following three forms:

9.871 $\mathfrak{L}[\mathfrak{D}(\overline{A})]$,
9.872 $\mathfrak{L}[\mathfrak{Cnv}_k(\overline{A})]$ for $k = 1, 2, 3$,
9.873 $\mathfrak{L}[L \cdot (V \times \overline{A})]$.
9.88 $\mathfrak{L}(\overline{A} \times \overline{B})$

Proof: By 4.871, $\overline{A} \times \overline{B} = (V \times \overline{B}) \cdot (\overline{A} \times V) = L \cdot (V \times \overline{B}) \cdot L \cdot (\overline{A} \times V)$, because $\overline{A} \times \overline{B} \subseteq L$, by 9.62. By 9.873 and 9.872, $\mathfrak{L}[L \cdot (V \times \overline{B})]$ and $\mathfrak{L}[L \cdot (\overline{A} \times V)]$. Hence, by 9.84 $\mathfrak{L}(\overline{A} \times \overline{B})$.

9.89 $\mathfrak{L}[\mathfrak{W}(\overline{A})]$

Proof: $\mathfrak{W}(\overline{A}) = \mathfrak{D}(\breve{\overline{A}})$, hence the result follows from 9.871 and 9.872.

9.90 $\mathfrak{L}[\overline{A} \restriction \overline{B}]$

Proof: $\overline{A} \restriction \overline{B} = \overline{A} \cdot (V \times \overline{B}) = \overline{A} \cdot L \cdot (V \times \overline{B})$, hence the theorem, by 9.873 and 9.84.

9.91 $\mathfrak{L}[\overline{A}``\overline{B}]$

Proof: $\overline{A}``\overline{B} = \mathfrak{W}(\overline{A} \restriction \overline{B})$, hence the theorem, by 9.89 and 9.90.

9.92 $\mathfrak{L}(\{\overline{X}, \overline{Y}\})$

Proof: By Definition 3.1 $\{\overline{X}, \overline{Y}\}$ is either 0 or $\{\overline{X}\}$ or $\{\overline{Y}\}$ or $\{\overline{X}, \overline{Y}\}$, where now only sets can appear within the braces. Hence the theorem, by 9.6, 9.65, 9.8.

Not all operations on constructible classes give necessarily constructible classes. For example, it cannot be shown that $\mathfrak{L}[\mathfrak{P}(\overline{X})]$.

Now consider *the model* Δ obtained as follows:

1. Class is construed as constructible class.
2. Set is construed as constructible set.
3. ϵ_l, the membership relation, is to be the ϵ-relation | confined to constructible classes, i.e, $\overline{X} \epsilon_l \overline{Y} . \equiv . \overline{X} \epsilon \overline{Y}$.

42

The operations, notions and special classes defined so far can be *relativized* for this model Δ by replacing in their definition or defining postulate the variables $X, Y, \ldots$ by $\overline{X}, \overline{Y}, \ldots$; the variables $x, y, \ldots$ by $\overline{x}, \overline{y}, \ldots$; ϵ by ϵ_l; and the previously defined concepts and variables by the corresponding relativized ones, leaving the logical symbols (in particular also $=$, which is considered as a logical concept) as they stand. The relativized of a variable $\mathfrak{x}$ is a variable whose range is obtained by relativizing the notion which defines the range of $\mathfrak{x}$. Note that for an operation or special class the relativized need not exist a priori, because the theorem which states existence and unicity (cf. page 12) may not hold in the model Δ; furthermore the relativized concept may depend on the particular definition which we chose, since equivalent definitions need not be equivalent in Δ. (However, as soon as we have proved that the axioms of Σ hold for Δ, we know that the relativized always does exist and does not depend on the particular definition.) If the relativized of a defined class A, operation $\mathfrak{A}$, notion $\mathfrak{B}$, variable $\mathfrak{x}$ exists (which presupposes that also the relativized of any symbol occurring in its definition exists), we denote it by A_l, $\mathfrak{A}_l$, $\mathfrak{B}_l$, $\mathfrak{x}_l$ (hence x_l, X_l have the same range as $\overline{x}$, $\overline{X}$). $\mathfrak{A}_l$ and $\mathfrak{B}_l$ are defined for constructible classes as arguments only, and we have the theorem:

10.1 If A_l and $\mathfrak{A}_l$ exist, then A_l is constructible and $\mathfrak{A}_l(\overline{X}_1, \ldots, \overline{X}_n)$ is constructible for any $\overline{X}_1, \ldots, \overline{X}_n$.

Evidently the relativized classes, notions and operations are at the same time classes, notions and operations of the system Σ, if the requirement on page 11, that they be defined for any classes as arguments, is met, e.g., by stipulating that $\mathfrak{A}_l\,(X_1, \ldots, X_n) = 0$ and $\mathfrak{B}_l\,(X_1, \ldots, X_n)$ is false, if $X_1, \ldots, X_n$ are not all constructible.

10. Dfn A special class A or operation $\mathfrak{A}$ or notion $\mathfrak{B}$ is called *absolute* if A_l, $\mathfrak{A}_l$, or $\mathfrak{B}_l$ exists, respectively, and $A_l = A$, $\mathfrak{A}_l\,(\overline{X}_1, \ldots, \overline{X}_n) = \mathfrak{A}\,(\overline{X}_1, \ldots, \overline{X}_n)$, or $\mathfrak{B}_l\,(\overline{X}_1, \ldots, \overline{X}_n) . \equiv . \mathfrak{B}\,(\overline{X}_1, \ldots, \overline{X}_n)$, respectively, for any $\overline{X}_1, \ldots, \overline{X}_n$. A variable $\mathfrak{x}$ is called *absolute* if the range of $\mathfrak{x}_l$ is the same as the range of $\mathfrak{x}$.

By Theorem 10.1 we have:

10.11 If A (the operation $\mathfrak{A}$) is absolute, then A is constructible ($\mathfrak{A}(\overline{X}_1, \ldots, \overline{X}_n)$ is constructible for any $\overline{X}_1, \ldots, \overline{X}_n$).

Concerning the meaning and purpose of the metamathematical notions of relativization and absoluteness, cf. page 1. The relativized of a propositional function ϕ or a proposition ψ | is denoted by ϕ_l, ψ_l, respectively, 43 and obtained by replacing any concept and variable occurring in it by the relativized one (presupposing that they all exist). In particular also the relativized of a theorem is quoted by putting a subscript l to its number.

10.12 ϵ is absolute.

This is true by definition of ϵ_l.

10.13 "$\subseteq$" is absolute.

Proof: $\overline{X} \subseteq_l \overline{Y} . \equiv . (\overline{u})[\overline{u} \,\epsilon_l\, \overline{X} . \supset . \overline{u} \,\epsilon_l\, \overline{Y}] . \equiv . (\overline{u})[\overline{u} \,\epsilon\, \overline{X} . \supset . \overline{u} \,\epsilon\, \overline{Y}]$. Also $\overline{X} \subseteq \overline{Y} . \equiv . (u)[u \,\epsilon\, \overline{X} . \supset . u \,\epsilon\, \overline{Y}]$. If $(u)[u \,\epsilon\, \overline{X} . \supset . u \,\epsilon\, \overline{Y}]$, then in particular $(\overline{u})[\overline{u} \,\epsilon\, \overline{X} . \supset . \overline{u} \,\epsilon\, \overline{Y}]$. On the other hand, the reverse implication holds, since, if u is not in L, the condition holds vacuously, because the hypothesis $u \,\epsilon\, \overline{X}$ is false. Therefore $\overline{X} \subseteq_l \overline{Y} . \equiv . \overline{X} \subseteq \overline{Y}$.

10.131 $\overline{X} \subseteq_l \overline{Y} . \overline{Y} \subseteq_l \overline{X} : \supset . \overline{X} = \overline{Y}$, i.e., the relativized axiom of extensionality holds.

Proof by 10.13 and the axiom of extensionality.

10.14 "$\subset$" is absolute.

Proof: $\overline{X} \subset_l \overline{Y} . \equiv : \overline{X} \subseteq_l \overline{Y} . \overline{X} \neq \overline{Y} . \equiv . \overline{X} \subset \overline{Y}$, by 10.13.
Similarly:

10.15 $\mathfrak{Ex}$ is absolute, i.e., $\mathfrak{Ex}_l(\overline{X}, \overline{Y}) . \equiv . \mathfrak{Ex}(\overline{X}, \overline{Y})$.
10.16 $\mathfrak{Em}$ is absolute, i.e., $\mathfrak{Em}_l(\overline{X}) . \equiv . \mathfrak{Em}(\overline{X})$.
10.17 The operation $\{\overline{X}, \overline{Y}\}$ is absolute.

Proof: By 3.1 $\{\overline{X}, \overline{Y}\}_l$ is the constructible class $\overline{Z}$ such that

$$(\overline{u})[\overline{u} \,\epsilon\, \overline{Z} . \equiv : \overline{u} = \overline{X} . \vee . \overline{u} = \overline{Y}].$$

$\{\overline{X}, \overline{Y}\}$ satisfies this condition on $\overline{Z}$ because it satisfies it even with (u) instead of $(\overline{u})$. Moreover, $\{\overline{X}, \overline{Y}\}$ is constructible, by 9.92. Also $\{\overline{X}, \overline{Y}\}$ is the only constructible class satisfying the condition (by 10.131). Hence the relativized of the operation $\{X,Y\}$ exists and $\{\overline{X},\overline{Y}\}_l = \{\overline{X},\overline{Y}\}$ for any $\overline{X}$, $\overline{Y}$, i.e., $\{X,Y\}$ is absolute.

10.18 If $\mathfrak{C}$ is defined by $\mathfrak{C}(\overline{X}) = \mathfrak{A}(\mathfrak{B}(\overline{X}))$ and $\mathfrak{A}$ and $\mathfrak{B}$ are absolute, then $\mathfrak{C}$ is absolute.

Proof: $\mathfrak{A}(\mathfrak{B}(\overline{X})) = \mathfrak{A}(\mathfrak{B}_l(\overline{X}))$, but $\mathfrak{B}_l(\overline{X})$ is constructible by 10.1, hence $\mathfrak{A}(\mathfrak{B}_l(\overline{X})) = \mathfrak{A}_l(\mathfrak{B}_l(\overline{X})) = \mathfrak{C}_l(\overline{X})$.
This principle holds also for operations with more than one argument.

44 | 10.19 The operation $\langle X,Y\rangle$ is absolute.

This is an immediate consequence of 10.17 and 10.18. Similarly:

10.20 The operation $\langle X,Y,Z\rangle$ is absolute.
10.21 $\mathfrak{Un}$ is absolute.

Proof: $\mathfrak{Un}_l(\overline{X}) . \equiv . (\overline{u},\overline{v},\overline{w})[\langle\overline{v},\overline{u}\rangle_l \,\epsilon_l\, \overline{X} . \langle\overline{w},\overline{u}\rangle_l \,\epsilon_l\, \overline{X} :\supset . \overline{v} = \overline{w}]$.
By 10.12 and 10.19 the subscript l can be dropped wherever it appears on the right. The condition is now equivalent to that obtained by replacing $\overline{u},\overline{v},\overline{w}$, by u, v, w, respectively, as in the proof of 10.13 (using 9.62).

10.22 $\mathfrak{M}$ is absolute and $\mathfrak{Pr}$ is absolute.

Proof: $\mathfrak{M}_l(\overline{X}) . \equiv . \overline{X} \,\epsilon\, L$, by definition of the model Δ on page 41, therefore $\mathfrak{M}_l(\overline{X}) . \equiv . \mathfrak{M}(\overline{X})$, by 9.64 and Axiom A2. Hence also $\sim\mathfrak{M}_l(\overline{X}) . \equiv . \sim\mathfrak{M}(\overline{X})$.

Not all concepts can be proved to be absolute; for example, $\mathfrak{P}$ and V cannot be proved to be absolute.

10.23 $V_l = L$

Proof: V_l is defined by the postulate $(\overline{x})[\overline{x} \,\epsilon\, V_l]$. L satisfies the condition, hence $L = V_l$, because of the relativized axiom of extensionality and because $\mathfrak{L}(L)$ by 9.81.

10.24 0 is absolute.

Proof: $(\overline{x})[\sim\overline{x} \,\epsilon\, 0]$, and 0 is the only constructible class satisfying this postulate.

| 45

Chapter VI
Proof of the axioms of groups A–D for the model Δ

Every notion and operation appearing in the *axioms* has now been shown to be absolute. This facilitates the proofs of the relativized axioms, since in forming the relativized of a proposition all absolute notions and operations can be left as they stand, because by 10.1 only constructible classes can appear as their arguments, so that the relativized axioms may be formed merely by replacing X by $\overline{X}$ and x by $\overline{x}$. For convenience we list the axioms in their relativized form:

A1_l $\mathfrak{L}(\overline{x})$,
2_l $\overline{X} \in \overline{Y} . \supset . \mathfrak{M}(\overline{X})$,
3_l $(\overline{u})[\overline{u} \in \overline{X} . \equiv . \overline{u} \in \overline{Y}] . \supset . \overline{X} = \overline{Y}$,
4_l $(\overline{x}, \overline{y})(\exists \overline{z})(\overline{u})[\overline{u} \in \overline{z} . \equiv : \overline{u} = \overline{y} . \vee . \overline{u} = \overline{x}]$;
B1_l $(\exists \overline{A})(\overline{x}, \overline{y})[\langle \overline{x}, \overline{y} \rangle \in \overline{A} . \equiv . \overline{x} \in \overline{y}]$,
2_l $(\overline{A}, \overline{B})(\exists \overline{C})(\overline{x})[\overline{x} \in \overline{C} . \equiv : \overline{x} \in \overline{A} . \overline{x} \in \overline{B}]$,
3_l $(\overline{A})(\exists \overline{B})(\overline{x})[\overline{x} \in \overline{B} . \equiv . \sim \overline{x} \in \overline{A}]$,
4_l $(\overline{A})(\exists \overline{B})(\overline{x})[\overline{x} \in \overline{B} . \equiv . (\exists \overline{y})[\langle \overline{y}, \overline{x} \rangle \in \overline{A}]]$,
5_l $(\overline{A})(\exists \overline{B})(\overline{x}, \overline{y})[\langle \overline{y}, \overline{x} \rangle \in \overline{B} . \equiv . \overline{x} \in \overline{A}]$,
6_l $(\overline{A})(\exists \overline{B})(\overline{x}, \overline{y})[\langle \overline{x}, \overline{y} \rangle \in \overline{B} . \equiv . \langle \overline{y}, \overline{x} \rangle \in \overline{A}]$,
7_l $(\overline{A})(\exists \overline{B})(\overline{x}, \overline{y}, \overline{z})[\langle \overline{x}, \overline{y}, \overline{z} \rangle \in \overline{B} . \equiv . \langle \overline{y}, \overline{z}, \overline{x} \rangle \in \overline{A}]$,
8_l $(\overline{A})(\exists \overline{B})(\overline{x}, \overline{y}, \overline{z})[\langle \overline{x}, \overline{y}, \overline{z} \rangle \in \overline{B} . \equiv . \langle \overline{x}, \overline{z}, \overline{y} \rangle \in \overline{A}]$;
C1_l $(\exists \overline{a})\{\sim \mathfrak{Em}(\overline{a}) . (\overline{x})[\overline{x} \in \overline{a} . \supset . (\exists \overline{y})(\overline{y} \in \overline{a} . \overline{x} \subset \overline{y})]\}$,
2_l $(\overline{x})(\exists \overline{y})(\overline{u}, \overline{v})[\overline{u} \in \overline{v} . \overline{v} \in \overline{x} : \supset . \overline{u} \in \overline{y}]$,
3_l $(\overline{x})(\exists \overline{y})(\overline{u})[\overline{u} \subseteq \overline{x} . \supset . \overline{u} \in \overline{y}]$,
4_l $(\overline{x}, \overline{A})\{\mathfrak{Un}(\overline{A}) . \supset . (\exists \overline{y})(\overline{u})[\overline{u} \in \overline{y} . \equiv . (\exists \overline{v})(\overline{v} \in \overline{x} . \langle \overline{u}, \overline{v} \rangle \in \overline{A})]\}$;
D_l $\sim \mathfrak{Em}(\overline{A}) . \supset . (\exists \overline{x})[\overline{x} \in \overline{A} . \mathfrak{Ex}(\overline{x}, \overline{A})]$.

A1_l is Theorem 9.65, A2_l is immediate from A2, A3_l holds by 10.131, A4_l is satisfied for $\overline{z} = \{\overline{x}, \overline{y}\}$, which is constructible by 9.6. Now we prove B1–8_l by exhibiting in each case a constructible class satisfying the conditions, as follows:

B1_l Take $\overline{A} = E \cdot L$. The class $E \cdot L$ is constructible by 9.82 and satisfies $\langle \overline{x}, \overline{y} \rangle \in E \cdot L . \equiv . \overline{x} \in \overline{y}$, because $\langle \overline{x}, \overline{y} \rangle \in E . \equiv . \overline{x} \in \overline{y}$ and $\langle \overline{x}, \overline{y} \rangle \in L$.

2_l Take $\overline{C} = \overline{A} \cdot \overline{B}$. This class is constructible by 9.84 and satisfies B2.

3_l Take $\overline{B} = L - \overline{A}$. This class is constructible by 9.83, 9.81 and satisfies B3.

4_l Take $\overline{B} = \mathfrak{D}(\overline{A})$. By 9.871 $\mathfrak{D}(\overline{A})$ is constructible.
$x \in \overline{B} . \equiv . (\exists y)[\langle y, x \rangle \in \overline{A}]$. Therefore, in particular,

| $$\overline{x} \in \overline{B} . \equiv . (\exists y)[\langle y, \overline{x} \rangle \in \overline{A}] . \equiv . (\exists \overline{y})[\langle \overline{y}, \overline{x} \rangle \in \overline{A}].$$ 46

The last equivalence holds, because, if there exists a y it must be constructible by 9.62.

5_l Take $\overline{B} = L \cdot (V \times \overline{A})$. $\overline{B}$ is constructible, by 9.873.
$\langle x, y\rangle \,\epsilon\, \overline{B} \,.\!\equiv.\, \langle x, y\rangle \,\epsilon\, L \,.\, y \,\epsilon\, \overline{A}$; therefore $\langle \overline{x}, \overline{y}\rangle \,\epsilon\, \overline{B} \,.\!\equiv.\, \langle \overline{x}, \overline{y}\rangle \,\epsilon\, L \,.\, \overline{y} \,\epsilon\, \overline{A}$ so that $\langle \overline{x}, \overline{y}\rangle \,\epsilon\, \overline{B} \,.\!\equiv.\, \overline{y} \,\epsilon\, \overline{A}$, since $\langle \overline{x}, \overline{y}\rangle \,\epsilon\, L$, by 9.62.

6_l Take $\overline{B} = \mathfrak{Cnv}(\overline{A})$. $\overline{B}$ is constructible, by 9.872.
$\langle x, y\rangle \,\epsilon\, \mathfrak{Cnv}(\overline{A}) \,.\!\equiv.\, \langle y, x\rangle \,\epsilon\, \overline{A}$; therefore, in particular,

$$\langle \overline{x}, \overline{y}\rangle \,\epsilon\, \mathfrak{Cnv}(\overline{A}) \,.\!\equiv.\, \langle \overline{y}, \overline{x}\rangle \,\epsilon\, \overline{A}.$$

Axioms B7–8_l are proved in the same manner. Now consider Axioms C1–4_l:

1_l $\mathrm{C}1_l$ is satisfied by $\overline{a} = F`\omega$.

Proof: $\omega \,\epsilon\, \mathfrak{W}(J_0)$ by 9.27, hence $F`\omega = F``\omega$. If $\overline{x} \,\epsilon\, \overline{a}$ (i.e., $\overline{x} = F`\alpha$, $\alpha < \omega$), let $\beta$ be an integer $\epsilon\, \mathfrak{W}(J_0)$ and $> \alpha$ (e.g., $\beta = J_0`\langle 0, \alpha \dotplus 1\rangle$ by 9.25 and 9.26) and put $\overline{y} = F`\beta$; then $\overline{y} \,\epsilon\, \overline{a}$ and $\overline{y} \supset \overline{x}$ because $F`\beta = F``\beta$ and $F`\alpha \subseteq F``\beta$.
Moreover: $F`\alpha \,\epsilon\, F``\beta$ but $\sim(F`\alpha \,\epsilon\, F`\alpha)$ so that $F`\alpha \subset F`\beta$.

2_l Consider $\mathfrak{S}(\overline{x})$; this is a set of constructible sets by 5.122 and 9.51. Therefore, by 9.63, there is a $\overline{y}$ such that $\mathfrak{S}(\overline{x}) \subseteq \overline{y}$. Hence $(u, v)[u \,\epsilon\, v \,.\, v \,\epsilon\, \overline{x} :\supset. u \,\epsilon\, \overline{y}]$, therefore

$$(\overline{u}, \overline{v})[\overline{u} \,\epsilon\, \overline{v} \,.\, \overline{v} \,\epsilon\, \overline{x} :\supset. \overline{u} \,\epsilon\, \overline{y}],$$

that is, $\overline{y}$ satisfies the condition of $\mathrm{C}2_l$.

3_l Consider $L \cdot \mathfrak{P}(\overline{x})$ (which is a set by 5.121) and take $\overline{y}$ such that $L \cdot \mathfrak{P}(\overline{x}) \subseteq \overline{y}$, by 9.63. Then $u \,\epsilon\, L \,.\, \mathfrak{P}(\overline{x}) \,.\!\supset.\, u \,\epsilon\, \overline{y}$. Therefore $\overline{u} \,\epsilon\, L \cdot \mathfrak{P}(\overline{x}) \,.\!\supset.\, \overline{u} \,\epsilon\, \overline{y}$, so that $\overline{u} \,\epsilon\, \mathfrak{P}(\overline{x}) \,.\!\supset.\, \overline{u} \,\epsilon\, \overline{y}$, that is, $\overline{u} \subseteq \overline{x} \,.\!\supset.\, \overline{u} \,\epsilon\, \overline{y}$.

4_l Take $\overline{y} = \overline{A}``\overline{x}$. $\overline{y}$ is constructible, by 9.91.

$$u \,\epsilon\, \overline{y} \,.\!\equiv.\, (\exists v)[v \,\epsilon\, \overline{x} \,.\, \langle u, v\rangle \,\epsilon\, \overline{A}],$$

therefore, in particular,

$$\overline{u} \,\epsilon\, \overline{y} \,.\!\equiv.\, (\exists v)[v \,\epsilon\, \overline{x} \,.\, \langle \overline{u}, v\rangle \,\epsilon\, \overline{A}].$$

Now, if there is a constructible v, there is a v satisfying the condition; on the other hand, if there is a v, v will be constructible, since $v \,\epsilon\, \overline{x}$. Therefore

$$\overline{u} \,\epsilon\, \overline{y} \,.\!\equiv.\, (\exists \overline{v})[\overline{v} \,\epsilon\, \overline{x} \,.\, \langle \overline{u}, \overline{v}\rangle \,\epsilon\, \overline{A}].$$

Finally, consider Axiom D_l. By Axiom D, $(\exists x)[x \,\epsilon\, \overline{A} \,.\, \mathfrak{Ex}(x, \overline{A})]$. But x is constructible, since $x \,\epsilon\, \overline{A}$. Hence there is an $\overline{x}$ satisfying the condition.

Since all axioms of Σ hold in Δ, it follows now that all theorems proved so far also hold in the model Δ, except perhaps those based on the axiom of choice. Therefore the existence and unicity theorems necessary for the definition of the special classes and the operations introduced so far also will hold in Δ, and, as a result, the relativized of every concept introduced so far exists (except those definitions marked by $*$, which depend on the axiom of choice); in particular also $\mathfrak{L}_l$ and L_l exist.

| 47

Chapter VII
Proof that $V = L$ holds in the model Δ

In order to prove that the axiom of choice and the generalized continuum hypothesis hold for the model Δ, we shall show: (1) that both of them follow from the axioms of Σ and the additional axiom $V = L$ (which says that every set is constructible) and (2) that $V = L$ holds in the model Δ, i.e., $V_l = L_l$. We begin with item (2). Since $V_l = L$ by 10.23, it is sufficient to prove $L_l = L$, that is, *the class of constructible sets is absolute.* To that end, it will be shown that all operations, etc. used in the construction of L are absolute.

A general remark for proofs of absoluteness will be useful. In order for the operation $\mathfrak{A}(X_1, \ldots, X_n)$ to be absolute it is sufficient to show that

(1) $\mathfrak{A}$ gives constructible classes when applied to constructible classes, and

(2) $\mathfrak{A}$ satisfies the relativized defining postulate, i.e., if $\mathfrak{A}$ is defined by $\phi(\mathfrak{A}(X_1, \ldots, X_n), X_1, \ldots, X_n)$, then $\phi_l(\mathfrak{A}(\overline{X}_1, \ldots, \overline{X}_n), \overline{X}_1, \ldots, \overline{X}_n)$.

It is easily verified that (1) and (2) are sufficient, namely, as follows: $\mathfrak{A}_l$ exists, since the model satisfies the axioms of Σ. Hence ϕ_l has the property that for any $\overline{X}_1, \ldots, \overline{X}_n$ there exists at most one $\overline{Y}$ such that $\phi_l(\overline{Y}, \overline{X}_1, \ldots, \overline{X}_n)$.
But

$$\phi_l(\mathfrak{A}_l(\overline{X}_1, \ldots, \overline{X}_n), \overline{X}_1, \ldots, \overline{X}_n)$$

by definition of $\mathfrak{A}_l$ and

$$\phi_l(\mathfrak{A}(\overline{X}_1, \ldots, \overline{X}_n), \overline{X}_1, \ldots, \overline{X}_n)$$

by assumption (2). Therefore

$$\mathfrak{A}_l(\overline{X}_1, \ldots, \overline{X}_n) = \mathfrak{A}(\overline{X}_1, \ldots, \overline{X}_n).$$

Similarly for the particular class A it is sufficient to show that it is constructible and satisfies the relativized postulate. Remember also that by

10.18 operations defined by substituting absolute operations into absolute operations are absolute.

11.1 "$\times$" is absolute.

Proof: $\overline{A} \times \overline{B}$ is constructible, by 9.88.
$u \,\epsilon\, \overline{A} \times \overline{B} \,.\!\equiv\!.\, (\exists v, w)[v \,\epsilon\, \overline{A} \,.\, w \,\epsilon\, \overline{B} \,.\, u = \langle v, w\rangle]$ by Definition 4.1. Therefore $\overline{u} \,\epsilon\, \overline{A} \times \overline{B} \,.\!\equiv\!.\, (\exists v, w)[v \,\epsilon\, \overline{A} \,.\, w \,\epsilon\, \overline{B} \,.\, \overline{u} = \langle v, w\rangle]$. Now, in the usual manner, the condition on the right is equivalent to that obtained by replacing v, w by $\overline{v}, \overline{w}$ respectively. Therefore $\overline{A} \times \overline{B}$ satisfies the relativized postulate; hence "$\times$" is absolute, by the remark made above.

11.11 The operations $A^2, A^3, \ldots$ are absolute.

48 | This follows from 10.18 and 11.1.

11.12 $\mathfrak{Rel}$ and $\mathfrak{Rel}_3$ are absolute.

Proof: $\mathfrak{Rel}(\overline{X}) \,.\equiv.\, \overline{X} \subseteq V^2$ and $\mathfrak{Rel}_l(\overline{X}) \,.\equiv.\, \overline{X} \subseteq L^2$, by 10.23; but $\overline{X} \subseteq L^2 \,.\!\equiv\!.\, \overline{X} \subseteq V^2$, by 9.62. Hence $\mathfrak{Rel}_l(\overline{X}) \,.\!\equiv\!.\, \mathfrak{Rel}(\overline{X})$. Similarly for $\mathfrak{Rel}_3$.

11.13 $\mathfrak{D}$ is absolute.

Proof: $\mathfrak{D}(\overline{A})$ is constructible, by 9.871. $x \,\epsilon\, \mathfrak{D}(\overline{A}) \,.\equiv.\, (\exists y)[\langle y, x\rangle \,\epsilon\, \overline{A}]$, therefore $\overline{x} \,\epsilon\, \mathfrak{D}(\overline{A}) \,.\!\equiv\!.\, (\exists y)[\langle y, \overline{x}\rangle \,\epsilon\, \overline{A}]$. In the usual way, the last condition is equivalent to that obtained by replacing y by $\overline{y}$, so that $\mathfrak{D}(A)$ satisfies the relativized postulate.

11.14 "$\cdot$" is absolute.

Proof: $\overline{A} \cdot \overline{B}$ is constructible, by 9.84. $x \,\epsilon\, \overline{A} \cdot \overline{B} \,.\!\equiv\!:\, x \,\epsilon\, \overline{A} \,.\, x \,\epsilon\, \overline{B}$; therefore $\overline{x} \,\epsilon\, \overline{A} \cdot \overline{B} \,.\!\equiv\!:\, \overline{x} \,\epsilon\, \overline{A} \,.\, \overline{x} \,\epsilon\, \overline{B}$, that is, $\overline{A} \cdot \overline{B}$ satisfies the relativized postulate.

11.15 $\mathfrak{Cnv}_k$ is absolute ($k = 1, 2, 3$).

Proof: $\mathfrak{Cnv}_k(\overline{A})$ is constructible, by 9.872. Consider, e.g., $\mathfrak{Cnv}_1(\overline{A})$. It satisfies the condition

$$\mathfrak{Rel}(\mathfrak{Cnv}_1(\overline{A})) \,.\, (x, y)[\langle x, y\rangle \,\epsilon\, \mathfrak{Cnv}_1(\overline{A}) \,.\!\equiv\!.\, \langle y, x\rangle \,\epsilon\, \overline{A}]$$

by definition. This condition implies the relativized statement by 11.12. Similarly for $\mathfrak{Cnv}_k(\overline{A})$.

11.16 "$\upharpoonright$" is absolute.

Proof: $\overline{A} \upharpoonright \overline{B} = \overline{A} \cdot (V \times \overline{B})$ and $\overline{A} \upharpoonright_l \overline{B} = \overline{A} \cdot (L \times \overline{B})$. But $\overline{A} \cdot (V \times \overline{B}) \subseteq L \times L$ by 9.62; therefore $\overline{A} \upharpoonright \overline{B} = \overline{A} \cdot (V \times \overline{B}) \cdot (L \times L) = \overline{A} \cdot (L \times \overline{B})$, by 4.87. Therefore $\overline{A} \upharpoonright \overline{B} = \overline{A} \upharpoonright_l \overline{B}$.

11.17 "$\mathfrak{W}$" is absolute.

Proof: $\mathfrak{W}(A) = \mathfrak{D}(\mathfrak{Cnv}(A))$ by definition. Hence the theorem by 10.18, 11.13, and 11.15.

11.18 The operation A"B is absolute.

Proof: A"$B = \mathfrak{W}(A \upharpoonright B)$, by definition. Hence the theorem by 10.18.

11.181 The relativized operation of the complement is $L - \overline{X}$.

Proof: $L - \overline{X}$ is constructible by 9.81, 9.83 and $\overline{y} \in L - \overline{X} . \equiv . \sim \overline{y} \in \overline{X}$.

11.19 The operation $A - B$ is absolute.

| Proof: $\overline{A} -_l \overline{B} = \overline{A} \cdot (L - \overline{B}) = \overline{A} \cdot L \cdot (-\overline{B}) = \overline{A} \cdot (-\overline{B}) = \overline{A} - \overline{B}$. 49

11.20 "+" is absolute.

Proof: $\overline{A} +_l \overline{B} = L - [(L - \overline{A}) \cdot (L - \overline{B})] = L - [L - (\overline{A} + \overline{B})] = \overline{A} + \overline{B}$ (since $\overline{A} + \overline{B} \subseteq L$).

11.21 $E_l = E \cdot L$

Proof: $E \cdot L$ is constructible by 9.82. Also

$$\mathfrak{Rel}_l\,(E \cdot L) \,.\, (\overline{x}, \overline{y})[\langle \overline{x}, \overline{y} \rangle \in E \cdot L \,.\!\equiv\!.\, \overline{x} \in \overline{y}],$$

since $\mathfrak{Rel}(E \cdot L)$ and since $\langle \overline{x}, \overline{y} \rangle \in L$ and $\langle \overline{x}, \overline{y} \rangle \in E \,.\!\equiv\!.\, \overline{x} \in \overline{y}$. Therefore $E \cdot L$ satisfies the relativized postulate.

11.22 $\mathfrak{F}_2$ is absolute.

Proof: $\mathfrak{F}_{2_l}(\overline{X}, \overline{Y}) = \overline{X} \cdot_l E_l = \overline{X} \cdot L \cdot E = \overline{X} \cdot E = \mathfrak{F}_2(\overline{X}, \overline{Y})$, by 11.14, 11.21.

11.221 All the fundamental operations $\mathfrak{F}_i$ $(i = 1, 2, \ldots, 8)$ are absolute.

The proof follows from 10.17, 11.22, 11.19, 11.16, 11.13, 11.15, respectively, using 10.18 and 11.14.

11.23 The binary operation $A`X$ is absolute.

Proof: Since any y satisfying $\langle y, \overline{X}\rangle \in \overline{A}$ is constructible by 9.51, we have: if there is exactly one *constructible* set y such that $\langle y, \overline{X}\rangle \in \overline{A}$, there is exactly one set, and vice versa. Therefore $(\overline{A}`)_l\overline{X} = \overline{A}`\overline{X}$ in this case; in the contrary case both are 0.

11.3 $\mathfrak{Comp}$ is absolute.

Proof:

$$\begin{aligned}\mathfrak{Comp}\,(\overline{X}) \,.\!\equiv.\, &(u)[u \in \overline{X} \,.\!\supset.\, u \subseteq \overline{X}].\\ \equiv.\, &(\overline{u})[\overline{u} \in \overline{X} \,.\!\supset.\, \overline{u} \subseteq \overline{X}] \,.\!\equiv.\, \mathfrak{Comp}_l\,(\overline{X}).\end{aligned}$$

11.31 $\mathfrak{Ord}$ is absolute.

Proof:

$$\begin{aligned}\mathfrak{Ord}\,(\overline{X}) \,.\!\equiv:\, &\mathfrak{Comp}\,(\overline{X}) \,.\, (u,v)[u,v \in \overline{X} \;:\supset\\ &\qquad :.\, u = v \,.\!\vee.\, u \in v \,.\!\vee.\, v \in u].\\ \equiv:\, &\mathfrak{Comp}_l\,(\overline{X}) \,.\, (\overline{u},\overline{v})[\overline{u},\overline{v} \in \overline{X} \;:\supset\\ &\qquad :.\, \overline{u} = \overline{v} \,.\!\vee.\, \overline{u} \in \overline{v} \,.\!\vee.\, \overline{v} \in \overline{u}].\\ \equiv:\, &\mathfrak{Ord}_l\,(\overline{X}).\end{aligned}$$

The first and last equivalences follow immediately from the definition of $\mathfrak{Ord}$ and $\mathfrak{Ord}_l$.

11.32 $\mathfrak{O}$ is absolute.

Proof:

$$\begin{aligned}\mathfrak{O}\,(\overline{X}) \,.\!\equiv:\, &\mathfrak{Ord}\,(\overline{X}) \,.\, \mathfrak{M}\,(\overline{X}) \;:\equiv:\; \mathfrak{Ord}_l\,(\overline{X}) \,.\, \mathfrak{M}_l\,(\overline{X}).\\ \equiv:\, &\mathfrak{O}_l\,(\overline{X}), \quad \text{by } 11.31, 10.22.\end{aligned}$$

50 | 11.31 says that the ordinals of the model Δ are the same as the ordinals which belong to the model Δ. This does not mean that the ordinals of the model are the same as the ordinals of the original system, since nothing is said of those ordinals which may not belong to the model (i.e., may not be constructible). Cf. however 11.42.

11.4 "$\mathfrak{Fnc}$" is absolute.

Proof:

$$\begin{aligned}\mathfrak{Fnc}_l(\overline{Y}) \,.\!\equiv:\, &\mathfrak{Rel}_l(\overline{Y}) \,.\, \mathfrak{Un}_l(\overline{Y}) \,.\\ \equiv:\, &\mathfrak{Rel}(\overline{Y}) \,.\, \mathfrak{Un}(\overline{Y}) \text{ by } 11.12, 10.21, 4.61.\end{aligned}$$

11.41 "$\mathfrak{Fn}$" is absolute.

Proof:

$$\overline{Y}\mathfrak{Fn}_l(\overline{X}) \,.\!\equiv\!: \mathfrak{Fnc}_l(\overline{Y}) \,.\, \mathfrak{D}_l(\overline{Y}) = \overline{X}\,.$$
$$\equiv: \mathfrak{Fnc}(\overline{Y}) \,.\, \mathfrak{D}(\overline{Y}) = \overline{X} \text{ by } 11.4, 11.13, 4.63.$$

11.42 *On* is absolute.

Proof: $\mathfrak{Ord}_l\,(On_l)$ by 7.16_l and $\mathfrak{Pr}_l(On_l)$ by 7.17_l. But On_l is constructible by 10.1, hence $\mathfrak{Ord}\,(On_l)$ and $\mathfrak{Pr}\,(On_l)$ because $\mathfrak{Ord}$ and $\mathfrak{Pr}$ are absolute by 11.31 and 10.22. Hence $On_l = On$ by 7.2.

By 10.11, it follows from 11.42 that $On \subseteq L$; in other words, every ordinal number is constructible. Furthermore 11.42 implies:

11.421 The variables $\alpha, \beta, \ldots$ are absolute.
11.43 "$<$" is absolute.

Proof: "$<$" is by definition the same as "ϵ".

11.44 "$\leq$" is absolute.

Proof: $X \leq Y$ is by definition $X \,\epsilon\, Y \,.\!\vee\!.\, X = Y$.

11.45 "$\dot{+}1$" is absolute.

Proof: 7.4, 10.17, 11.20 and 10.18.

11.451 Each of the symbols $0, 1, 2, 3, \ldots$, etc. is absolute.

Proof by 10.24 and 11.45.

11.46 $\mathfrak{S}$ (and therefore $\mathfrak{Max}$ and $\mathfrak{Lim}$) is absolute.

Proof:

$$z \,\epsilon\, \mathfrak{S}\,(\overline{X}) \,.\!\equiv\!.\, (\exists v)[z \,\epsilon\, v \,.\, v \,\epsilon\, \overline{X}] \,.\!\equiv\!.\, (\exists \overline{v})[z \,\epsilon\, \overline{v} \,.\, \overline{v} \,\epsilon\, \overline{X}] \,.\!\equiv\!.\, z \,\epsilon\, \mathfrak{S}_l\,(\overline{X}).$$

Therefore $\mathfrak{S}\,(\overline{X}) = \mathfrak{S}_l\,(\overline{X})$ by the axiom of extensionality.

| What is left now is to show that the special classes R, S, J, K_1, K_2, F, and finally L, are absolute, where R is the ordering for pairs defined in 7.81, S is the ordering of the triples $\langle i, \alpha, \beta \rangle$ defined by 9.2, and F is the function introduced by 9.3 which defines L. For each of these the proof of absoluteness will be based on the following lemma: 51

If the class A is defined by the postulate $\phi(A)$ and if all defined classes, operations, notions, and variables appearing in ϕ are absolute, then A is absolute.

Proof: If ϕ satisfies the condition above, then $\phi_l(\overline{X}) . \equiv . \phi(\overline{X})$. Also $\phi_l(A_l)$ and $\phi(A)$ by definition of A_l and A. Since by 10.1 A_l is constructible, $\phi_l(A_l)$ implies $\phi(A_l)$, hence $A_l = A$, because both $\phi(A_l)$ and $\phi(A)$.

11.5 "R" is absolute.

Proof: By Definition 7.81 we have

$$R \subseteq (On^2)^2 . (\alpha, \beta, \gamma, \delta)[\langle\langle\alpha, \beta\rangle, \langle\gamma, \delta\rangle\rangle \in R . \equiv \\ ::. \mathfrak{Max}\{\alpha, \beta\} < \mathfrak{Max}\{\gamma, \delta\} . \vee :: \mathfrak{Max}\{\alpha, \beta\} = \mathfrak{Max}\{\gamma, \delta\} :. \\ \beta < \delta . \vee : \beta = \delta . \alpha < \gamma].$$

The following concepts appear in the defining postulate: $\subseteq$, On, ^{2}, $\langle\ \rangle$, $\mathfrak{Max}$, $\{\ \}$, $<$, ϵ, and variables $\alpha, \beta, \ldots$, all of which have been proved absolute by 10.13, 11.42, 11.11, 10.19, 11.46, 10.17, 11.43, 10.12, 11.421, respectively.

11.51 "S" is absolute.

Proof: By Definition 9.2 we have

$$S \subseteq (9 \times On^2)^2 . (\alpha, \beta, \gamma, \delta, \mu, \nu)\{\mu < 9 . \nu < 9 : \supset \\ :: \langle\langle\mu, \alpha, \beta\rangle, \langle\nu, \gamma, \delta\rangle\rangle \in S . \equiv :. \langle\langle\alpha, \beta\rangle, \langle\gamma, \delta\rangle\rangle \in R . \vee \\ : \langle\alpha, \beta\rangle = \langle\gamma, \delta\rangle . \mu < \nu\}.$$

In the postulate for S the following concepts appear, other than those appearing previously in the postulate for R: $\times$, R, 9, which are absolute by 11.1, 11.5, 11.451, respectively.

11.52 "J" is absolute.

Proof: By Definition 9.21 we have

$$J \mathfrak{Fn} (9 \times On^2) . \mathfrak{W}(J) = On . (\alpha, \beta, \gamma, \delta, \mu, \nu)[\mu, \nu < 9 . \supset \\ : \langle\mu, \alpha, \beta\rangle S \langle\nu, \gamma, \delta\rangle . \supset . J`\langle\mu, \alpha, \beta\rangle < J`\langle\nu, \gamma, \delta\rangle].$$

The only additional symbols in this postulate are: $\mathfrak{Fn}$, $\mathfrak{W}$ and ‘, all of which have been proved absolute by 11.41, 11.17, 11.23, respectively.

11.53 Each "J_i" is absolute, $i = 0, 1, 2, \ldots, 8$.

Proof: $J_0^{\text{‘}}\langle\alpha,\beta\rangle = J^{\text{‘}}\langle 0,\alpha,\beta\rangle$. $J_0\,\mathfrak{Fn}\,On^2$. Here there are no symbols but those mentioned before. Similarly for $J_1,\ldots,J_8$.

| 11.54 K_1 and K_2 are absolute. 52

Proof: In the defining postulate 9.24 there are no symbols but those mentioned before.

11.6 "F" is absolute.

Proof: The only additional symbols appearing in the defining postulate 9.3 are $\upharpoonright$ and $\mathfrak{F}_1,\ldots,\mathfrak{F}_8$, which are absolute by 11.16, 11.221, respectively.

11.7 "L" is absolute.

Proof: $L = \mathfrak{W}(F)$, and $\mathfrak{W}$, F are absolute by 11.6, 11.17.

It has now been demonstrated from the axioms of Σ that $L_l = L$, hence also that $V_l = L_l$, i.e., that the proposition $V = L$ holds in the model Δ . This proves that, if there exists a model for the axioms of groups A, B, C, D, there exists also a model for the augmented set of axioms obtained by adding as an axiom the proposition $V = L$, namely, the model consisting of the classes and sets "constructible" in the given model for Σ. Thus if the system A, B, C, D is consistent, the augmented system is consistent. Another way of putting this argument is as follows: If a contradiction were obtained from $V = L$ and the axioms of Σ (i.e., the axioms of groups A, B, C, D), then the same contradiction could be derived also from $V_l = L_l$ and the relativized axioms A_l, B_l, C_l, D_l. But $V_l = L_l$ and A_l, B_l, C_l, D_l can be proved in Σ as shown before; hence Σ would be contradictory, and a contradiction in Σ could actually be constructed if a contradiction from Σ and $V = L$ were given.

| 53

Chapter VIII
Proof that $V = L$ implies the axiom of choice and the generalized continuum hypothesis

Now it remains only to be shown that the axiom of choice and the generalized continuum hypothesis follow from $V = L$ and Σ.

For the axiom of choice this is immediate since the relation As defined in 11.8, which singles out the element of least order in any non-vacuous constructible set, evidently satisfies Axiom E if $V = L$.

11.8 Dfn $\langle y, x\rangle \,\epsilon\, As \,.\!\equiv:\, y \,\epsilon\, x \,.\, (z)[Od`z < Od`y \,.\!\supset.\, {\sim} z \,\epsilon\, x] \,.\, \mathfrak{Rel}(As)$

$As`x$ is what may be called the "designated" element of x.

11.81 Dfn $C`\alpha = Od`[As`(F`\alpha)] \,.\, C\,\mathfrak{Fn}\,On$

$C`\alpha$ is the order of the "designated" element of $F`\alpha$. Hence $C`\alpha \leq \alpha$ by 9.52.

The rest of these lectures is devoted to the derivation of the generalized continuum hypothesis from $V = L$ and the axioms of Σ. Since we have just derived the axiom of choice from these assumptions, we are justified in using all starred theorems and definitions in this derivation. The theorems which follow from now on are only claimed to be consequences of Σ and $V = L$. However, only 12.2 really depends on $V = L$; in all the others $V = L$ is not used, and even the axiom of choice could be avoided in their proofs, if one wanted to.

12.1 $\overline{\overline{F``\omega_\alpha}} = \omega_\alpha$

Proof: $\overline{\overline{F``\omega_\alpha}} \leq \overline{\overline{\omega}}_\alpha = \omega_\alpha$ by 8.31. On the other hand, there exists a subset of ω_α, namely $\omega_\alpha \cdot \mathfrak{W}(J_0)$, such that the values of F over this subset are all different. For if $\gamma \neq \delta$ and $\gamma, \delta \,\epsilon\, \omega_\alpha \cdot \mathfrak{W}(J_0)$, assume $\gamma < \delta$; then $F`\gamma \,\epsilon\, F`\delta$, by 9.3, so that $F`\gamma \neq F`\delta$. But $\overline{\overline{\omega_\alpha \cdot \mathfrak{W}(J_0)}} \geq \omega_\alpha$, because $J_0``(\omega_\alpha^2) \subseteq \omega_\alpha \cdot \mathfrak{W}(J_0)$ by 9.26 and $J_0$ is one-to-one. Hence $\overline{\overline{F``\omega_\alpha}} \geq \omega_\alpha$.

By 12.1 the generalized continuum hypothesis follows immediately from the following theorem:

54 | 12.2 $\mathfrak{P}(F``\omega_\alpha) \subseteq F``\omega_{\alpha+1}$.

This theorem is proved by means of the following lemma:

12.3 If $m \subseteq On$ and m is closed with respect to C, K_1, K_2 and with respect to $J_0, \ldots, J_8$ as triadic relations and if G is an isomorphism from m to an ordinal o with respect to E, then G is also an isomorphism with respect to $\hat{\alpha}\hat{\beta}[F`\alpha \,\epsilon\, F`\beta]$, i.e., $\alpha, \beta \,\epsilon\, m \supset [F`\alpha \,\epsilon\, F`\beta \,.\!\equiv.\, F`G`\alpha \,\epsilon\, F`G`\beta]$.

We show first that 12.3 implies 12.2.

Proof: Consider $u \,\epsilon\, \mathfrak{P}(F``\omega_\alpha)$, that is, $u \subseteq F``\omega_\alpha$. By $V = L$ there is a δ such that $u = F`\delta$; form the closure of the set $\omega_\alpha + \{\delta\}$ with respect to C, K_1, K_2, and with respect to J_i, $i = 0, 1, \ldots, 8$, as triadic relations, according to 8.73, and let the closure be denoted by m. Now, by 8.73, m is a set and $\overline{\overline{m}} = \omega_\alpha$. Furthermore, m is a set of ordinals; hence m is well-ordered by E by 7.161 and is isomorphic to some ordinal number o by 7.7. Let the isomorphism be denoted by G, so that $G``m = o$. For brevity, let α' denote $G`\alpha$. By Lemma 12.3 we have:

$$\alpha, \beta \,\epsilon\, m \,.\supset: F`\alpha \,\epsilon\, F`\beta \,.\equiv.\, F`\alpha' \,\epsilon\, F`\beta'.$$

Now consider δ', the image of δ by G. $\delta' \,\epsilon\, o$, that is, $\delta' < o$. Since G is one-to-one as an isomorphism, $\overline{\overline{o}} = \overline{\overline{m}} = \omega_\alpha$, from which it follows that $o < \omega_{\alpha+1}$, hence $\delta' < \omega_{\alpha+1}$. Also, for any $\beta \,\epsilon\, m$,

$$F`\beta \,\epsilon\, F`\delta \,.\equiv.\, F`\beta' \,\epsilon\, F`\delta'.$$

$\omega_\alpha \subseteq m$, by definition and ω_α is complete (as an ordinal number). Therefore ω_α is an E-section of m; hence ω_α is mapped by G on an E-section of o, i.e., by 7.21 on an ordinal number. But, by 7.62, this can be only the identical mapping of ω_α onto itself. Therefore, if $\beta \,\epsilon\, \omega_\alpha$, then $\beta' = \beta$. Hence $F`\beta \,\epsilon\, F`\delta \,.\equiv.\, F`\beta \,\epsilon\, F`\delta'$, for $\beta \,\epsilon\, \omega_\alpha$; that is, $F`\delta$ and $F`\delta'$ have exactly the same elements with $F``\omega_\alpha$ in common, i.e., $F`\delta \cdot F``\omega_\alpha = F`\delta' \cdot F``\omega_\alpha$; but $F`\delta \subseteq F``\omega_\alpha$, by assumption; therefore $F`\delta = F`\delta' \cdot F``\omega_\alpha$. But $\omega_\alpha \,\epsilon\, \mathfrak{W}(J_0)$ by 9.27, therefore, by 9.35, $F``\omega_\alpha = F`\omega_\alpha$, hence $u = F`\delta = F`\delta' \cdot F`\omega_\alpha$. Therefore by 9.611 $Od`u < \omega_{\alpha+1}$, in other words, $u \,\epsilon\, F``\omega_{\alpha+1}$, q.e.d.

In order to prove 12.3, we prove at first the following auxiliary theorem:

12.4 From the hypothesis of 12.3 (leaving out closure with respect to C) it follows that
(1) G is an isomorphism for the triadic relations J_i $(i = 0, \ldots, 8)$, i.e. (if $G`\alpha$ is abbreviated by $\alpha'$): $J_i`\langle\alpha', \beta'\rangle = [J_i`\langle\alpha, \beta\rangle]'$ for α, $\beta \,\epsilon\, m$, $i < 9$, and
(2) o is closed with respect to the triadic relations J_i.

In outline, the proof runs as follows: By definition of J and the closure property of m, J establishes an isomorphism with respect to S and E between the class of triples $\langle i, \alpha, \beta\rangle$, $i < 9$, $\alpha, \beta \,\epsilon\, m$, and m. By G this isomorphism is carried over to an isomorphism be|tween the set t of triples $\langle i, \alpha, \beta\rangle$, $i < 9$, $\alpha, \beta \,\epsilon\, o$, and o. But J likewise defines an isomorphic correspondence between t and some ordinal γ, also with respect to S and E; from this it is inferred by 7.62 that $\gamma = o$ and that J confined to t coincides with the image by G of J confined to $9 \times m^2$. But this is what the assertion of the theorem says. The detailed proof is as follows: 55

Set $j = J \restriction (9 \times m^2)$. Then we have $\mathfrak{D}(j) = 9 \times m^2$. Now $\mathfrak{W}(j) \subseteq m$, since m is closed with respect to all the J_i. But also $m \subseteq \mathfrak{W}(j)$; for suppose $\gamma \epsilon m$, then $\gamma = J`\langle i, \alpha, \beta\rangle$ for some i, α, β where $\alpha, \beta \epsilon m$, since m is closed with respect to K_1 and K_2; hence $\gamma \epsilon \mathfrak{W}(j)$. Therefore $\mathfrak{W}(j) = m$. Moreover,

$$i, k < 9 \,.\, \alpha, \beta, \gamma, \delta \epsilon m \,.\, \langle i, \alpha, \beta\rangle S \langle k, \gamma, \delta\rangle :\supset . \, j`\langle i, \alpha, \beta\rangle < j`\langle k, \gamma, \delta\rangle,$$

since J has this property, and since for this domain J coincides with j. Therefore

$$j\, \mathfrak{Isom}_{SE}(9 \times m^2, m).$$

Now denote by $\bar{j}$ the function into which j is carried over by G, that is, $\bar{j}$ is defined by:

$$\bar{j}\, \mathfrak{Fn}(9 \times o^2) \,.\, \bar{j}`\langle i, \alpha', \beta'\rangle = [j`\langle i, \alpha, \beta\rangle]',$$

for $\alpha, \beta \epsilon m$ and $i < 9$. This may be written $\bar{j}`\langle i, \alpha, \beta\rangle = [j`\langle i, \alpha_1, \beta_1\rangle]'$, for $\alpha, \beta \epsilon o$, $i < 9$, where $\breve{G}`\alpha$ is denoted by α_1. We want to show that $\bar{j}\, \mathfrak{Isom}_{SE}(9 \times o^2, o)$. Now: $\mathfrak{D}(\bar{j}) = 9 \times o^2$ and $\mathfrak{W}(\bar{j}) = o$, because j has the corresponding properties. Since G is an isomorphism with respect to E, it follows by Definition 7.8 that $\langle \alpha, \beta\rangle Le \langle \gamma, \delta\rangle . \equiv . \langle \alpha', \beta'\rangle Le \langle \gamma', \delta'\rangle$ for α, β, γ, δ ϵ m. Likewise, by Definition 7.81, $\langle \alpha, \beta\rangle R \langle \gamma, \delta\rangle . \equiv . \langle \alpha', \beta'\rangle R \langle \gamma', \delta'\rangle$ for $\alpha, \beta, \gamma, \delta \epsilon m$. It follows then by Definition 9.2 that

$$\langle i, \alpha_1, \beta_1\rangle S \langle k, \gamma_1, \delta_1\rangle . \equiv . \langle i, \alpha, \beta\rangle S \langle k, \gamma, \delta\rangle$$

for α, β, γ, δ ϵ o and $i, k < 9$. Now suppose α, β, γ, δ ϵ o, $i, k < 9$ and $\langle i, \alpha, \beta\rangle S \langle k, \gamma, \delta\rangle$. We have then $\langle i, \alpha_1, \beta_1\rangle S \langle k, \gamma_1, \delta_1\rangle$, which implies, since $j\, \mathfrak{Isom}_{SE}(9 \times m^2, m)$, that $j`\langle i, \alpha_1, \beta_1\rangle E j`\langle k, \gamma_1, \delta_1\rangle$. Now, since G is an isomorphism with respect to E, we conclude that

$$[j`\langle i, \alpha_1, \beta_1\rangle]' E [j`\langle k, \gamma_1, \delta_1\rangle]',$$

that is, $\bar{j}`\langle i, \alpha, \beta\rangle E \bar{j}`\langle k, \gamma, \delta\rangle$. Therefore $\bar{j}\, \mathfrak{Isom}_{SE}(9 \times o^2, o)$.

Now define $j_o = J \upharpoonright (9 \times o^2)$. Then $\mathfrak{D}(j_o) = 9 \times o^2$ and $\mathfrak{W}(j_o)$ is some ordinal number γ, since $9 \times o^2$ is an S-section of $9 \times On^2$. Therefore under J the image must be an E-section of On, i.e., an ordinal by 7.21. Hence both $j_o\, \mathfrak{Isom}_{SE}(9 \times o^2, \gamma)$ and $\bar{j}\, \mathfrak{Isom}_{SE}(9 \times o^2, o)$; but there can exist but one isomorphism of this kind of a set on an ordinal number, by 7.62, hence $\gamma = o$ and $j_o = \bar{j}$. Therefore

$$j_o` \langle i, \alpha', \beta'\rangle = \bar{j}`\langle i, \alpha', \beta'\rangle = [j`\langle i, \alpha, \beta\rangle]',$$

for $\alpha, \beta \epsilon m$, $i < 9$, which is equivalent, by the construction of j_o and j, to the statement: $J`\langle i, \alpha', \beta'\rangle = [J`\langle i, \alpha, \beta\rangle]'$, for $\alpha, \beta \epsilon m$, $i < 9$, which, in turn, is the same as: $J_i`\langle \alpha', \beta'\rangle = [J_i`\langle \alpha, \beta\rangle]'$, for $i = 0, \ldots, 8$, $\alpha, \beta \epsilon m$, which is what we set out to prove. That o is closed with respect to the J_i follows immediately from the last equality.

12.4 can be stated symmetrically as follows:

12.5 If $m \subseteq On$, $m' \subseteq On$, m, m' both closed with respect to K_1, K_2 and the J_i as triadic relations and if $| \, G \,\mathfrak{Isom}_{EE}\,(m, m')$, then G is an isomorphism for the triadic relations J_i. 56

The proof is obtained by mapping m and m' on the same ordinal o by 7.7 and then applying 12.4.

12.51 The hypothesis of 12.5 implies, furthermore,
$\alpha \,\epsilon\, \mathfrak{W}\,(J_i) \,.\!\supset\!.\, G`\alpha \,\epsilon\, \mathfrak{W}(J_i)$ for $\alpha \,\epsilon\, m$, $i = 0, \ldots, 8$.

Proof: $\alpha \,\epsilon\, \mathfrak{W}\,(J_i)$ implies $\alpha = J_i`\langle\beta, \gamma\rangle$, $\beta, \gamma \,\epsilon\, m$, since $m$ is closed with respect to $K_1, K_2$. Hence $\alpha' = J_i`\langle\alpha', \beta'\rangle$ by 12.5; hence $\alpha' \,\epsilon\, \mathfrak{W}\,(J_i)$.

Next it will be shown that:

12.6 If m, m', G satisfy the hypothesis of 12.5 and in addition m and m' are also closed with respect to C, then G is an isomorphism for the relations $\hat{\alpha}\hat{\beta}(F`\alpha \,\epsilon\, F`\beta)$ and $\hat{\alpha}\hat{\beta}(F`\alpha = F`\beta)$. In other words,

$$\text{(a)} \qquad \alpha, \beta \,\epsilon\, m \,.\!\supset\!:\! F`\alpha \,\epsilon\, F`\beta \,.\!\equiv\!.\, F`\alpha' \,\epsilon\, F`\beta' \,. \\ F`\alpha = F`\beta \,.\!\equiv\!.\, F`\alpha' = F`\beta',$$

where again $G`\alpha$ is abbreviated by α'.

The scheme of the proof will be to carry out an induction on $\eta = \mathfrak{Max}\{\alpha, \beta\}$. We will assume as the hypothesis of the induction that (a) is true for $\alpha, \beta \,\epsilon\, m$ and $\alpha, \beta < \eta$, and prove it for $\alpha, \beta \,\epsilon\, m$, $\mathfrak{Max}\{\alpha, \beta\} = \eta$. (Hence the property which is shown by induction to belong to all ordinals η is given by the propositional function:

$$(\alpha, \beta)[\alpha, \beta \,\epsilon\, m \cdot \eta = \mathfrak{Max}\{\alpha, \beta\} \,:\!\supset\!:\, (F`\alpha \,\epsilon\, F`\beta \,.\!\equiv\!.\, F`G`\alpha \,\epsilon\, F`G`\beta) \,. \\ (F`\alpha = F`\beta \,.\!\equiv\!.\, F`G`\alpha = F`G`\beta)].$$

This expression is normal; therefore we can apply induction by 7.161.) If $\mathfrak{Max}\,\{\alpha, \beta\} = \eta$ there are 3 possible cases, namely

1) $$\alpha = \beta = \eta.$$

In this case the equivalences (a) both hold, since, in the first, both members are false, and, in the second, both are true. The remaining two cases are $\alpha = \eta$, $\beta < \eta$ and $\alpha < \eta$, $\beta = \eta$. Hence what has to be proved is:

$$\left.\begin{array}{l} F`\alpha \,\epsilon\, F`\eta \,.\!\equiv\!.\, F`\alpha' \,\epsilon\, F`\eta', \\ F`\eta \,\epsilon\, F`\beta \,.\!\equiv\!.\, F`\eta' \,\epsilon\, F`\beta', \\ F`\eta = F`\beta \,.\!\equiv\!.\, F`\eta' = F`\beta', \end{array}\right\} \quad \text{for } \alpha, \beta < \eta, \quad \alpha, \beta \,\epsilon\, m,$$

under the hypothesis that $\eta \,\epsilon\, m$ and

$$\left.\begin{array}{l} \text{I. } F`\alpha \,\epsilon\, F`\beta \,.\!\equiv\!.\, F`\alpha' \,\epsilon\, F`\beta', \\ \text{II. } F`\alpha = F`\beta \,.\!\equiv\!.\, F`\alpha' = F`\beta', \end{array}\right\} \quad \text{for } \alpha, \beta \,\epsilon\, m \cdot \eta.$$

Everything which follows from now on up to the end of the proof of Theorem 12.6 (in particular the theorems (1)–(9) on pages 57–9) depends on this inductive hypothesis in addition to the hypothesis of Theorem 12.6.

57 The following abbreviations will be convenient: $r = F``m$, | $r_\eta = F``(m \cdot \eta)$, $r' = F``m'$, and $r'_\eta = F``(m' \cdot \eta')$. Hence $r_\eta \subseteq r$ and $r'_\eta \subseteq r'$. Now we can define a one-to-one mapping H of r_η on r'_η by $H = F|G|F^{-1}$. Because of the inductive hypothesis II, H is one-to-one and $H`x = F`\alpha'$ if $x = F`\alpha$, $\alpha \,\epsilon\, m \cdot \eta$. Because of inductive hypothesis I, H is an isomorphism with respect to E. Note that the hypotheses of Theorem 12.6 and the inductive hypothesis are perfectly symmetric in m, m' and η, η', so that whatever is proved from them will also hold if m, η, r, r_η, G, H are interchanged respectively with m', η', r', r'_η, $\breve{G}$, $\breve{H}$.

The next step will be to show that H is an isomorphism for the triadic relation $\hat{z}\hat{x}\hat{y}[z = \langle x, y\rangle]$ and the tetradic relation $\hat{z}\hat{u}\hat{v}\hat{w}[z = \langle u, v, w\rangle]$, and for the Q_i. In order to establish this some preliminary results are needed.

(1) r is closed with respect to the fundamental operations.

Proof: Take $x, y \,\epsilon\, r$; then $x = F`\alpha$, $y = F`\beta$, $\alpha, \beta \,\epsilon\, m$, so that $\mathfrak{F}_i(x, y) \,\epsilon\, r$, by 9.31–9.34, since m is closed with respect to the J_i. Therefore $x - y$, $\{x, y\}$, $\langle x, y\rangle$, $\langle x, y, z\rangle$, and $x \cdot Q_i``y$ are in $r$ if $x, y, z \,\epsilon\, r$. In particular, it follows that $x \cdot Q_i``\{y\} \,\epsilon\, r$ if $x, y \,\epsilon\, r$.

(2) $x \,\epsilon\, r \,.\!\supset\!.\, Od`x \,\epsilon\, m$

Proof: $\{x\} \,\epsilon\, r$ by (1), hence there is an $\alpha \,\epsilon\, m$ such that $\{x\} = F`\alpha$. Set $\beta = C`\alpha$. Then $\beta \,\epsilon\, m$, since m is closed with respect to C and $\beta = Od`x$, by definition of C (11.81).

(3) $(x \in r) . (x \neq 0) :\supset . x \cdot r \neq 0$

Proof: There is an $\alpha \in m$ such that $x = F`\alpha$. $F`C`\alpha \in x$, by Definition 11.81; but $F`C`\alpha \in r$, since m is closed with respect to C; hence $x \cdot r \neq 0$.

(3.1) $\{x, y\} \in r . \supset . x, y \in r; \quad \langle x, y \rangle \in r . \supset . x, y \in r;$
$\langle x, y, z \rangle \in r . \supset . x, y, z \in r.$

Proof: It follows from (3) that $\{x\} \in r . \supset . x \in r$, because x is the only element of $\{x\}$, also $\{x, y\} \in r . \supset . x, y \in r$, for, either x or $y \in r$ by (3); suppose $x \in r$, then $\{x\} \in r$ by (1), hence $\{x, y\} - \{x\} \in r$ by (1), so that $\{y\} \in r$ if $x \neq y$, hence $y \in r$. By iteration, $\langle x, y \rangle \in r . \supset . x,\ y \in r$ and $\langle x, y, z \rangle \in r . \supset . x, y, z \in r$. It follows then that

(4) $y \in r . \langle y, x \rangle \in Q_i :\supset . x \in r \quad$ for $i \neq 5$.

Proof: Consider Q_6, the permutation of the ordered pair: If $\langle y, x \rangle \in Q_6$, then $y = \langle u, v \rangle$, $\langle v, u \rangle = x$ for some u, v. $\langle u, v \rangle \in r$ by assumption, hence $u, v \in r$ by (3.1) so that $\langle v, u \rangle \in r$, by (1), | that is, $x \in r$. Similarly for the other permutations, i.e., Q_7, Q_8. Now consider $Q_4 = P_2^{-1}$: assume $y \in r$, $\langle y, x \rangle \in P_2^{-1}$; then $\langle x, y \rangle \in P_2$, i.e., y is an ordered pair and x its second member, hence $x \in r$ by (3.1). 58

There is a weak completeness theorem for r_η:

(5) $x \in r_\eta . y \in x \ :\supset:\ y \in r . \supset . y \in r_\eta.$

Proof: Set $\alpha = Od`y$. Now $\alpha \in m$, by (2); $Od`y < Od`x < \eta$, by 9.52; hence $\alpha \in m \cdot \eta$, that is, $y \in r_\eta$.

(6) $y \in F`\eta . y \in r :\supset . y \in r_\eta$

Proof: $Od`y < \eta$ by 9.52. $Od`y \in m$ by (2); hence $Od`y \in m \cdot \eta$, i.e., $y \in r_\eta$.

(7) $\{x, y\} \in r_\eta . \supset . x, y \in r_\eta$ and $\langle x, y \rangle \in r_\eta . \supset . x, y \in r_\eta;$
$\langle x, y, z \rangle \in r_\eta . \supset . x, y, z \in r_\eta.$

Proof: $\{x, y\} \in r$, therefore $x, y \in r$ by (3.1); hence the result follows, by (5). By iteration it follows that $\langle x, y \rangle \in r_\eta . \supset . x, y \in r_\eta$, and similarly for triples.

(8) H is an isomorphism with respect to $\hat{z}\hat{x}\hat{y}[z = \{x, y\}]$, $\hat{z}\hat{x}\hat{y}[z = \langle x, y \rangle]$, $\hat{z}\hat{x}\hat{y}\hat{t}[z = \langle x, y, t \rangle]$, and the Q_i $(i = 4, 5, \ldots, 8)$.

(In the sequel $H`x$ is abbreviated by x'. So the prime is an abbreviation for G or H according as to whether it occurs with a Greek or a Latin letter.)

Proof: Consider $\{x, y\}$; we wish to show that

$$x, y, z \in r_\eta . \supset : z = \{x, y\} . \equiv . z' = \{x', y'\}.$$

Recalling the symmetry of the hypotheses, and that $x, y, z \in r_\eta$ is equivalent to $x', y', z' \in r'_\eta$, it is obvious that it is sufficient to prove implication in one direction, in order to establish the equivalence. We prove implication from right to left; $z' = \{x', y'\}$ implies $x' \in z'$ and $y' \in z'$, hence, since H is an isomorphism with respect to E, $x \in z$ and $y \in z$, i.e., $\{x, y\} \subseteq z$. We have then only to show that $z - \{x, y\} = 0$. Since $x, y, z \in r$, $z - \{x, y\} \in r$ by (1); hence by (3), if $z - \{x, y\} \neq 0$, there is a $u \in r$ such that $u \in [z - \{x, y\}]$. So $u \in z$, and $z \in r_\eta$, hence, by (5), $u \in r_\eta$. Then $u \in z, u \neq x$ and $u \neq y$, hence $u' \in z'$, $u' \neq x'$ and $u' \neq y'$, because H is one-to-one and isomorphic for E. But this means $z' \neq \{x', y'\}$, contrary to assumption.

To establish that H is an isomorphism for $z = \langle x, y \rangle$, it must be shown that

$$x, y, z \in r_\eta . \supset : z = \langle x, y \rangle . \equiv . z' = \langle x', y' \rangle.$$

Again it is sufficient to establish implication in one direction. Assume $z = \langle x, y \rangle$. It follows that $z = \{u, v\}$, where $u = \{x, x\}$ and $v = \{x, y\}$. By 59 (7), $u, v \in r_\eta$; hence, forming z', u', v', x', y', it follows | that $v' = \{x', y'\}$, $u' = \{x', x'\}$, and $z' = \{u', v'\}$, that is, $z' = \langle x', y' \rangle$.

For the ordered triple, assume $z = \langle x, y, t \rangle$; then $z = \langle x, s \rangle$, where $s = \langle y, t \rangle$; $t, s \in r_\eta$, by (7), since $z \in r_\eta$; therefore $z' = \langle x', s' \rangle$, $s' = \langle y', t' \rangle$, that is, $z' = \langle x', y', t' \rangle$.

Consider now $Q_5 = P_2$; we must show that

$$x, y \in r_\eta . \supset : \langle x, z \rangle \in P_2 . \equiv . \langle x', z' \rangle \in P_2.$$

As usual, only the implication in one direction is necessary. Assume $\langle x, z \rangle \in P_2$; then there is a y such that $z = \langle y, x \rangle$; by (7) $y \in r_\eta$, therefore $z' = \langle y', x' \rangle$ by (8), that is, $\langle x', z' \rangle \in P_2$. Now since H is an isomorphism with respect to P_2, H must be an isomorphism also with respect to $Q_4 = P_2^{-1}$.

There remain only the permutations Q_6, Q_7, Q_8. Consider Q_6, for example. Assume $\langle x, y \rangle \in Q_6$; then there exist u and v such that $x = \langle u, v \rangle$ and $y = \langle v, u \rangle$. Since $x, y \in r_\eta$, it follows by (7) that $u, v \in r_\eta$, hence $x' = \langle u', v' \rangle$ and $y' = \langle v', u' \rangle$ by (8), that is, $\langle x', y' \rangle \in Q_6$. The proofs are similar for Q_7 and Q_8.

Now consider the three relations which must be proved to establish the induction, namely,

$$(9) \quad \left.\begin{array}{l} 1.\ F`\alpha \in F`\eta \,.\!\equiv\!.\, F`\alpha' \in F`\eta', \\ 2.\ F`\eta \in F`\beta \,.\!\equiv\!.\, F`\eta' \in F`\beta', \\ 3.\ F`\eta = F`\beta \,.\!\equiv\!.\, F`\eta' = F`\beta', \end{array}\right\} \quad \text{for } \alpha, \beta \in m \cdot \eta.$$

We shall show now that it is sufficient to prove the first of these three relations. Let us assume then that the first is true, and prove the third. Assume that $F`\eta \neq F`\beta$. Then either $F`\eta - F`\beta \neq 0$ or $F`\beta - F`\eta \neq 0$, and $[F`\eta - F`\beta] \in r$, $[F`\beta - F`\eta] \in r$ by (1). Hence by (1) and (3) there is a $u \in r$ such that either $u \in [F`\eta - F`\beta]$ or $u \in [F`\beta - F`\eta]$. Therefore $u \in F`\eta$ or $u \in F`\beta$, hence in both cases $u \in r_\eta$, by (6) and (5), since $F`\beta \in r_\eta$. Let us now assume $u \in [F`\eta - F`\beta]$; then $u \in F`\eta$ and $\sim(u \in F`\beta)$. Hence by the inductive hypothesis I, we have $\sim(u' \in F`\beta')$; but also $u' \in F`\eta'$, because we have assumed (9) 1 to be true. Therefore $F`\eta' - F`\beta' \neq 0$. Suppose, on the other hand, that $u \in [F`\beta - F`\eta]$; then $u \in F`\beta$ and $\sim(u \in F`\eta)$. Exactly as above, we have $u' \in F`\beta'$ and $\sim(u' \in F`\eta')$, that is, $F`\eta' \neq F`\beta'$. Thus we have shown

$$F`\eta \neq F`\beta \,.\!\supset\!.\, F`\eta' \neq F`\beta',$$

and the inverse follows by symmetry as usual.

We have now established that the third relation of (9) follows from the first. Now we derive the second from the first and third. Assume that $F`\eta \in F`\beta$; set $\alpha = Od`F`\eta$. By 9.52 $\alpha < \beta < \eta$ and by (2) $\alpha \in m \cdot \eta$. So $F`\eta = F`\alpha$ and therefore $F`\alpha \in F`\beta$; from $F`\eta = F`\alpha$ it follows by (9) 3 that $F`\eta' = F`\alpha'$; moreover $F`\alpha' \in F`\beta'$, by the inductive hypothesis I, hence $F`\eta' \in F`\beta'$, that is, $F`\eta \in F`\beta \,.\!\supset\!.\, F`\eta' \in F`\beta'$ and the inverse implication by reasons of symmetry. Therefore it is sufficient to show (9) 1 and by symmetry it is sufficient to show:

$$F`\alpha \in F`\eta \,.\!\supset\!.\, F`\alpha' \in F`\eta' \quad \text{for } \alpha \in m \cdot \eta.$$

| 60

So we assume $F`\alpha \in F`\eta$, and consider separate cases according to the index i such that $\eta \in \mathfrak{W}(J_i)$.

1. Suppose $\eta \in \mathfrak{W}(J_0)$; by 12.51, $\eta' \in \mathfrak{W}(J_0)$, hence $F`\eta = F``\eta$ and $F`\eta' = F``\eta'$, by 9.35, so that both members of the equivalence (9) 1 are true, hence trivially equivalent.

2. Suppose $\eta \in \mathfrak{W}(J_1)$. Then $\eta = J_1`\langle\beta,\gamma\rangle$, where $\beta, \gamma \in m$ (by the closure property of $m$) and $\beta, \gamma < \eta$, by 9.25. Also $\eta' = J_1`\langle\beta',\gamma'\rangle$ by 12.5, so that 9.31 gives: $F`\eta = \{F`\beta, F`\gamma\}$ and $F`\eta' = \{F`\beta', F`\gamma'\}$. Suppose $F`\alpha \in F`\eta$; then $F`\alpha = F`\beta$ or $F`\alpha = F`\gamma$; therefore, by the inductive hypothesis II, $F`\alpha' = F`\beta'$ or $F`\alpha' = F`\gamma'$, that is,

$$F`\alpha' \,\epsilon\, \{F`\beta', F`\gamma'\};$$

in other words, $F`\alpha' \,\epsilon\, F`\eta'$.

3. If $\eta \,\epsilon\, \mathfrak{W}(J_2)$, then we have, as before, $\eta = J_2`\langle\beta, \gamma\rangle$ and $\eta' = J_2`\langle\beta', \gamma'\rangle$, $\beta, \gamma \,\epsilon\, m \cdot \eta$. By 9.32, $F`\eta = E \cdot F`\beta$ and $F`\eta' = E \cdot F`\beta'$. If $F`\alpha \,\epsilon\, F`\eta$, then $F`\alpha \,\epsilon\, F`\beta$ and $F`\alpha \,\epsilon\, E$. It follows that $F`\alpha' \,\epsilon\, F`\beta'$, by the hypothesis I of the induction. From $F`\alpha \,\epsilon\, E$ it follows that $F`\alpha = \langle x, y\rangle$ and $x \,\epsilon\, y$ for some $x, y$; $F`\alpha \,\epsilon\, r_\eta$, hence $x, y \,\epsilon\, r_\eta$, by (7), therefore $F`\alpha' = \langle x', y'\rangle$, by (8), and $x' \,\epsilon\, y'$, that is, $F`\alpha' \,\epsilon\, E$. Hence

$$F`\alpha' \,\epsilon\, E \cdot F`\beta';$$

in other words, $F`\alpha' \,\epsilon\, F`\eta'$.

4. If $\eta \,\epsilon\, \mathfrak{W}(J_3)$, we get in the same fashion, by 9.33, $F`\eta = F`\beta - F`\gamma$ and $F`\eta' = F`\beta' - F`\gamma'$, $\beta, \gamma \,\epsilon\, m \cdot \eta$. Assume $F`\alpha \,\epsilon\, F`\eta$, and the inductive hypothesis I applied to $F`\alpha$, with $F`\beta$ and $F`\gamma$, gives $F`\alpha' \,\epsilon\, F`\eta'$ immediately.

5. Suppose $\eta \,\epsilon\, \mathfrak{W}(J_i)$, $i = 4, 6, 7, 8$. As above, $\eta = J_i`\langle\beta, \gamma\rangle$, $\eta' = J_i`\langle\beta', \gamma'\rangle$, $\beta, \gamma \,\epsilon\, m \cdot \eta$, so that $F`\eta = F`\beta \cdot Q_i``F`\gamma$ and $F`\eta' = F`\beta' \cdot Q_i``F`\gamma'$, by 9.34. Now assume $F`\alpha \,\epsilon\, F`\eta$, that is, $F`\alpha \,\epsilon\, F`\beta$ and $F`\alpha \,\epsilon\, Q_i``F`\gamma$. It follows that $F`\alpha' \,\epsilon\, F`\beta'$; also by Definition 4.52 there is an $x \,\epsilon\, F`\gamma$ such that $\langle F`\alpha, x\rangle \,\epsilon\, Q_i$. Now $x \,\epsilon\, r$ by (4) and $x \,\epsilon\, F`\gamma \,\epsilon\, r_\eta$, hence $x \,\epsilon\, r_\eta$, by (5), so that, by (8), $\langle F`\alpha', x'\rangle \,\epsilon\, Q_i$; in addition $x' \,\epsilon\, F`\gamma'$, hence

$$F`\alpha' \,\epsilon\, Q_i``F`\gamma';$$

hence $F`\alpha' \,\epsilon\, F`\eta'$.

6. There remains now only the case $\eta \,\epsilon\, \mathfrak{W}(J_5)$. As before, $\eta = J_5`\langle\beta, \gamma\rangle$ and $\eta' = J_5`\langle\beta', \gamma'\rangle$, that is, $F`\eta = F`\beta \cdot P_2``(F`\gamma)$ and $F`\eta' = F`\beta' \cdot P_2``(F`\gamma')$. Note that $x \,\epsilon\, P_2``y$ is equivalent to $y \cdot P_2``\{x\} \neq 0$. Suppose $F`\alpha \,\epsilon\, F`\eta$; then $F`\alpha \,\epsilon\, F`\beta$, and $F`\alpha \,\epsilon\, P_2``F`\gamma$, that is, $F`\gamma \cdot \breve{P}_2``\{F`\alpha\} \neq 0$. $F`\alpha \,\epsilon\, r$ and $F`\gamma \,\epsilon\, r$; hence by (1) $[F`\gamma \cdot \breve{P}_2``\{F`\alpha\}] \,\epsilon\, r$, therefore by (3) there is a $u \,\epsilon\, r$ such that $u \,\epsilon\, F`\gamma \,.\, u \,\epsilon\, \breve{P}_2``\{F`\alpha\}$. Then by (5) $u \,\epsilon\, r_\eta$; since $u \,\epsilon\, F`\gamma$ and $\langle u, F`\alpha\rangle \,\epsilon\, \breve{P}_2$, it follows that $u' \,\epsilon\, F`\gamma'$ and $\langle u', F`\alpha'\rangle \,\epsilon\, \breve{P}_2$ by (8), that is,

$$F`\alpha' \,\epsilon\, P_2``(F`\gamma');$$

therefore, since $F`\alpha' \,\epsilon\, F`\beta'$, it follows that $F`\alpha' \,\epsilon\, F`\eta'$. This concludes the proof of 12.6.

61 | 12.3 follows immediately from 12.6, since if m, o satisfy the hypothesis of 12.3, o must be closed with respect to J_i by 12.4 and with respect to C, K_1, K_2 (because $K_1`\alpha \leq \alpha$, $K_2`\alpha \leq \alpha$ by 9.25 and $C`\alpha \leq \alpha$ by definition). Hence m, o satisfy the hypothesis of 12.6.

But on page 54 it was shown that 12.2 follows from 12.3. So it is proved that the generalized continuum hypothesis is a consequence of Σ and the additional axiom $V = L$, q.e.d.[18–22]

[18][*Note added in 1951*: The above given consistency proof can easily be extended for the case that stronger axioms of infinity are added (e.g., the axiom of the existence of unaccessible numbers, or others given by P. Mahlo (*1911, 1913*)), for the simple reason that all these axioms of infinity imply their own relativized form. A similar remark also applies to extensions of the system Σ by other axioms suggested by the intuitive meaning of the primitive terms. Note added in 1966: This holds for the axioms of infinity and other additional axioms known at that time (1951).]

[19][*Note added in August 1965*: In the past few years decisive progress in the foundations of set theory has been achieved by Paul J. Cohen, who invented a powerful method for constructing denumerable models. This method yields answers to several most important consistency questions. In particular Paul J. Cohen (*1963, 1964, 1966*) has proved that Cantor's continuum hypothesis is unprovable from the axioms of set theory (including Mahlo or Levy type axioms of infinity), provided these axioms are consistent. The value that can consistently be assigned to $2^{\aleph_\alpha}$ turns out to be almost completely arbitrary. See *Cohen 1963*, *1964*, *1966*, *Solovay 1963* and *Easton 1964*, *1964a*.]

[20][*Note added in August 1965*: Other quite important progress has been made in the area of axioms of infinity, namely:

1. Mahlo's axioms of infinity have been derived from a general principle regarding the totality of sets, that was first introduced by A. Levy (*1960*). It gives rise to a hierarchy of different precise formulations. One, given by P. Bernays, implies all of Mahlo's axioms (see *Bernays 1961*).

2. Propositions which, if true, are extremely strong axioms of infinity of an entirely new kind have been formulated and investigated as to their consequences and mutual implications in *Tarski 1962*, *Keisler and Tarski 1964* and the papers cited there. In contradistinction to Mahlo's axioms the truth (or consistency) of these axioms does not immediately follow from "the basic intuitions underlying abstract set theory" (*Tarski 1962*, page 134), nor can it, as of now, be derived from them. However, the new axioms are supported by rather strong arguments from analogy, such as the fact that they are implied by the existence of generalizations of Stone's representation theorem to Boolean algebras with operations on infinitely many elements. As was conjectured in a general way in *Gödel 1947*, page 520, one of the new axioms implies the existence of non-constructible sets (see *Scott 1961*). Whether one of them implies the negation of the generalised continuum hypothesis has not yet been determined.]

[21][*Note added in August 1965*: A general discussion of Cantor's continuum problem and its relationship to the foundations of set theory is given in *Gödel 1947* and *1964*.]

[22][*Note added in August 1965*: A slightly different version of the consistency proof given in these lectures, which exhibits more clearly the basic idea of it, is outlined in *Gödel 1939a*.]

62 |

Appendix

The following list is a continuation of the list of page 13 and shows by the method explained there that all notions and operations for which special symbols are introduced in these lectures (except only $\simeq$ and $\mathfrak{We}$) are normal.[23]

4.1 $Z \epsilon X \times Y . \equiv . (\exists u, v)[\langle u, v\rangle = Z . u \epsilon X . v \epsilon Y]$

4.11 $Z \epsilon X^2 . \equiv . Z \epsilon X \times X$ (similarly for X^3)

4.2 $\mathfrak{Rel}(X) . \equiv . X \subseteq V^2$ (similarly for $\mathfrak{Rel}_3$)

4.4, 4.41, 4.411 $Z \epsilon \mathfrak{Cnv}(X) . \equiv . (\exists u, v)[\langle u, v\rangle = Z . \langle v, u\rangle \epsilon X]$ (similarly for $\mathfrak{Cnv}_2, \mathfrak{Cnv}_3$)

4.42 $Z \epsilon X + Y . \equiv : Z \epsilon X . \vee . Z \epsilon Y$

4.43 $Z \epsilon X - Y . \equiv : Z \epsilon X . \sim (Z \epsilon Y)$

4.45 $Z \epsilon \mathfrak{W}(X) . \equiv . Z \epsilon \mathfrak{D}(\mathfrak{Cnv}(X))$

4.5 $Z \epsilon X \restriction Y . \equiv . Z \epsilon X \cdot (V \times Y)$ (similarly for 1 (4.512))

4.52 $Z \epsilon X``Y . \equiv . Z \epsilon \mathfrak{W}(X \restriction Y)$

4.53 $Z \epsilon X|Y . \equiv . (\exists u, v, w)[Z = \langle u, w\rangle . \langle u, v\rangle \epsilon X . \langle v, w\rangle \epsilon Y]$

4.6 $\mathfrak{Un}_2(X) . \equiv : \mathfrak{Un}(X) . \mathfrak{Un}(\mathfrak{Cnv}(X))$

4.61 $\mathfrak{Fnc}(X) . \equiv : \mathfrak{Rel}(X) . \mathfrak{Un}(X)$

4.63 $X \mathfrak{Fn} Y . \equiv : \mathfrak{Fnc}(X) . \mathfrak{D}(X) = Y$

4.65 $Z \epsilon X`Y . \equiv . (\exists u)[Z \epsilon u . (v)(\langle v, Y\rangle \epsilon X . \equiv . v = u)]$

4.8 $Z \epsilon \mathfrak{S}(X) . \equiv . (\exists u)[Z \epsilon u . u \epsilon X]$ (the same proposition holds for $\mathfrak{Max}$ and $\mathfrak{Lim}$)

4.84 $Z \epsilon \mathfrak{P}(X) . \equiv : \mathfrak{M}(Z) . Z \subseteq X$

6.1 $X \mathfrak{Con} Y . \equiv . Y^2 \subseteq X + \mathfrak{Cnv}(X) + I$

6.3 $X \mathfrak{Sect}_Z Y \ . \equiv : Y \cdot Z``X \subseteq X . X \subseteq Y$

6.31 $Z \epsilon \mathfrak{Seg}_T(X, Y) . \equiv . Z \epsilon X \cdot T``\{Y\}$

6.4 $Z \mathfrak{Isom}_{P,Q}(X, Y) . \equiv :: . \mathfrak{Rel}(Z) . \mathfrak{Un}_2(Z) . \mathfrak{D}(Z) = X .$
$\mathfrak{W}(Z) = Y . (u, v)[u, v \epsilon X . \supset . (\langle u, v\rangle \epsilon P . \equiv . \langle Z`u, Z`v\rangle \epsilon Q)]$

6.5 $\mathfrak{Comp}(X) . \equiv . \mathfrak{S}(X) \subseteq X$

6.6 $\mathfrak{Ord}(X) . \equiv : \mathfrak{Comp}(X) . E \mathfrak{Con} X$

6.61 $\mathfrak{O}(X) . \equiv : \mathfrak{Ord}(X) . \mathfrak{M}(X)$
$\alpha, \beta, \gamma, \ldots$ are normal variables since their range is the class On.

6.63, 6.64 $X < Y . \equiv . X \epsilon Y; \quad X \leq Y . \equiv : X \epsilon Y . \vee . X = Y$

7.4 $Z \epsilon X \dotplus 1 . \equiv : Z \epsilon X . \vee . [Z = X . \mathfrak{M}(Z)]$

8.12 $X \simeq' Y . \equiv : . (\exists u)[\mathfrak{Rel}(u) . \mathfrak{Un}_2(u) . X = \mathfrak{D}(u) . Y = \mathfrak{W}(u)]$

8.2 $Z \epsilon \overline{\overline{X}} . \equiv : Z \epsilon On . (\alpha)[\alpha \simeq' X . \supset . Z \epsilon \alpha]$

8.48, 8.49 $\mathfrak{Fin}(X) . \equiv . (\exists \alpha)[\alpha \epsilon \omega . X \simeq' \alpha]$, $\mathfrak{Inf}(X) . \equiv . \sim \mathfrak{Fin}(X)$
9.1 $Z \epsilon \mathfrak{F}_1(X,Y) . \equiv . Z \epsilon \{X,Y\}$, $Z \epsilon \mathfrak{F}_2(X,Y) . \equiv . Z \epsilon E \cdot X$
(similarly for $\mathfrak{F}_3, \ldots, \mathfrak{F}_8$)
9.41 $\mathfrak{L}(X) . \equiv : X \subseteq L . (u)[u \epsilon L \supset u \cdot X \epsilon L]$

Index

I. Special Symbols

II. Letters and Combinations of Letters

(Note that the letters C, F, R, S also occur as variables before their respective definitions as constants. Operations and notions are denoted in general by German letters, classes by Latin letters.)

64 |

Variables:

$X, Y, Z, \ldots,$ $A, B, C, \ldots,$ for classes
$x, y, z, \ldots,$ $a, b, c, \ldots,$ for sets
$\alpha, \beta, \gamma, \ldots,$ for ordinal numbers
$i, k, \ldots,$ for integers
$\overline{X}, \overline{Y}, \ldots,$ $\overline{A}, \overline{B}, \ldots,$ for constructible classes
$\overline{x}, \overline{y}, \ldots,$ $\overline{a}, \overline{b}, \ldots,$ for constructible sets

| 65

III. Technical Terms

Introductory note to *1944*

This paper was written for *The philosophy of Bertrand Russell*, a volume of Paul Arthur Schilpp's series the Library of living philosophers. In his letter of invitation of 18 November 1942, Schilpp proposed the title of the paper and also wrote: "In talking the matter over last night with Lord Russell in person, I learned that he too would not only very greatly appreciate your participation in this project, but that he considers you the scholar *par excellence* in this field." Gödel sent in the manuscript on 17 May 1943. There followed a lengthy correspondence about stylistic editing proposed by Schilpp and Gödel's own deliberation concerning revision. Before Gödel had submitted the final version, Russell had completed his reply to the other papers and decided that under the circumstances he would not reply to Gödel's. When Gödel finally sent in the revised version on 28 September, he wrote to Russell attempting to change his mind about not replying. He undertook to disabuse Russell of his impression that what Gödel said would not be controversial, and emphasized his criticisms of Russell. However, Russell confined himself to the following brief note:

> Dr. Gödel's most interesting paper on my mathematical logic came into my hands after my replies had been completed, and at a time when I had no leisure to work on it. As it is now about eighteen years since I last worked on mathematical logic, it would have taken me a long time to form a critical estimate of Dr. Gödel's opinions. His great ability, as shown in his previous work, makes me think it highly probable that many of his criticisms of me are justified. The writing of *Principia Mathematica* was completed thirty-three years ago, and obviously, in view of subsequent advances in the subject, it needs amending in various ways. If I had the leisure, I should be glad to attempt a revision of its introductory portions, but external circumstances make this impossible. I must therefore ask the reader to give Dr. Gödel's work the attention that it deserves, and to form his own critical judgment on it. (*Schilpp 1944*, page 741)

Gödel subsequently contributed *1949a* to the volume on Einstein in Schilpp's series and also accepted an invitation to contribute to the Carnap volume (*Schilpp 1963*). Several drafts of this paper on Carnap survive in Gödel's *Nachlass*, but it was never actually submitted. Still later, Gödel declined an invitation to contribute to the Popper volume.

The paper *1944* was reprinted twice (*1964a, 1972b*), with only editorial changes in the text. In an opening footnote added in *1964a*, Gödel clarifies the difference between the use of the term "constructivistic" in that paper and the more usual uses; this remark is revised and expanded in *1972b*.

The paper is notable as Gödel's first and most extended philosophical statement. It holds mainly to the form of a commentary on Russell; as such it has been quite influential. However, Gödel is not reticent in expressing his own views. In the present note I shall give more emphasis to what it reveals about the thought of its author.

The organization of the paper is difficult for the present commentator to analyze. He cannot but endorse Hermann Weyl's remark that the paper "is the work of a pointillist: a delicate pattern of partly disconnected, partly interrelated, critical remarks and suggestions" (*1946*, page 210). Nonetheless, Gödel's paper might be divided as follows:

1. Introductory remarks (125–128).
2. Russell's theory of descriptions (128–131).
3. The paradoxes and the vicious-circle principle (131–137).
4. Gödel's own realistic view of classes and "concepts" (137–141).
5. Contrast with Russell's "no-classes theory" and the ramified theory of types; limitations of the latter (141–147).
6. The simple theory of types (147–150).
7. The analyticity of the axioms of *Principia* (150–152).
8. Concluding remarks on mathematical logic and Leibniz' project of a universal characteristic (152–153).

I follow this division in the remainder of this note, where I use a number in parentheses without a date to indicate a page number in *Gödel 1944*.

The Gödel archive contains reprints of *1944* (designated below as A–D) and a loose page (designated E) containing annotations to it. All changes on A–E are listed in the textual notes for *1944* at the end of this volume. Many of the annotations are textual emendations, for the most part either of a stylistic nature or for greater explicitness. A few indicate changes of view on specific points. There is no way of knowing whether any of these changes represents a final position for Gödel. They are hardly reflected at all in the reprints *1964a* and *1972b*. E contains some remarks on Bernays' review *1946* of the paper. I will comment below on only a few of the annotations.

1. Introductory remarks

Early on, Gödel remarks on Russell's "pronouncedly realistic attitude" and the analogy with natural science expressed by Russell's remark, "Logic is concerned with the real world just as truly as zoology, though with its more abstract and general features" (*1919*, page 169). He also mentions another, epistemological analogy in Russell's view that the axioms of logic and mathematics do not have to be evident in themselves but can obtain justification from the fact that their consequences agree with what has been found evident in the course of the history of mathematics (127). Gödel remarks: "This view has been largely justified by subsequent development, and it is to be expected that it will be still more so in the future."

The essay as a whole might be seen as a defense of these attitudes of Russell against the reductionism prominent in his philosophy and implicit in much of his actual logical work. It was perhaps the most robust defense of realism about mathematics and its objects since the paradoxes had come to the consciousness of the mathematical world after 1900. Bernays' earlier defense of realism (for example, in *1935*) was more cautious. Gödel begins to develop this theme when he turns to Russell's approach to the paradoxes (see especially §§3–5 below).

2. The theory of descriptions

This discussion is noteworthy. Gödel indicates (128–129 and note 5) a formal argument for Frege's thesis that the signification (his translation of Frege's *Bedeutung*) of two sentences is the same if they agree in truth value. The argument collapses intensional distinctions on the basis of simple assumptions about signification. A similar argument to the same conclusion, from somewhat different assumptions, occurs in *Church 1943*, pages 299–300. Such collapsing arguments have been prominent since in philosophical discussions of meaning and reference, modality and propositional attitudes. Gödel concedes that Russell's theory of descriptions avoids Frege's conclusion and allows a sentence to signify a fact or a proposition.[a] He expresses the suspicion that it only evades the problem (130).

[a]Church (*1942*) observes that Russell's theory of descriptions eliminates apparent violations of the substitutivity of identity in intensional contexts. The universal substitutivity of identity is one of the assumptions on which Gödel's and Church's collapsing arguments turn. Church's observation is the basis for the reply to Quine's criticism of modal logic in *Smullyan 1948*. Quine's use of the argument in criticizing modal logic is its most widely known and influential use. However, it does not occur in either of his two early papers on this theme (*1943* and *1947*). The earliest occurrence of a collapsing argument of the Gödel–Church type that I have been able to find in Quine's writings is *1953*, p. 159; see also *1953a* (*1976*, pp. 163–164). Both these arguments are essentially the same as Church's.

3. The paradoxes and the vicious-circle principle

Gödel says that Russell "freed them ⟦the paradoxes⟧ from all mathematical technicalities, thus bringing to light the amazing fact that our logical intuitions (i.e., intuitions concerning such notions as: truth, concept, being, class, etc.) are self-contradictory" (131). Many readers have been puzzled by the contrast between this statement and the defense of the concept of set in *1947*, where he says that the set-theoretical paradoxes "are a very serious problem, but not for Cantor's set theory" (page 518), revised in *1964* to "... problem, not for mathematics, however, but rather for logic and epistemology" (page 262).[b] Closer examination makes the passages not difficult to reconcile. In the later paper Gödel says that the concept of set in contemporary mathematics, including Cantor's set theory, can be taken to be what we call the iterative conception of set, according to which sets are obtained by iterated application of the formation of sets of previously given objects, beginning with some well-defined objects such as the integers. It is this conception that has "never led to any antinomy whatsoever" and whose "perfectly 'naive' and uncritical working" has "so far proved completely self-consistent" (*1964*, page 263).

In the present paper, Gödel is following Russell in being concerned with the foundations of logic in a larger sense; note that in the above quotation our logical intuitions are said to be "concerning such notions as: truth, concept, being, class, etc." Though it is mentioned (144), the iterative conception of set is kept in the background, perhaps more so than was optimal for the purpose of defending realism. But, as we shall see, Gödel's purpose was not limited to defending realism about sets or the objects of classical mathematics.

After short remarks about two proposals that Russell discussed briefly in *1906*, the "theory of limitation of size" and the "zig-zag theory", Gödel turns to the vicious-circle principle. The masterly analysis of the ambiguities of this principle and the criticism of the principle itself constitute probably the most influential piece of direct commentary on Russell in the essay and also supply a major argument for Gödel's own position. Gödel's main criticism, which had already been intimated by Ramsey in *1926* (*1931*, page 41), is that the strongest form of the principle, that no totality can contain members *definable only in terms of* this totality, is true only if the entities whose totality is in question are "constructed by ourselves". Gödel adds: "If, however, it is a question of

[b]Perhaps Gödel thought it necessary to clarify the difference of his concerns from those of *1944*.

objects that exist independently of our constructions, there is nothing in the least absurd in the existence of totalities containing members, which can be described ... only by reference to the totality" (136). Earlier Gödel observed that this form of the principle is not satisfied by classical mathematics or even by the system of *Principia mathematica* with the axiom of reducibility. He considered this "rather as a proof that the vicious-circle principle is false than that classical mathematics is false" (135).

These remarks lead Gödel into the declaration of his realistic point of view. But before discussing this we should note his characterization of the position that would justify the strong vicious-circle principle, which he calls constructivistic or nominalistic. He seems to regard this viewpoint as involving the eliminability of reference to such objects as classes and propositions (136–137). His model is clearly Russell's no-class theory.

4. Gödel's realism

Classes and concepts, according to Gödel, may be understood as real objects "existing independently of our definitions and constructions" (137). It should be stressed, as it has not been in previous commentaries on this paper, that Gödel's realism extends not only to *sets* as described in axiomatic set theory, but also to what he calls *concepts*: "the properties and relations of things existing independently of our definitions and constructions". He is clearly referring to both in his often-quoted remark that "the assumption of such objects is quite as legitimate as the assumption of physical bodies" and that "they are in the same sense necessary to obtain a satisfactory system of mathematics as physical bodies are necessary for a satisfactory theory of our sense perceptions" (137). These remarks have excited critical comment.[c] To deal in an adequate way with the questions they raise would be quite beyond the scope of a note of this kind. However, we should state some of these questions.

(i) What does Gödel mean by "real objects" and "existing independently of our definitions and constructions"? A question of this kind arises about any form of philosophical realism. If realism about sets and concepts is to go beyond what would be asserted by a non-

[c]For example, *Chihara 1973*, pp. 61, 75–81; *Chihara 1982*, part I; *Dummett 1978*, p. 204.

subjectivist form of constructivism, this reality and independence will have to amount to more than objective existence. In the present paper, Gödel does not undertake very directly to clarify his meaning, though something can be learned from his discussion of the ramified theory of types (see below).[d] Of course, the general question "What is realism?" has been much debated in quite recent times, largely through the stimulation of the writings of Michael Dummett.[e]

A point that needs to be stressed, however, is that Gödel saw his realism in the context of concrete problems and as motivating mathematical research programs. This is perhaps most evident in *1947/1964*, with its defense of the view that the continuum hypothesis is definitely true or false even though probably (and by 1963 certainly) independent of the established axioms of set theory. Further reflection shows that it is very much present in the paper under discussion, where Gödel criticizes ideas of Russell that obstructed the transfinite extension of the hierarchies of simple and ramified type theories. The remarks about his own theory of constructible sets (146–147; see below) are an illustration of the "cash value" of realism for Gödel.[f]

(ii) How does Gödel understand the parallel between the objects of mathematics and "physical bodies"? It would be tempting to suppose that Gödel views sets and concepts as postulated in a theory to *explain* certain data.[g] This is suggested by the parallel itself between the necessity of sets and concepts for a "satisfactory system of mathematics" and the necessity of physical bodies for a "satisfactory theory of our sense perceptions"; it is also in line with Gödel's approval, noted above, of Russell's suggestion that mathematical axioms can be justified by

[d]In *1949a* Gödel argues that the general theory of relativity calls in question the objectivity of time and change. He sees this as a confirmation of idealistic views, in particular Kant's. He does not attempt to draw any parallel with a possible anti-realist view of mathematics.

Of course, Kant did not "deny the objectivity of change" (*Gödel 1949a*, p. 557) if what is meant by the latter is the existence of an objective temporal order that is the same for all observers with our forms of intuition. Evidently Gödel thinks of the dependence of the temporal ordering of events on the position and state of motion of the observer according to relativity theory as parallel to the dependence of the very temporality of the experienced world on the constitution of our cognitive faculties according to Kant.

[e]For example, the essays in *Dummett 1978*, especially "Truth", "Realism", and "The reality of the past".

[f]Cf. note 48a of *Gödel 1931*, which Burton Dreben called to my attention.

[g]This interpretation is assumed by Dummett (*1978*, p. 204) and in some of Chihara's criticisms (for example, *1982*, pp. 214–215).

their consequences. But more direct evidence for this interpretation is lacking.[h]

Gödel does, however, use the notion of "data" with reference to mathematics. Indeed, one point of parallelism between mathematics and physics is simply that "in both cases it is impossible to interpret the propositions one wants to assert about these entities as propositions about the 'data'" (137). But at this point he refrains from saying what plays the role of data in the case of mathematics. We shall return to this matter in connection with his discussion of the ramified theory.

(iii) By "concepts" Gödel evidently means objects signified in some way by predicates. The notion "property of set", which he counts among the primitives of set theory (*1947*, note 17, or *1964*, note 18), is clearly a special case of this notion. Why he should have considered "property of set" a primitive of set theory is clear enough from the role of classes in set theory and from the generalization with respect to predicates contained in the axioms of separation and replacement. Gödel therefore did not lack mathematical motivation for adding something like concepts to his ontology.[i]

But what sort of theory of concepts did Gödel envisage? What consequences for the theory does realism about concepts have, once the existence of sets as "real objects" is granted? The use of the notion of class that is standard in set theory is predicative relative to the universe of sets. In the above-cited note, he seems to envisage impredicative theories of properties based on the simple theory of types, which he also mentions as a theory of concepts in the present paper (140). In the note he remarks that such theories are not deductively stronger than extensions of the axioms referring to sets.

It is clear that Gödel takes his realism about concepts to justify an impredicative theory, and he suggests strongly that he would prefer a stronger theory than the simple theory of types. He claims (139) that impredicative specifications of properties do not themselves lead to absurdity and that a property might "involve" a totality of properties to

[h]Gödel does not use the language of explanation in the two passages where he is most explicit about the justification of mathematical axioms through their consequences (*1964*, pp. 265, 272), although in the former he does describe a (hypothetical) situation in which axioms "would have to be accepted in at least the same sense as any well-established physical theory." In spite of its lack of direct support, the interpretation in terms of explanation is difficult to refute.

In these passages in *1964*, Gödel seems to me to be considerably more cautious about justification of axioms by their consequences than he appears to be in the above-noted passage (127).

[i]This view may be reflected in Gödel's choice of a theory with class variables as the framework for *1940*. Note the remark on page 2, "Classes are what appear in Zermelo's formulation ... as 'definite Eigenschaften'."

which it belongs, thus contradicting the second of the three forms of the vicious-circle principle he has earlier distinguished.[j] He also remarks: "Nor is it self-contradictory that a proper part should be identical (not merely equal) to the whole, as is seen in the case of structures in the abstract sense" (139). Of course, on the set-theoretic conception of structure, such identity (as opposed to isomorphism) does not obtain and would indeed be self-contradictory, at least if one takes "proper part" in its obvious meaning of a substructure whose domain is a proper subset of the whole. Gödel is evidently thinking in terms of an informal notion of structure according to which isomorphic "structures" (in the set-theoretic sense) are *the same* structure. But it is a problem to construct a theory in which this sameness is interpreted as identity.[k]

Gödel seems to regard the simple theory of types as the best presently available solution to the paradoxes for a theory of concepts, but "such a solution may be found ... in the future perhaps in the development of the ideas sketched on pages 132 and 150" (140). The former refers to his remarks on Russell's "zig-zag theory", the latter to the frequently quoted but enigmatic suggestion that a concept might be assumed "significant

[j]Whether the simple theory of types conflicts with this form of the principle depends on how it is interpreted. It seems clear, for example from the emphasis Gödel places on the claim (136) that classical mathematics, and in particular *Principia* with reducibility, does not satisfy the *first* form, that Gödel thinks that the extensional simple theory of types with its higher-order variables interpreted to range over sets does satisfy the second form (and the third as well). Whether an intensional form of the simple theory, in which the variables range over properties and relations, satisfies the second form will depend on the underlying notion of intension. Gödel's remark (139) that "the totality of all properties (or of all those of a given type) does lead to situations of this kind", in which the second form is violated, makes it clear that he envisages a notion of property that would lead to an interpretation of the simple theory of types where the second form of the principle *is* violated. Gödel's annotations to reprint A, however, call this into question. In *1944*, the violation of the second form of the principle is said to arise because a universal quantification over properties of a given type contains these properties as constituents of their content (139). This is questioned in A on the ground that universal quantification "does not mean in the same way as conjunction does."

It should be noted that Gödel's first remark in E on *Bernays 1946* is "Misunderstanding of my interpretation of type theory for concepts." Presumably he is attributing such a misunderstanding to Bernays. I am unable to determine in what the misunderstanding consists.

[k]In the language of category theory, we could say that an object A is a proper part of an object B if there is a monomorphism of A into B that is not epi; this does not exclude identity. The alternative, to say that A is a proper part of B if there is a monomorphism of A into B but there is not an epimorphism, does of course exclude identity but does not fit what Gödel says.

Of course, the relevant difference between the construal of structures as tuples of a domain and relations on it, and the language of categories, is that the former forces, while the latter does not, a distinction between isomorphism and identity. This is a much more superficial difference than the matter of the self-applicability of categories (see below).

everywhere except for certain 'singular points' or 'limiting points', so that the paradoxes would appear as something analogous to dividing by zero." Evidently he has in mind a type-free theory; he mentions as attempts the work of the early 1930s on theories based on the λ-calculus (in particular *Church 1932, 1933*), which he views as having had a negative outcome, in view of *Kleene and Rosser 1935*. Gödel does not return to this theme in later publications, except for the brief remark that "the spirit of the modern abstract disciplines of mathematics, in particular the theory of categories, transcends this ⟦iterative⟧ concept of set, as becomes apparent, e.g., by the self-applicability of categories" (*1964*, page 262, footnote 12; not in *1947*). However, he evidently thought that Mac Lane's distinction between large and small categories captured "the mathematical content of the theory" as it then stood. But the program of constructing a strong type-free theory has attracted others, with inconclusive results so far.[l]

Gödel's remarks about realistic theories of concepts in the present paper have an inconclusive character; no available theory satisfies him. In later publications, as we have noted, he is virtually silent on the subject. The question arises whether Gödel himself worked on the project of constructing a theory that would answer to his conception. Whether he did is not known, but the absence of more definite information would suggest the conjecture that he never formulated such a theory to his own satisfaction. It is to be hoped that transcriptions of Gödel's shorthand notebooks will shed light on these questions.

5. The ramified theory of types

Returning to Russell, Gödel begins his discussion of the ramified theory by remarking on Russell's "pronounced tendency to build up logic as far as possible without the assumption of the objective existence of such entities as classes and concepts" (141). He reads the contextual definitions of locutions involving classes in *Principia* as a reduction of classes to concepts,[m] but reasonably enough finds matters not so clear when it comes to concepts and propositions. Influenced especially by

[l]For a survey see *Feferman 1984* and its sequel.

[m]Gödel's reading is misleading in that he clearly understands concepts to be *objects*, while the "ambiguity" that Russell attributes to propositional functions is close to Frege's "unsaturatedness". But, from his later comment on the notion of propositional function (147–148), it is clear that this does not result from misunderstanding but is rather a conscious assimilation of Russell's conceptual scheme to his own. See note t below.

the introduction to the second edition, Gödel finds in *Principia* a program according to which all concepts and propositions except logically simple ones are to "appear as something constructed (i.e., as something not belonging to the 'inventory' of the world)" (142). This program offers an intrinsic motivation for the ramification of the type hierarchy, but does not yield a theory strong enough for classical mathematics, for well-known reasons: the impredicative character of standard arguments in analysis, and the question whether, to construct number theory, one can replace the Frege–Russell definition of the predicate "natural number" by one in which the second-order quantifier is restricted to a definite order (145–146).[n]

It is not as clear as it might be how Gödel sees the realization of the program he attributes to Russell even to obtain ramified type theory without reducibility. The introduction to the second edition of *Principia* proceeds on the basis of the Wittgensteinian idea that "functions of propositions are always truth functions, and that a function can only occur in a proposition through its values" (*Whitehead and Russell 1925*, page xiv). But it is hard to see how propositions involving quantifiers are to be interpreted as truth functions of atomic propositions unless infinitary propositional combinations are allowed, as Ramsey in effect proposed; Gödel's comment on that is that one might as well adopt the iterative conception of sets as pluralities (144). Gödel says that Russell "took a less metaphysical course by confining himself to such truth-functions as can actually be constructed" (145). But what are the allowed means of construction? Gödel apparently has in mind an interpretation of the ramified theory in which the higher-order variables range over *predicates*, that is, the linguistic expressions that "express" the propositional functions that the quantifiers range over on the naive reading. This is indicated by Gödel's later remark that for propositional functions to be "defined (as in the second edition of *Principia*) to be certain finite ... combinations (of quantifiers, propositional connectives, etc.)" (146) would presuppose the notion of finiteness and therefore arithmetic, and by the earlier characterization of the constructivist view he is trying to explicate as a form of nominalism (136–137).

Such an interpretation is certainly possible and well known, provided that at the level of individuals one has elementary syntax. But this reading of Gödel still leaves some puzzles. Translating a statement about propositional functions of order n as one about predicates, namely predicates containing quantifiers only for propositional functions of order

[n]Gödel leaves the latter question open. A negative answer is claimed without proof in *Wang 1959* (see *1962*, p. 642). A full treatment, with a proof, is given in *Myhill 1974*.

$< n$, requires the notion of satisfaction, or at least truth, for the language with quantifiers of order $< n$. It is surprising that Gödel nowhere remarks on this fact, particularly since in this context we would have to suppose satisfaction or truth introduced by an inductive definition, another obstacle to applying the idea in reducing arithmetic to logic. Moreover, such an interpretation entirely eliminates quantification over the sort of entities Gödel calls concepts, at least in the absence of such locutions as propositional attitudes, and therefore does not leave "the primitive predicates and relations such as 'red' or 'colder'" as "real objects" (142). But to suppose that Gödel would have remarked on the latter point if he had had the present interpretation in mind may be to attribute to him a Quinean distinction of "ontology" and "ideology" that is foreign to him.

The difficulties faced by this last interpretation suggest to me that Gödel did not distinguish clearly in his own mind between a nominalist theory of concepts in which such entities are *eliminated*, and a theory in which every concept in the range of a quantifier is "signified" by an expression for it that is antecedently understood, but in which reference to concepts is not actually eliminated, because, presumably, one does not give a contextual definition of quantifiers over them. The latter sort of theory might now be realized by a substitutional interpretation of quantifiers over concepts; given such an interpretation, the most that can be asked in establishing the "existence" of a concept is the construction of a meaningful expression "signifying" it.

Something more should be said about Gödel's use of the term "data". In the discussion of the ramified theory it refers to that on the basis of which classes and concepts are constructed, or perhaps to what is allowed as primitive in a theory in which reference to classes and concepts is eliminated. (See for example note 33, page 142.) Gödel does not say here what epistemological force this might have. The analogy with sense perception may be limited to the context of interpreting Russell, who was interested in a program that would represent the objects of physics as "logical constructions" from sense-data. In Gödel's own epistemological view of mathematics, what corresponds most closely to sense perception is something quite different, namely elementary arithmetical evidence (see 128).

Like many other commentators, Gödel found that in the *first* edition of *Principia* the constructivistic attitude was fatally compromised by the axiom of reducibility, but his description (143) of what survived is worth noting.[o] However, Gödel's remarks about the axiom of reducibility show

[o]Gödel mentions here the treatment of propositional connectives as applied to propositions containing quantifiers, presumably in *9, which he says "proved its

lack of sensitivity to the essentially intensional character of Russell's logic; the fact that every propositional function is *coextensive* with one of lowest order does not imply "the existence in the data of the kind of objects to be constructed" (141), if the objects in question are concepts or propositional functions rather than classes. It is similarly misleading to say that, "owing to the axiom of reducibility, there always exist real objects in the form of primitive predicates, or combinations of such, corresponding to each defined symbol" (143). Russell himself was closer to the mark in saying that the axiom accomplishes "what common sense effects by the admission of *classes*" (*1908*, page 167 of *van Heijenoort 1967*). The ramified theory with reducibility would fit well with a conception according to which classes are admitted as "real objects", but the conception of propositional functions (concepts) is constructivistic. This insensitivity is quite common in commentators on Russell, but is somewhat surprising in Gödel, since his own "concepts" are evidently intensions of a kind, and his *1958* shows a very subtle and fruitful handling of intensional notions.[p]

Gödel concludes his discussion of the ramified theory with well known remarks in which he views his own theory of constructible sets as an extension of the hierarchy of orders, now within the framework of ordinary (impredicative) mathematics, to arbitrary transfinite orders (146–147). After what I have said about Gödel's treatment of the axiom of reducibility, I should call attention to his characterization of his theorem that every constructible set of integers has order $< \omega_1$ as a "transfinite theorem of reducibility"[q]—thus, when set against Russell, a striking application of his realistic point of view.

Gödel remarks that even from the predicative standpoint an extension of the hierarchy of orders is possible and, moreover, demanded by the theory. This remark may be the first suggestion of a program that

fecundity in a consistency proof for arithmetic". The connection between *9 and Herbrand's work suggests that he has in mind *Herbrand 1931*, which, however, covered only first-order arithmetic with quantifier-free induction. A proof within Herbrand's framework that covers full first-order arithmetic became known only some years later; see *Dreben and Denton 1970* and *Scanlon 1973*. Could Gödel have seen at this time how to extend Herbrand's proof? The errors in the proof of the fundamental theorem in *Herbrand 1930* would have been an obstacle. At the time Gödel was at least aware of a difficulty; see *van Heijenoort 1967*, p. 525.

Another possibility, suggested by van Heijenoort, is that Gödel was thinking of the proof he published in *1958*, which he had discovered not long before writing the present paper. This now seems to me somewhat more likely.

[p]Of course, the theory of intensional equality in *1958* is very different from more usual constructions in intensional logic, and in the present paper he suggests that concepts might obey extensionality (137).

[q]In reprint A, Gödel amended the text in a way that omits this phrase.

was pursued by several logicians from 1950 on, of extending and analyzing the resources of predicative mathematics by means of ramified theories with transfinite orders. The first work of this kind (*Lorenzen 1951a*, *1951b*, *1955*; *Wang 1954*) sought to give a better reconstruction of classical mathematics than earlier predicative work by constructing transfinite ramified theories. It left open the question what ordinal levels can be admitted in such a construction.

Gödel already offers a hint in remarking that one can extend the hierarchy of orders "to such transfinite ordinals as can be constructed within the framework of finite orders" (147). It seems evident that such a procedure might be iterated; this gives rise to the notion of autonomous iteration that is prominent in later analyses of predicativity. The first such proposal is made in *Wang 1954*: given an interpreted language Σ_α (in his setting, ramified set theory with ordinal levels $< \alpha$), one extends it by admitting as new levels ordinals β such that a well-ordering of type β is *definable* in Σ_α (*1962*, page 579). In this case, since Wang's theory Σ_0 could define all recursive orderings, the iteration closes after one step. This was shown in *Spector 1957*. Spector constructed a sequence of systems similar to Wang's, indexed by the recursive ordinal notations, and showed that any well-ordering definable in one of the systems has as order type a recursive ordinal; he also showed that the sets of natural numbers definable in some system of the sequence are exactly the hyperarithmetic sets.

Kleene's work on the hyperarithmetic hierarchy related it to a transfinite ramified hierarchy, and a number of technical results suggested the thesis, advanced with some reservations by G. Kreisel (*1960a*, page 373), that a set of natural numbers is predicatively definable if and only if it is hyperarithmetic. Kreisel seems to have been more confident of the "only if" than of the "if" part of this thesis (*1960a*, pages 387–388; see also *1962*, page 318, and *Feferman 1964*, page 10). The main reservation about the latter concerned the use in the definition of a hyperarithmetic set of the notion of a recursive ordinal; should one not demand, for an ordinal to count as predicatively obtained, that an ordering of that type be *predicatively recognized* to be a well-ordering (*1960a*, page 387)?[r]

[r]Accepting this demand seems to involve giving up the attempt to characterize predicative definability independently of predicative provability. In fact the independent discussion of predicative definability seems to have petered out when Feferman's and Schütte's results became known.

That a definite meaning can be given to the notion "predicatively definable set of natural numbers" without using impredicative concepts seems very doubtful, and Kreisel already stressed in *1960a* that he was approaching the analysis of predicativity with the help of impredicative notions, that of a well-ordering in particular. (See also Lorenzen's reasons for rejecting the question how far his iterative construction

This question leads into the analysis of predicative *provability*, in which another type of autonomous iteration (first suggested in *Kreisel 1960*) plays a role. In the context of ramified theories, one constructs a progression of theories (in the precise sense of *Feferman 1962*) such that levels up to an ordinal given by a primitive recursive ordering are admitted if at an earlier stage that ordering has been *proved* to be well-founded. This led to a precise characterization of the predicatively provable statements of ramified analysis (*Feferman 1964*; *Schütte 1965, 1965a*). The same idea is applied to the admission of stages in the progression of theories, and it is therefore applicable to unramified theories, which allowed Feferman to extend his characterization to the usual language of analysis (*1964*) and to set theory (*1966*, *1974*). The results of Feferman and Schütte all point to the conclusion that an ordering that can be predicatively proved to be a well-ordering is of type less than a certain recursive ordinal Γ_0, the first strongly critical number.[s]

6. The simple theory of types

What is of greatest interest in Gödel's discussion of the simple theory is his questioning of it as a theory of concepts and the hint for a possible type-free theory commented on above (§4). It should be noted that he motivates his suggestion by means of Russell's idea that propositional functions have limited ranges of significance (149–150).

7. Analyticity

Gödel now turns to "the question whether (and in which sense) the axioms of *Principia* can be considered to be analytic". In a first sense—roughly, reducibility by explicit or contextual definitions to instances of the law of identity—he says that even arithmetic is demonstrably non-analytic because of its undecidability. This sense is of interest because it seems to be directly inspired by Leibniz. If infinite reduction, with intermediary sentences of infinite length, is allowed (as would be suggested by Leibniz's theory of contingent propositions), then all the axioms of *Principia* can be proved analytic, but the proof would require "the whole of mathematics ... e.g., the axiom of choice can be proved to be analytic

of language strata can be carried, *1955*, p. 189.) Kreisel's attitude was in line with Gödel's realism and may have been influenced by it. But it should be noted that the notions involved can still be understood constructively.

[s] *Feferman 1964*, part II; see also *Schütte 1977*, chapter VIII.

only if it is assumed to be true" (151). This remark anticipates later arguments criticizing the thesis that mathematics is analytic, such as *Quine 1960*. Gödel continues: "In a second sense a proposition is called analytic if it holds 'owing to the meaning of the concepts occurring in it', where this meaning may perhaps be undefinable (i.e., irreducible to anything more fundamental)." In this sense, Gödel affirms the analyticity of the axioms of the first edition of *Principia* other than the axiom of infinity for two interpretations, "namely if the term 'predicative function' is replaced either by 'class' (in the extensional sense) or (leaving out the axiom of choice) by 'concept'" (151). The first is the sort of interpretation suggested by remarks of Russell such as that in *1908* quoted above.[t] The second prompts the remark that "the meaning of the term 'concept' seems to imply that every propositional function defines a concept". Gödel's intuitive notion of concept seems in that respect to have resembled Frege's notion of extension.[u]

For analyticity in this sense, Gödel sees the difficulty that "we don't perceive the concepts of 'concept' and 'class' with sufficient distinctness, as is shown by the paradoxes" (151). But, rather than following Russell's reductionism, the actual development of logic (even by Russell in much of his work) has consisted in "trying to make the meaning of the terms 'class' and 'concept' clearer, and to set up a consistent theory of classes and concepts as objectively existing entities" (152). In spite of the success of the simple theory of types and of axiomatic set theory, "many symptoms show only too clearly, however, that the primitive concepts need further elucidation". But surely Gödel must have seen matters differently with respect to the two notions of class and concept: in the former, he seems to insist in *1947*, there is a well-motivated theory that is quite satisfactory as far as it goes; what is lacking in our "perception" of the notion of set is intuition regarding the truth of axioms that would decide such questions as the continuum problem. In the latter, he had no theory to offer that answered to his intuitive notion, and it does not appear that such a theory has been constructed since.

[t]Gödel's remarks about the axiom of reducibility commented on above might suggest that he had in mind an extensional interpretation that would collapse the ramification of the hierarchy. But that in the absence of extensionality such collapse does not occur is made clear in *Church 1976*. It is straightforward to construct possible-worlds models of Church's formulation where reducibility holds but orders do not collapse.

[u]Except for Frege's commitment to an extensional language, Gödel's "concepts" closely resemble the *objects* that Frege says are signified by such phrases as "the concept *horse*". In the Fregean context, they are hardly distinguishable from the extensions. Indeed, Gödel's suggestion (150; see above) that one might assume a concept significant everywhere except for certain "singular points" recalls Frege's unsuccessful proposal for a way out of the paradoxes. (See *Frege 1903*, p. 262, and *Quine 1955*.)

The seriousness with which Gödel takes the notion of analyticity in this section has not attracted the attention of commentators. It seems to show a greater engagement with the ideas of the Vienna Circle than has usually been attributed to Gödel, although there is no doubt that the disagreement is deep. With the Vienna Circle and many other analytic philosophers before the impact of Quine's criticism, Gödel believes firmly that mathematical propositions are true by virtue of "the meaning of the concepts occurring" in them, though others might have said "words" instead of "concepts", but he denies that mathematics is true by convention (as perhaps it would be if it were analytic in his first sense) or that its truth is constituted by linguistic rules that we lay down or embody in our usage.[v] Thus, he says, this position does not contradict his view that "mathematics is based on axioms with a real content", since the *existence* of the concepts involved would have to be an axiom of this kind (151, note 47), presumably if one were undertaking to derive the truth of the axioms from their being in some way implied by the concepts.[w]

8. Concluding remarks

A sort of *coda* to this intricate paper is formed by Gödel's closing remarks noting that mathematical logic had not yet come close to fulfilling the hopes of "Peano and others" that it would contribute to the solution of problems in mathematics, and attributing this to "incomplete understanding of the foundations": "For how can one hope to solve mathematical problems by mere analysis of the concepts occurring, if our analysis so far does not even suffice to set up the axioms?" (152). His suggestion that the hopes expressed by Leibniz for his *characteristica universalis* might, after all, be realistic is one of his most striking

[v]Gödel's annotation in reprint A to note 47 seems to reject the possibility that by virtue of meaning every mathematical proposition can be "reduced to a special case of $a = a$". This thesis is mentioned in the remarks in E on *Bernays 1946*. Bernays appears to argue against it, using Frege's distinction of sense and signification (Gödel's translation of *Bedeutung*): since the transformations that reduce a mathematical proposition to an identity will in general not preserve sense, the "meaning" by virtue of which the reduction proceeds can only be signification (*1946*, p. 78). On the latter reading the thesis reduces to triviality.

There is no way of knowing whether this argument influenced Gödel. It seems to me to be unconvincing. That P can be reduced to Q by virtue of the sense of P need not imply that the sense of Q is the same as that of P.

[w]Gödel evidently elaborated further on this issue in the unfinished paper, "Is mathematics syntax of language?" undertaken in the mid-1950s for *Schilpp 1963*. (See *Wang 1981*, p. 658.) From a very brief examination, it seems to reinforce the points of agreement and disagreement with the Vienna Circle noted in the text.

and enigmatic utterances. It is known that Gödel was for a time much occupied with the study of Leibniz and that he regarded Leibniz as the greatest influence on him of the philosophers of the past. But the strongly Leibnizian flavor of the last pages of the present paper recedes in his later writings (except perhaps *1946*), and about the substance of his reflections on Leibniz little is known.[x]

Charles Parsons[y]

[x]But see also *Wang 1981*, p. 657, n. 8.

[y]I am much indebted to John W. Dawson, Jr., for assistance, and to Burton Dreben, Wilfried Sieg, Hao Wang, and the editors (especially Solomon Feferman) for comments and suggestions. Without the work of the late Jean van Heijenoort on Gödel's annotations to A–E, my own brief comments on them would not have been possible.

Russell's mathematical logic*
(*1944*)

Mathematical logic, which is nothing else but a precise and complete formulation of formal logic, has two quite different aspects. On the one hand, it is a section of mathematics treating of classes, relations, combinations of symbols, etc., instead of numbers, functions, geometric figures, etc. On the other hand, it is a science prior to all others, which contains the ideas and principles underlying all sciences. It was in this second sense that mathematical logic was first conceived by Leibniz in his *Characteristica universalis*, of which it would have formed a central part. But it was almost two centuries after his death before his idea of a logical calculus really sufficient for the kind of reasoning occurring in the exact sciences was put into effect (in some form at least, if not the one Leibniz had in mind) by Frege and Peano.[1] Frege was chiefly interested in the analysis of thought and used his calculus in the first place for deriving arithmetic from pure logic. Peano, on the other hand, was more interested in its applications within mathematics and created an elegant and flexible symbolism, which permits expressing even the most complicated mathematical theorems in a perfectly precise and often very concise manner by single formulas.

It was in this line of thought of Frege and Peano that Russell's work set in. Frege, in consequence of his painstaking analysis of the proofs, had not gotten beyond the most elementary properties of the series of integers, while Peano had accomplished a big collection of mathematical theorems expressed | in the new symbolism, but without proofs. It was only in *Principia mathematica* that full use was made of the new method for actually deriving large parts of mathematics from a very few logical concepts and 126

*[*Author's addition of 1964, expanded in 1972*: The author wishes to note (1) that since the original publication of this paper advances have been made in some of the problems discussed and that the formulations given could be improved in several places, and (2) that the term "constructivistic" in this paper is used for a strictly nominalistic kind of constructivism, such as that embodied in Russell's "no class theory". Its meaning, therefore, is very different from that used in current discussions on the foundations of mathematics, i.e., from both "intuitionistically admissible" and "constructive" in the sense of the Hilbert School. Both these schools base their constructions on a mathematical intuition whose avoidance is exactly one of the principal aims of Russell's constructivism (see the first alternative in the last sentence of footnote 23 below). What, in Russell's own opinion, can be obtained by his constructivism (which might better be called fictionalism) is the system of finite orders of the ramified hierarchy without the axiom of infinity for individuals. The explanation of the term constructive given in footnote 22 below is to be replaced by the remarks just made.]

[1]Frege has doubtless the priority, since his first publication about the subject, which already contains all the essentials, appeared ten years before Peano's.

axioms. In addition, the young science was enriched by a new instrument, the abstract theory of relations. The calculus of relations had been developed before by Peirce and Schröder, but only with certain restrictions and in too close analogy with the algebra of numbers. In *Principia* not only Cantor's set theory but also ordinary arithmetic and the theory of measurement are treated from this abstract relational standpoint.

It is to be regretted that this first comprehensive and thorough-going presentation of a mathematical logic and the derivation of mathematics from it so greatly lacking in formal precision in the foundations (contained in *1–*21 of *Principia*) that it presents in this respect a considerable step backwards as compared with Frege. What is missing, above all, is a precise statement of the syntax of the formalism. Syntactical considerations are omitted even in cases where they are necessary for the cogency of the proofs, in particular in connection with the "incomplete symbols". These are introduced not by explicit definitions, but by rules describing how sentences containing them are to be translated into sentences not containing them. In order to be sure, however, that (or for what expressions) this translation is possible and uniquely determined and that (or to what extent) the rules of inference apply also the new kind of expressions, it is necessary to have a survey of all possible expressions, and this can be furnished only by syntactical considerations. The matter is especially doubtful for the rule of substitution and of replacing defined symbols by their *definiens*. If this latter rule is applied to expressions containing other defined symbols it requires that the order of elimination of these be indifferent. This however is by no means always the case ($\phi!\hat{u} = \hat{u}[\phi!u]$, e.g., is a counter-example). In *Principia* such eliminations are always carried out by substitutions in the theorems corresponding to the definitions, so that it is chiefly the rule of substitution which would have to be proved.

127 I do not want, however, to go into any more details about | either the formalism or the mathematical content of *Principia*,[2] but want to devote the subsequent portion of this essay to Russell's work concerning the analysis of the concepts and axioms underlying mathematical logic. In this field Russell had produced a great number of interesting ideas some of which are presented most clearly (or are contained only) in his earlier writings. I shall therefore frequently refer also to these earlier writings, although their content may partly disagree with Russell's present standpoint.

What strikes one as surprising in this field is Russell's pronouncedly realistic attitude, which manifests itself in many passages of his writings. "Logic is concerned with the real world just as truly as zoology, though with its more abstract and general features", he says, e.g., in his *Introduction to mathematical philosophy* (edition of 1920, page 169). It is true, however,

[2]Cf. in this respect *Quine 1941*.

that this attitude has been gradually decreasing in the course of time[3] and also that it always was stronger in theory than in practice. When he started on a concrete problem, the objects to be analyzed (e.g., the classes or propositions) soon for the most part turned into "logical fictions". Though perhaps this need not necessarily mean (according to the sense in which Russell uses this term) that these things do not exist, but only that we have no direct perception of them.

The analogy between mathematics and a natural science is enlarged upon by Russell also in another respect (in one of his earlier writings). He compares the axioms of logic and mathematics with the laws of nature and logical evidence with sense perception, so that the axioms need not necessarily be evident in themselves, but rather their justification lies (exactly as in physics) in the fact that they make it possible for these "sense perceptions" to be deduced; which of course would not exclude that they also have a kind of intrinsic plausibility similar to that in physics. I think that (provided "evidence" is understood in a sufficiently strict sense) this view has been largely justified by subsequent developments, and it is to be expected that it will be still more so in the future. It has turned out that (under the | assumption that modern mathematics is consistent) the solution of certain arithmetical problems requires the use of assumptions essentially transcending arithmetic, i.e., the domain of the kind of elementary indisputable evidence that may be most fittingly compared with sense perception. Furthermore it seems likely that for deciding certain questions of abstract set theory and even for certain related questions of the theory of real numbers new axioms based on some hitherto unknown idea will be necessary. Perhaps also the apparently unsurmountable difficulties which some other mathematical problems have been presenting for many years are due to the fact that the necessary axioms have not yet been found. Of course, under these circumstances mathematics may lose a good deal of its "absolute certainty"; but, under the influence of the modern criticism of the foundations, this has already happened to a large extent. There is some resemblance between this conception of Russell and Hilbert's "supplementing the data of mathematical intuition" by such axioms as, e.g., the law of excluded middle which are not given by intuition according to Hilbert's view; the borderline, however, between data and assumptions would seem to lie in different places according to whether we follow Hilbert or Russell.

128

An interesting example of Russell's analysis of the fundamental logical concepts is his treatment of the definite article "the". The problem is: what do the so-called descriptive phrases (i.e., phrases as, e.g., "the author

[3]The above quoted passage was left out in the later editions of the *Introduction*. ⟦Blackwell (*1976*) has observed that Gödel was mistaken on this factual matter.⟧

of *Waverley*" or "the king of England") denote or signify[4] and what is the meaning of sentences in which they occur? The apparently obvious answer that, e.g., "the author of *Waverley*" signifies Walter Scott, leads to unexpected difficulties. For, if we admit the further apparently obvious axiom, that the signification of a composite expression, containing constituents which have themselves a signification, depends only on the signification of these constituents (not on the manner in which this signification is expressed), then it follows that the sentence "Scott is the author of *Waverley*" 129 signifies the same thing as "Scott is Scott"; and this again leads | almost inevitably to the conclusion that all true sentences have the same signification (as well as all false ones).[5] Frege actually drew this conclusion; and he meant it in an almost metaphysical sense, reminding one somewhat of the Eleatic doctrine of the "One". "The True"—according to Frege's view—is analyzed by us in different ways in different propositions, "the True" being the name he uses for the common signification of all true propositions.[6]

Now, according to Russell, what corresponds to sentences in the outer world is facts. However, he avoids the term "signify" or "denote" and uses "indicate" instead (in his earlier papers he uses "express" or "being a symbol for"), because he holds that the relation between a sentence and a fact is quite different from that of a name to the thing named. Furthermore, he uses "denote" (instead of "signify") for the relation between things and names, so that "denote" and "indicate" together would correspond to Frege's "*bedeuten*". So, according to Russell's terminology and view, true sentences "indicate" facts and, correspondingly, false ones indicate nothing.[7] Hence Frege's theory would in a sense apply to false sentences, since they all indicate the same thing, namely nothing. But

[4]I use the term "signify" in the sequel because it corresponds to the German word "*bedeuten*" which Frege, who first treated the question under consideration, used in this connection.

[5]The only further assumptions one would need in order to obtain a rigorous proof would be (1) that "$\phi(a)$" and the proposition "a is the object which has the property ϕ and is identical with a" means the same thing and (2) that every proposition "speaks about something", i.e., can be brought to the form $\phi(a)$. Furthermore one would have to use the fact that for any two objects a, b, there exists a true proposition of the form $\phi(a,b)$ as, e.g., $a \neq b$ or $a = a.b = b$.

[6]Cf. *Frege 1892*, p. 35.

[7]From the indication (*Bedeutung*) of a sentence is to be distinguished what Frege called its meaning (*Sinn*), which is the conceptual correlate of the objectively existing fact (or "the True"). This one should expect to be in Russell's theory a possible fact (or rather the possibility of a fact), which would exist also in the case of a false proposition. But Russell, as he says, could never believe that such "curious shadowy" things really exist. Thirdly, there is also the psychological correlate of the fact which is called "signification" and understood to be the corresponding belief in Russell's latest book *An inquiry into meaning and truth* [*1940*]. "Sentence", in contradistinction to "proposition", is used to denote the mere combination of symbols.

different true sentences may indicate many different things. Therefore this view concerning sentences makes it necessary either to drop the above mentioned principle about the signification (i.e., in Rus|sell's terminology the corresponding one about the denotation and indication) of composite expressions or to deny that a descriptive phrase denotes the object described. Russell did the latter[8] by taking the viewpoint that a descriptive phrase denotes nothing at all but has meaning only in context; for example, the sentence "the author of *Waverley* is Scotch" is defined to mean: "There exists exactly one entity who wrote *Waverley* and whoever wrote *Waverley* is Scotch." This means that a sentence involving the phrase "the author of *Waverley*" does not (strictly speaking) assert anything about Scott (since it contains no constituent denoting Scott), but is only a roundabout way of asserting something about the concepts occurring in the descriptive phrase. Russell adduces chiefly two arguments in favor of this view, namely (1) that a descriptive phrase may be meaningfully employed even if the object described does not exist (e.g., in the sentence: "The present king of France does not exist."); (2) that one may very well understand a sentence containing a descriptive phrase without being acquainted with the object described, whereas it seems impossible to understand a sentence without being acquainted with the objects about which something is being asserted. The fact that Russell does not consider this whole question of the interpretation of descriptions as a matter of mere linguistic conventions, but rather as a question of right and wrong, is another example of his realistic attitude, unless perhaps he was aiming at a merely psychological investigation of the actual processes of thought. As to the question in the logical sense, I cannot help feeling that the problem raised by Frege's puzzling conclusion has only been evaded by Russell's theory of descriptions and that there is something behind it which is not yet completely understood. 130

There seems to be one purely formal respect in which one may give preference to Russell's theory of descriptions. By defining the meaning of sentences involving descriptions in the above manner, he avoids in his logical system any axioms about the particle "the", i.e., the analyticity of the theorems about "the" is made explicit; they can be shown to follow from | the explicit definition of the meaning of sentences involving "the". Frege, on the contrary, has to assume an axiom about "the", which of course is also analytic, but only in the implicit sense that it follows from the meaning of the undefined terms. Closer examination, however, shows that this advantage of Russell's theory over Frege's subsists only as long as one interprets definitions as mere typographical abbreviations, not as 131

[8]He made no explicit statement about the former; but it seems it would hold for the logical system of *Principia*, though perhaps more or less vacuously.

introducing names for objects described by the definitions, a feature which is common to Frege and Russell.

I pass now to the most important of Russell's investigations in the field of the analysis of the concepts of formal logic, namely those concerning the logical paradoxes and their solution. By analyzing the paradoxes to which Cantor's set theory had led, he freed them from all mathematical technicalities, thus bringing to light the amazing fact that our logical intuitions (i.e., intuitions concerning such notions as: truth, concept, being, class, etc.) are self-contradictory. He then investigated where and how these common-sense assumptions of logic are to be corrected and came to the conclusion that the erroneous axiom consists in assuming that for every propositional function there exists the class of objects satisfying it, or that every propositional function exists "as a separate entity";[9] by which is meant something separable from the argument (the idea being that propositional functions are abstracted from propositions which are primarily given) and also something distinct from the combination of symbols expressing the propositional function; it is then what one may call the notion or concept defined by it.[10] The existence of this concept already suffices for the paradoxes in their 132 "intensional" form, where the concept of | "not applying to itself" takes the place of Russell's paradoxical class.

Rejecting the existence of a class or concept in general, it remains to determine under what further hypotheses (concerning the propositional function), these entities do exist. Russell pointed out (loc. cit.) two possible directions in which one may look for such a criterion, which he called the zig-zag theory and the theory of limitation of size, respectively, and which might perhaps more significantly be called the intensional and the extensional theory. The second one would make the existence of a class or concept depend on the extension of the propositional function (requiring that it be not too big), the first one on its content or meaning (requiring a certain kind of "simplicity", the precise formulation of which would be the problem).

The most characteristic feature of the second (as opposed to the first) would consist in the non-existence of the universal class or (in the intensional interpretation) of the notion of "something" in an unrestricted sense.

[9] In Russell's first paper about the subject (*1906*). If one wants to bring such paradoxes as "the liar" under his viewpoint, universal (and existential) propositions must be considered to involve the class of objects to which they refer.

[10] "Propositional function" (without the clause "as a separate entity") may be understood to mean a proposition in which one or several constituents are designated as arguments. One might think that the pair consisting of the proposition and the argument could then for all purposes play the role of the "propositional function as a separate entity", but it is to be noted that this pair (as one entity) is again a set or a concept and therefore need not exist.

Axiomatic set theory as later developed by Zermelo and others can be considered as an elaboration of this idea as far as classes are concerned.[11] In particular the phrase "not too big" can be specified (as was shown by J. von Neumann[12]) to mean: not equivalent with the universe of all things, or to be more exact, a propositional function can be assumed to determine a class when and only when there exists no relation (in intension, i.e., a propositional function with two variables) which associates, in a one-to-one manner with each object, an object satisfying the propositional function and vice versa. This criterion, however, does not appear as the basis of the theory but as a consequence of the axioms and inversely can replace two of the axioms (the axiom of replacement and that of choice).

For the second of Russell's suggestions too, i.e., for the zig-zag theory, there has recently been set up a logical system which shares some essential features with this scheme, namely, | Quine's system.[13] It is, moreover, not unlikely that there are other interesting possibilities along these lines. 133

Russell's own subsequent work concerning the solution of the paradoxes did not go in either of the two afore-mentioned directions pointed out by himself, but was largely based on a more radical idea, the "no-class theory", according to which classes or concepts *never* exist as real objects, and sentences containing these terms are meaningful only to such an extent as they can be interpreted as a *façon de parler*, a manner of speaking about other things (cf. page 141). Since in *Principia* and elsewhere, however, he formulated certain principles discovered in the course of the development of this theory as general logical principles without mentioning any longer their dependence on the no-class theory, I am going to treat of these principles first.

I mean in particular the vicious circle principle, which forbids a certain kind of "circularity" which is made responsible for the paradoxes. The fallacy in these, so it is contended, consists in the circumstance that one defines (or tacitly assumes) totalities, whose existence would entail the existence of certain new elements of the same totality, namely elements definable only in terms of the whole totality. This led to the formulation of a principle which says that "no totality can contain members definable only in terms of this totality, or members involving or presupposing this totality" (vicious circle principle). In order to make this principle applicable to the intensional paradoxes, still another principle had to be assumed, namely that "every propositional function presupposes the totality of its

[11] The intensional paradoxes can be dealt with e.g. by the theory of simple types or the ramified hierarchy, which do not involve any undesirable restrictions if applied to concepts only and not to sets.

[12] Cf. *von Neumann 1929*.

[13] Cf. *Quine 1937*.

values" and therefore evidently also the totality of its possible arguments.[14] (Otherwise the concept of "not applying to itself" would presuppose no totality (since it involves no quantifications),[15] and the vicious circle principle would not prevent its application to itself.) A corresponding vicious circle principle | [134] for propositional functions which says that nothing defined in terms of a propositional function can be a possible argument of this function is then a consequence.[16] The logical system to which one is led on the basis of these principles is the theory of orders in the form adopted, e.g., in the first edition of *Principia*, according to which a propositional function which either contains quantifications referring to propositional functions of order n or can be meaningfully asserted of propositional functions of order n is at least of order $n+1$, and the range of significance of a propositional function as well as the range of a quantifier must always be confined to a definite order.

In the second edition of *Principia*, however, it is stated in the Introduction (pages xl and xli) that "in a limited sense" also functions of a higher order than the predicate itself (therefore also functions defined in terms of the predicate as, e.g., in $p‘\kappa \in \kappa$) can appear as arguments of a predicate of functions; and in Appendix B such things occur constantly. This means that the vicious circle principle for propositional functions is virtually dropped. This change is connected with the new axiom that functions can occur in propositions only "through their values", i.e., extensionally, which has the consequence that any propositional function can take as an argument any function of appropriate type, whose extension is defined (no matter what order of quantifiers is used in the definition of this extension). There is no doubt that these things are quite unobjectionable even from the constructive standpoint (see page 136), provided that quantifiers are always restricted to definite orders. The paradoxes are avoided by the theory of simple types,[17] which in | [135] *Principia* is combined with the theory of

[14]Cf. *Whitehead and Russell 1925*, p. 39.

[15]Quantifiers are the two symbols $(\exists x)$ and (x), meaning respectively "there exists an object x" and "for all objects x". The totality of objects x to which they refer is called their range.

[16]Cf. *Whitehead and Russell 1925*, p. 47, section IV.

[17]By the theory of simple types I mean the doctrine which says that the objects of thought (or, in another interpretation, the symbolic expressions) are divided into types, namely: individuals, properties of individuals, relations between individuals, properties of such relations, etc. (with a similar hierarchy for extensions), and that sentences of the form: "a has the property ϕ", "b bears the relation R to c", etc. are meaningless, if a, b, c, R, ϕ are not of types fitting together. Mixed types (such as classes containing individuals and classes as elements) and therefore also transfinite types (such as the class of all classes of finite types) are excluded. That the theory of simple types suffices for avoiding also the epistemological paradoxes is shown by a closer analysis of these. (Cf. *Ramsey 1926* and *Tarski 1935*, p. 399.

orders (giving as a result the "ramified hierarchy") but is entirely independent of it and has nothing to do with the vicious circle principle (cf. page 147).

Now as to the vicious circle principle proper, as formulated on page 133, it is first to be remarked that, corresponding to the phrases "definable only in terms of", "involving", and "presupposing", we have really three different principles, the second and third being much more plausible than the first. It is the first form which is of particular interest, because only this one makes impredicative definitions[18] impossible and thereby destroys the derivation of mathematics from logic, effected by Dedekind and Frege, and a good deal of modern mathematics itself. It is demonstrable that the formalism of classical mathematics does not satisfy the vicious circle principle in its first form, since the axioms imply the existence of real numbers definable in this formalism only by reference to all real numbers. Since classical mathematics can be built up on the basis of *Principia* (including the axiom of reducibility), it follows that even *Principia* (in the first edition) does not satisfy the vicious circle principle in the first form, if "definable" means "definable within the system" and no methods of defining outside the system (or outside other systems of classical mathematics) are known except such as involve still more comprehensive totalities than those occurring in the systems.

I would consider this rather as a proof that the vicious circle principle is false than that classical mathematics is false, and this is indeed plausible also on its own account. For, first of all one may, on good grounds, deny that reference to a totality necessarily implies reference to all single elements of it or, in other words, that "all" means the same as an infinite logical | conjunction. One may, e.g., follow Langford's and Carnap's[19] suggestion to interpret "all" as meaning analyticity or necessity or demonstrability. There are difficulties in this view; but there is no doubt that in this way the circularity of impredicative definitions disappears. 136

Secondly, however, even if "all" means an infinite conjunction, it seems that the vicious circle principle in its first form applies only if the entities involved are constructed by ourselves. In this case there must clearly exist a definition (namely the description of the construction) which does not refer to a totality to which the object defined belongs, because the construction of a thing can certainly not be based on a totality of things to which the thing to be constructed itself belongs. If, however, it is a question of

[18]These are definitions of an object α by reference to a totality to which α itself (and perhaps also things definable only in terms of α) belong. As, e.g., if one defines a class α as the intersection of all classes satisfying a certain condition ϕ and then concludes that α is a subset also of such classes u as defined in terms of α (provided they satisfy ϕ).

[19]See *Carnap 1931*, p. 103, and *1937*, p. 162, and *Langford 1927*, p. 599.

objects that exist independently of our constructions, there is nothing in the least absurd in the existence of totalities containing members which can be described (i.e., uniquely characterized)[20] only by reference to this totality.[21] Such a state of affairs would not even contradict the second form of the vicious circle principle, since one cannot say that an object described by reference to a totality "involves" this totality, although the description itself does; nor would it contradict the third form, if "presuppose" means "presuppose for the existence" not "for the knowability".

So it seems that the vicious circle principle in its first form applies only if one takes the constructivistic (or nominalistic) standpoint[22] toward the objects of logic and mathematics, in particular toward propositions, classes and notions, e.g., if one understands by a notion a symbol together with a rule for translating sentences containing the symbol into such sentences 137 as do | not contain it, so that a separate object denoted by the symbol appears as a mere fiction.[23]

Classes and concepts may, however, also be conceived as real objects, namely classes as "pluralities of things" or as structures consisting of a plurality of things and concepts as the properties and relations of things existing independently of our definitions and constructions.

It seems to me that the assumption of such objects is quite as legitimate as the assumption of physical bodies and there is quite as much reason to believe in their existence. They are in the same sense necessary to obtain a satisfactory system of mathematics as physical bodies are necessary for a satisfactory theory of our sense perceptions and in both cases it is impossible to interpret the propositions one wants to assert about these entities as propositions about the "data", i.e., in the latter case the actually occurring sense perceptions. Russell himself concludes in the last chapter of his book on *Meaning and truth* [[*1940*]], though "with hesitation", that there exist "universals", but apparently he wants to confine this statement to concepts of sense perceptions, which does not help the logician. I shall use the term "concept" in the sequel exclusively in this objective sense. One formal difference between the two conceptions of notions would be that any two different definitions of the form $\alpha(x) = \phi(x)$ can be assumed to define two

[20]An object a is said to be described by a propositional function $\phi(x)$ if $\phi(x)$ is true for $x = a$ and for no other object.

[21]Cf. *Ramsey 1926*.

[22]I shall use in the sequel "constructivism" as a general term comprising both these standpoints and also such tendencies as are embodied in Russell's "no class" theory.

[23]One might think that this conception of notions is impossible, because the sentences into which one translates must also contain notions so that one would get into an infinite regress. This, however, does not preclude the possibility of maintaining the above viewpoint for all the more abstract notions, such as those of the second and higher types, or in fact for all notions except the primitive terms which might be only a very few.

different notions α in the constructivistic sense. (In particular this would be the case for the nominalistic interpretation of the term "notion" suggested above, since two such definitions give different rules of translation for propositions containing α.) For concepts, on the contrary, this is by no means the case, since the same thing may be described in different ways. It might even be that the axiom of extensionality[24] or at least something near to it holds for | concepts. The difference may be illustrated by the following definition of the number two: "Two is the notion under which fall all pairs and nothing else." There is certainly more than one notion in the constructivistic sense satisfying this condition, but there might be one common "form" or "nature" of all pairs. 138

Since the vicious circle principle, in its first form, does apply to constructed entities, impredicative definitions and the totality of all notions or classes or propositions are inadmissible in constructivistic logic. What an impredicative definition would require is to construct a notion by a combination of a set of notions to which the notion to be formed itself belongs. Hence if one tries to effect a retranslation of a sentence containing a symbol for such an impredicatively defined notion it turns out that what one obtains will again contain a symbol for the notion in question.[25] At least this is so if "all" means an infinite conjunction; but Carnap's and Langford's idea (mentioned on page 136) would not help in this connection, because "demonstrability", if introduced in a manner compatible with the constructivistic standpoint towards notions, would have to be split into a hierarchy of orders, which would prevent one from obtaining the desired results.[26] As Chwistek has shown,[27] it is even possible under certain assumptions admissible within constructivistic logic to derive an actual contradiction from the unrestricted admission of impredicative definitions. To be more specific, he has shown that the system of simple types becomes contradictory if one adds the "axiom of intensionality" which says (roughly speaking) that to different definitions belong different notions. This axiom, however, as has just been pointed out, can be assumed to hold for notions in the constructivistic sense.

Speaking of concepts, the aspect of the question is changed completely. Since concepts are supposed to exist objectively, there seems to be objection neither to speaking of all of them | (cf. page 143) nor to describing 139

[24]I.e., that no two different properties belong to exactly the same things, which, in a sense, is a counterpart to Leibniz's *Principium identitatis indiscernibilium*, which says no two different things have exactly the same properties.

[25]Cf. *Carnap 1931*, p. 103, and *1937*, p. 162.

[26]Nevertheless the scheme is interesting because it again shows the constructibility of notions which can be meaningfully asserted of notions of arbitrarily high order.

[27]See *Chwistek 1933*.

some of them by reference to all (or at least all of a given type). But, one may ask, isn't this view refutable also for concepts because it leads to the "absurdity" that there will exist properties ϕ such that $\phi(a)$ consists in a certain state of affairs involving all properties (including ϕ itself and properties defined in terms of ϕ), which would mean that the vicious circle principle does not hold even in its second form for concepts or propositions? There is no doubt that the totality of all properties (or of all those of a given type) does lead to situations of this kind, but I don't think they contain any absurdity.[28] It is true that such properties ϕ (or such propositions $\phi(a)$) will have to contain themselves as constituents of their content (or of their meaning), and in fact in many ways, because of the properties defined in terms of ϕ; but this only makes it impossible to construct their meaning (i.e., explain it as an assertion about sense perceptions or any other non-conceptual entities), which is no objection for one who takes the realistic standpoint. Nor is it self-contradictory that a proper part should be identical (not merely equal) to the whole, as is seen in the case of structures in the abstract sense. The structure of the series of integers, e.g., contains itself as a proper part and it is easily seen that there exist also structures containing infinitely many different parts, each containing the whole structure as a part. In addition there exist, even within the domain of constructivistic logic, certain approximations to this self-reflexivity of impredicative properties, namely propositions which contain as parts of their meaning not themselves but their own formal demonstrability.[29] Now formal demonstrability of a proposition (in case the axioms and rules of inference are correct) implies this proposition and in many cases is equiva|lent to it. Furthermore, there doubtlessly exist sentences referring to a totality of sentences to which they themselves belong as, e.g., the sentence: "Every sentence (of a given language) contains at least one relation word."

140

Of course, this view concerning the impredicative properties makes it necessary to look for another solution of the paradoxes, according to which the fallacy (i.e., the underlying erroneous axiom) does not consist in the assumption of certain self-reflexivities of the primitive terms but in other assumptions about these. Such a solution may be found for the present in the simple theory of types and in the future perhaps in the development of the ideas sketched on pages 132 and 150. Of course, all this refers only

[28]The formal system corresponding to this view would have, instead of the axiom of reducibility, the rule of substitution for functions described, e.g., in *Hilbert and Bernays 1934*, p. 90, applied to variables of any type, together with certain axioms of intensionality required by the concept of property which, however, would be weaker than Chwistek's. It should be noted that this view does not necessarily imply the existence of concepts which cannot be expressed in the system, if combined with a solution of the paradoxes along the lines indicated on p. 149.

[29]Cf. my *1931*, p. 173, or *Carnap 1937*, §35.

to concepts. As to notions in the constructivistic sense, there is no doubt that the paradoxes are due to a vicious circle. It is not surprising that the paradoxes should have different solutions for different interpretations of the terms occurring.

As to classes in the sense of pluralities or totalities, it would seem that they are likewise not created but merely described by their definitions and that therefore the vicious circle principle in the first form does not apply. I even think there exist interpretations of the term "class" (namely as a certain kind of structures) where it does not apply in the second form either.[30] But for the development of all contemporary mathematics one may even assume that it does apply in the second form, which for classes as mere pluralities is, indeed, a very plausible assumption. One is then led to something like Zermelo's axiom system for set theory, i.e., the sets are split up into "levels" in such a manner that only sets of lower levels can be elements of sets of higher levels (i.e., $x \epsilon y$ is always false if x belongs to a higher level than y). There is no reason for classes in this sense to exclude mixtures of levels in one set and transfinite levels. The place of the axiom of reducibility is now taken by the axiom | of classes (Zermelo's *Aussonderungsaxiom*) which says that for each level there exists for an arbitrary propositional function $\phi(x)$ the set of those x of this level for which $\phi(x)$ is true, and this seems to be implied by the concept of classes as pluralities. 141

Russell adduces two reasons against the extensional view of classes, namely, the existence of (1) the null class, which cannot very well be a collection, and (2) the unit classes, which would have to be identical with their single elements. But it seems to me that these arguments could, if anything, at most prove that the null class and the unit classes (as distinct from their only element) are fictions (introduced to simplify the calculus like the points at infinity in geometry), not that all classes are fictions.

But in Russell the paradoxes had produced a pronounced tendency to build up logic as far as possible without the assumption of the objective existence of such entities as classes and concepts. This led to the formulation of the aforementioned "no class theory", according to which classes and concepts were to be introduced as a *façon de parler*. But propositions, too, (in particular those involving quantifications)[31] were later on largely included in this scheme, which is but a logical consequence of this standpoint, since e.g., universal propositions as objectively existing entities evidently belong to the same category of idealistic objects as classes and concepts and lead to the same kind of paradoxes, if admitted without restrictions. As regards

[30]Ideas tending in this direction are contained in *Mirimanoff 1917*, *1917a*, and *1920*. Cf. in particular *1917a*, p. 212.

[31]Cf. *Russell 1906a*.

classes, this program was actually carried out; i.e., the rules for translating sentences containing class names or the term "class" into such as do not contain them were stated explicitly; and the basis of the theory, i.e., the domain of sentences into which one has to translate, is clear, so that classes can be dispensed with (within the system *Principia*), but only if one assumes the existence of a concept whenever one wants to construct a class. When it comes to concepts and the interpretation of sentences containing this or some synonymous term, the state of affairs is by no means as clear. 142 First of all, some of them | (the primitive predicates and relations such as "red" or "colder") must apparently be considered as real objects;[32] the rest of them (in particular according to the second edition of *Principia*, all notions of a type higher than the first and therewith all logically interesting ones) appear as something constructed (i.e., as something not belonging to the "inventory" of the world); but neither the basic domain of propositions in terms of which finally everything is to be interpreted, nor the method of interpretation is as clear as in the case of classes (see below).

This whole scheme of the no-class theory is of great interest as one of the few examples, carried out in detail, of the tendency to eliminate assumptions about the existence of objects outside the "data" and to replace them by constructions on the basis of these data.[33] The result has been in this case essentially negative; i.e., the classes and concepts introduced in this way do not have all the properties required for their use in mathematics, unless one either introduces special axioms about the data (e.g., the axiom of reducibility), which in essence already mean the existence in the data of the kind of objects to be constructed, or makes the fiction that one can form propositions of infinite (and even non-denumerable) length,[34] i.e., operates with truth-functions of infinitely many arguments, regardless of whether or not one can construct them. But what else is such an infinite truth-function but a special kind of an infinite extension (or structure) and even a more complicated one than a class, endowed in addition with a hypothetical meaning, which can be understood only by an infinite mind? All this is only a verification of the view defended above that logic and mathematics (just as physics) are built up on axioms with a real content which cannot be "explained away".

What one can obtain on the basis of the constructivistic attitude is the 143 theory of orders (cf. page 134); only now (and this | is the strong point of

[32]In Appendix C of *Principia* a way is sketched by which these also could be constructed by means of certain similarity relations between atomic propositions, so that these latter would be the only ones remaining as real objects.

[33]The "data" are to be understood in a relative sense here; i.e., in our case as logic without the assumption of the existence of classes and concepts.

[34]Cf. *Ramsey 1926*.

the theory) the restrictions involved do not appear as ad hoc hypotheses for avoiding the paradoxes, but as unavoidable consequences of the thesis that classes, concepts, and quantified propositions do not exist as real objects. It is not as if the universe of things were divided into orders and then one were prohibited to speak of all orders; but, on the contrary, it is possible to speak of all existing things; only, classes and concepts are not among them; and if they are introduced as a *façon de parler*, it turns out that this very extension of the symbolism gives rise to the possibility of introducing them in a more comprehensive way, and so on indefinitely. In order to carry out this scheme one must, however, presuppose arithmetic (or something equivalent), which only proves that not even this restricted logic can be built up on nothing.

In the first edition of *Principia*, where it was a question of actually building up logic and mathematics, the constructivistic attitude was, for the most part, abandoned, since the axiom of reducibility for types higher than the first together with the axiom of infinity makes it absolutely necessary that there exist primitive predicates of arbitrarily high types. What is left of the constructive attitude is only: (1) The introduction of classes as a *façon de parler*; (2) the definition of $\sim$, $\vee$, etc., as applied to propositions containing quantifiers (which incidentally proved its fecundity in a consistency proof for arithmetic); (3) the step by step construction of functions of orders higher than 1, which, however, is superfluous owing to the axiom of reducibility; (4) the interpretation of definitions as mere typographical abbreviations, which makes every symbol introduced by definition an incomplete symbol (not one naming an object described by the definition). But the last item is largely an illusion, because, owing to the axiom of reducibility, there always exist real objects in the form of primitive predicates, or combinations of such, corresponding to each defined symbol. Finally also Russell's theory of descriptions is something belonging to the constructivistic order of ideas.

In the second edition of *Principia* (or, to be more exact, in the introduction to it) the constructivistic attitude is resumed again. The axiom of reducibility is dropped, and it is stated explicitly | that all primitive predicates belong to the lowest type and that the only purpose of variables (and evidently also of constants) of higher orders and types is to make it possible to assert more complicated truth-functions of atomic propositions,[35] which is only another way of saying that the higher types and orders are solely a *façon de parler*. This statement at the same time informs us of what kind of propositions the basis of the theory is to consist, namely of truth-functions of atomic propositions. 144

[35]I.e., propositions of the form $S(a)$, $R(a, b)$, etc., where S, R are primitive predicates and a, b individuals.

This, however, is without difficulty only if the number of individuals and primitive predicates is finite. For the opposite case (which is chiefly of interest for the purpose of deriving mathematics) Ramsey (*1926*) took the course of considering our inability to form propositions of infinite length as a "mere accident", to be neglected by the logician. This of course solves (or rather cuts through) the difficulties; but is to be noted that, if one disregards the difference between finite and infinite in this respect, there exists a simpler and at the same time more far reaching interpretation of set theory (and therewith of mathematics). Namely, in case of a finite number of individuals, Russell's *aperçu* that propositions about classes can be interpreted as propositions about their elements becomes literally true, since, e.g., "$x \epsilon m$" is equivalent to

$$``x = a_1 \vee x = a_2 \vee \cdots \vee x = a_k"$$

where the a_i are the elements of m; and "there exists a class such that ..." is equivalent to "there exist individuals $x_1, x_2, \ldots, x_n$ such that ...",[36] provided n is the number of individuals in the world and provided we neglect for the moment the null class which would have to be taken care of by an additional clause. Of course, by an iteration of this procedure one can obtain classes of classes, etc., so that the logical system obtained would resemble the theory of simple types except for the circumstance that mixture of types would be possible. Axiomatic set theory appears, then, as an extrapolation of this scheme for the case of infinitely many individuals or an infinite iteration of the process of forming sets.

145 | Ramsey's viewpoint is, of course, everything but constructivistic, unless one means constructions of an infinite mind. Russell, in the second edition of *Principia*, took a less metaphysical course by confining himself to such truth-functions as can actually be constructed. In this way one is again led to the theory of orders, which, however, appears now in a new light, namely as a method of constructing more and more complicated truth-functions of atomic propositions. But this procedure seems to presuppose arithmetic in some form or other (see next paragraph).

As to the question of how far mathematics can be built up on this basis (without any assumptions about the data i.e., about the primitive predicates and individuals except, as far as necessary, the axiom of infinity), it is clear that the theory of real numbers in its present form cannot be obtained.[37] As to the theory of integers, it is contended in the second edition of *Principia* that it can be obtained. The difficulty to be overcome

[36]The x_i may, of course, as always, be partly or wholly identical with each other.

[37]As to the question how far it is possible to build up the theory of real numbers presupposing the integers, cf. *Weyl 1918* or *1932*.

is that in the definition of the integers as "those cardinals which belong to every class containing 0 and containing $x+1$ if containing x", the phrase "every class" must refer to a given order. So one obtains integers of different orders, and complete induction can be applied to integers of order n only for properties of order n; whereas it frequently happens that the notion of integer itself occurs in the property to which induction is applied. This notion, however, is of order $n+1$ for the integers of order n. Now, in Appendix B of the second edition of *Principia*, a proof is offered that the integers of any order higher than 5 are the same as those of order 5, which of course would settle all difficulties. The proof as it stands, however, is certainly not conclusive. In the proof of the main lemma *89.16, which says that every subset α (of arbitrarily high order)[38] of an inductive class β of order 3 is itself an inductive class of order 3, induction is applied to a property of β involving α (namely $\alpha - \beta \neq \Lambda$, which, however, | should read $\alpha - \beta \sim \epsilon$ Induct$_2$, because (3) is evidently false). This property, however, is of an order > 3 if α is of an order > 3. So the question whether (or to what extent) the theory of integers can be obtained on the basis of the ramified hierarchy must be considered as unsolved at the present time. It is to be noted, however, that, even in case this question should have a positive answer, this would be of no value for the problem whether arithmetic follows from logic, if propositional functions of order n are defined (as in the second edition of *Principia*) to be certain finite (though arbitrarily complex) combinations (of quantifiers, propositional connectives, etc.), because then the notion of finiteness has to be presupposed, which fact is concealed only be taking such complicated notions as "propositional function of order n" in an unanalyzed form as primitive terms of the formalism and giving their definition only in ordinary language. The reply may perhaps be offered that in *Principia* the notion of a propositional function of order n is neither taken as primitive nor defined in terms of the notion of a finite combination, but rather quantifiers referring to propositional functions of order n (which is all one needs) are defined as certain infinite conjunctions and disjunctions. But then one must ask: Why doesn't one define the integers by the infinite disjunction:

$$x = 0 \vee x = 0+1 \vee x = 0+1+1 \vee \cdots \text{ ad infinitum},$$

saving in this way all the trouble connected with the notion of inductiveness? This whole objection would not apply if one understands by a propositional function of order n one "obtainable from such truth-functions of

146

[38]That the variable α is intended to be of undetermined order is seen from the later applications of *89.17 and from the note to *89.17. The main application is in line (2) of the proof of *89.24, where the lemma under consideration is needed for α's of arbitrarily high orders.

atomic propositions as presuppose for their definition no totalities except those of the propositional functions of order $< n$ and of individuals"; this notion, however, is somewhat lacking in precision.

The theory of orders proves more fruitful if considered from a purely mathematical standpoint, independently of the philosophical question whether impredicative definitions are admissible. Viewed in this manner, i.e., as a theory built up within the framework of ordinary mathematics, where impredicative definitions are admitted, there is no objection to extending it to arbitrarily high transfinite orders. Even if one rejects im-
147 predicative definitions, there would, I think, be no objection to | extend it to such transfinite ordinals as can be constructed within the framework of finite orders. The theory in itself seems to demand such an extension since it leads automatically to the consideration of functions in whose definition one refers to all functions of finite orders, and these would be functions of order ω. Admitting transfinite orders, an axiom of reducibility can be proved. This, however, offers no help to the original purpose of the theory, because the ordinal α—such that every propositional function is extensionally equivalent to a function of order α—is so great, that it presupposes impredicative totalities. Nevertheless, so much can be accomplished in this way, that all impredicativities are reduced to one special kind, namely the existence of certain large ordinal numbers (or well-ordered sets) and the validity of recursive reasoning for them. In particular, the existence of a well-ordered set, of order type ω_1 already suffices for the theory of real numbers. In addition this transfinite theorem of reducibility permits the proof of the consistency of the axiom of choice, of Cantor's continuum hypothesis and even of the generalized continuum hypothesis (which says that there exists no cardinal number between the power of any arbitrary set and the power of the set of its subsets) with the axioms of set theory as well as of *Principia.*

I now come in somewhat more detail to the theory of simple types which appears in *Principia* as combined with the theory of orders; the former is, however, (as remarked above) quite independent of the latter, since mixed types evidently do not contradict the vicious circle principle in any way. Accordingly, Russell also based the theory of simple types on entirely different reasons. The reason adduced (in addition to its "consonance with common sense") is very similar to Frege's, who, in his system, already had assumed the theory of simple types for functions, but failed to avoid the paradoxes, because he operated with classes, (or rather functions in extension) without any restriction. This reason is that (owing to the variable it contains) a propositional function is something ambiguous (or, as Frege says, something unsaturated, wanting supplementation) and therefore can occur in a meaningful proposition only in such a way that this ambiguity
148 is eliminated (e.g., by substituting a | constant for the variable or applying quantification to it). The consequences are that a function cannot replace

an individual in a proposition, because the latter has no ambiguity to be removed, and that functions with different kinds of arguments (i.e., different ambiguities) cannot replace each other; which is the essence of the theory of simple types. Taking a more nominalistic viewpoint (such as suggested in the second edition of *Principia* and in *Meaning and truth*) one would have to replace "proposition" by "sentence" in the foregoing considerations (with corresponding additional changes). But, in both cases, this argument clearly belongs to the order of ideas of the "no class" theory, since it considers the notions (or propositional functions) as something constructed out of propositions or sentences by leaving one or several constituents of them undetermined. Propositional functions in this sense are so to speak "fragments" of propositions, which have no meaning in themselves, but only in so far as one can use them for forming propositions by combining several of them, which is possible only if they "fit together", i.e., if they are of appropriate types. But, it should be noted that the theory of simple types (in contradistinction to the vicious circle principle) cannot in a strict sense follow from the constructive standpoint, because one might construct notions and classes in another way, e.g., as indicated on page 144, where mixtures of types are possible. If on the other hand one considers concepts as real objects, the theory of simple types is not very plausible since what one would expect to be a concept (such as, e.g., "transitivity" or the number two) would seem to be something behind all its various "realizations" on the different levels and therefore does not exist according to the theory of types. Nevertheless, there seems to be some truth behind this idea of realizations of the same concept on various levels, and one might, therefore, expect the theory of simple types to prove useful or necessary at least as a stepping stone for a more satisfactory system, a way in which it has already been used by Quine.[39] Also Russell's "typical ambiguity" is 49 a step in this direction. Since, however, it only adds certain simpli|fying symbolic conventions to the theory of types, it does not de facto go beyond this theory.

It should be noted that the theory of types brings in a new idea for the solution of the paradoxes, especially suited to their intensional form. It consists in blaming the paradoxes not on the axiom that every propositional function defines a concept or class, but on the assumption that every concept gives a meaningful proposition, if asserted for any arbitrary object or objects as arguments. The obvious objection that every concept can be extended to all arguments, by defining another one which gives a false proposition whenever the original one was meaningless, can easily be dealt with by pointing out that the concept "meaningfully applicable" need not itself be always meaningfully applicable.

[39] *Quine 1937.*

The theory of simple types (in its realistic interpretation) can be considered as a carrying through of this scheme, based, however, on the following additional assumption concerning meaningfulness: "Whenever an object x can replace another object y in one meaningful proposition, it can do so in every meaningful proposition."[40] This of course has the consequence that the objects are divided into mutually exclusive ranges of significance, each range consisting of those objects which can replace each other; and that therefore each concept is significant only for arguments belonging to one of those ranges, i.e., for an infinitely small portion of all objects. What makes the above principle particularly suspect, however, is that its very assumption makes its formulation as a meaningful proposition impossible,[41] because x and y must then be confined to definite ranges of significance which are either the same or different, and in both cases the statement does not express the principle or even part of it. Another consequence is that the fact that an object x is (or is not) of a given type also cannot be expressed by a meaningful proposition.

150 | It is not impossible that the idea of limited ranges of significance could be carried out without the above restrictive principle. It might even turn out that it is possible to assume every concept to be significant everywhere except for certain "singular points" or "limiting points", so that the paradoxes would appear as something analogous to dividing by zero. Such a system would be most satisfactory in the following respect: our logical intuitions would then remain correct up to certain minor corrections, i.e., they could then be considered to give an essentially correct, only somewhat "blurred", picture of the real state of affairs. Unfortunately the attempts made in this direction have failed so far;[42] on the other hand, the impossibility of this scheme has not been proved either, in spite of the strong inconsistency theorems of Kleene and Rosser.[43]

In conclusion I want to say a few words about the question whether (and in which sense) the axioms of *Principia* can be considered to be analytic. As to this problem, it is to be remarked that analyticity may be understood in two senses. First, it may have the purely formal sense that

[40] Russell formulates a somewhat different principle with the same effect, in *Principia*, vol. 1, p. 95.

[41] This objection does not apply to the symbolic interpretation of the theory of types, spoken of on p. 148, because there one does not have objects but only symbols of different types.

[42] A formal system along these lines is Church's (cf. his *1932* and *1933*), where, however, the underlying idea is expressed by the somewhat misleading statement that the law of excluded middle is abandoned. However, this system has been proved to be inconsistent. See footnote 43.

[43] Cf. *Kleene and Rosser 1935*.

the terms occurring can be defined (either explicitly or by rules for eliminating them from sentences containing them) in such a way that the axioms and theorems become special cases of the law of identity and disprovable propositions become negations of this law. In this sense even the theory of integers is demonstrably non-analytic, provided that one requires of the rules of elimination that they allow one actually to carry out the elimination in a finite number of steps in each case.[44] Leaving out this condition by admitting, e.g., sentences of infinite (and non-denumerable) length as intermediate steps of the process of reduction, all axioms of *Principia* | (including the axioms of choice, infinity and reducibility) could be proved to be analytic for certain interpretations (by considerations similar to those referred to on page 144).[45] But this observation is of doubtful value, because the whole of mathematics as applied to sentences of infinite length has to be presupposed in order to prove this analyticity, e.g., the axiom of choice can be proved to be analytic only if it is assumed to be true. 151

In a second sense a proposition is called analytic if it holds "owing to the meaning of the concepts occurring in it", where this meaning may perhaps be undefinable (i.e., irreducible to anything more fundamental).[46] It would seem that all axioms of *Principia*, in the first edition, (except the axiom of infinity) are in this sense analytic for certain interpretations of the primitive terms, namely if the term "predicative function" is replaced either by "class" (in the extensional sense) or (leaving out the axiom of choice) by "concept", since nothing can express better the meaning of the term "class" than the axiom of classes (cf. page 140) and the axiom of choice, and since, on the other hand, the meaning of the term "concept" seems to imply that every propositional function defines a concept.[47] The difficulty is only that we don't perceive the concepts of "concept" and of "class"

[44]Because this would imply the existence of a decision procedure for all arithmetical propositions. Cf. *Turing 1937*.

[45]Cf. also *Ramsey 1926*, where, however, the axiom of infinity cannot be obtained, because it is interpreted to refer to the individuals in the world.

[46]The two significations of the term *analytic* might perhaps be distinguished as tautological and analytic.

[47]This view does not contradict the opinion defended above that mathematics is based on axioms with a real content, because the very existence of the concept of e.g., "class" constitutes already such an axiom; since, if one defined e.g., "class" and "ϵ" to be "the concepts satisfying the axioms", one would be unable to prove their existence. "Concept" could perhaps be defined in terms of "proposition" (cf. p. 148), although I don't think that this would be a natural procedure; but then certain axioms about propositions, justifiable only with reference to the undefined meaning of this term, will have to be assumed. It is to be noted that this view about analyticity makes it again possible that every mathematical proposition could perhaps be reduced to a special case of $a = a$, namely if the reduction is effected not in virtue of the definitions of the terms occurring, but in virtue of their meaning, which can never be completely expressed in a set of formal rules.

with sufficient distinctness, as is shown by the paradoxes. In view of this 152 situation, Russell took the course of considering | both classes and concepts (except the logically uninteresting primitive predicates) as nonexistent and of replacing them by constructions of our own. It cannot be denied that this procedure has led to interesting ideas and to results valuable also for one taking the opposite viewpoint. On the whole, however, the outcome has been that only fragments of mathematical logic remain, unless the things condemned are reintroduced in the form of infinite propositions or by such axioms as the axiom of reducibility which (in case of infinitely many individuals) is demonstrably false unless one assumes either the existence of classes or of infinitely many "*qualitates occultae*". This seems to be an indication that one should take a more conservative course, such as would consist in trying to make the meaning of the terms "class" and "concept" clearer, and to set up a consistent theory of classes and concepts as objectively existing entities. This is the course which the actual development of mathematical logic has been taking and which Russell himself has been forced to enter upon in the more constructive parts of his work. Major among the attempts in this direction (some of which have been quoted in this essay) are the simple theory of types (which is the system of the first edition of *Principia* in an appropriate interpretation) and axiomatic set theory, both of which have been successful at least to this extent, that they permit the derivation of modern mathematics and at the same time avoid all known paradoxes. Many symptoms show only too clearly, however, that the primitive concepts need further elucidation.

It seems reasonable to suspect that it is this incomplete understanding of the foundations which is responsible for the fact that mathematical logic has up to now remained so far behind the high expectations of Peano and others who (in accordance with Leibniz's claims) had hoped that it would facilitate theoretical mathematics to the same extent as the decimal system of numbers has facilitated numerical computations. For how can one expect to solve mathematical problems systematically by mere analysis of the concepts occurring if our analysis so far does not even suffice to set up the axioms? But there is no need to give up hope. Leibniz did not in his writings about the *Characteristica universalis* speak of a utopian 153 project; if we are to | believe his words he had developed this calculus of reasoning to a large extent, but was waiting with its publication till the seed could fall on fertile ground.[48] He went even so far[49] as to estimate the time which would be necessary for his calculus to be developed by a few select scientists to such an extent "that humanity would have a new kind of an instrument increasing the powers of reason far more than any optical

[48] *Leibniz 1890*, p. 12. Cf. also *Vacca 1903*, p. 72, and the preface to *Leibniz 1923*.
[49] *Leibniz 1890*, p. 187.

instrument has ever aided the power of vision." The time he names is five years, and he claims that his method is not any more difficult to learn than the mathematics or philosophy of his time. Furthermore, he said repeatedly that, even in the rudimentary state to which he had developed the theory himself, it was responsible for all his mathematical discoveries; which, one should expect, even Poincaré would acknowledge as a sufficient proof of its fecundity.[50]

[50] I wish to express my thanks to Professor Alonzo Church of Princeton University, who helped me to find the correct English expressions in a number of places.

Courtesy Seeley G. Mudd Manuscript Library, Princeton University

Participants in the Princeton bicentennial conference on problems in mathematics, December 1946

Conference Photograph by Orren Jack Turner

THE PROBLEMS OF MATHEMATICS

1. Morse, M., Institute for Advanced Study
2. Ancochea, G., University of Salamanaca, Spain
3. Borsuk, K., University of Warsaw, Poland
4. Cramér, H., University of Stockholm, Sweden
5. Hlavaty, V., University of Prague, Czechoslovakia
6. Whitehead, J. H. C., University of Oxford, England
7. Garding, L. J., Princeton
8. Riesz, M., University of Lund, Sweden
9. Lefschetz, S., Princeton
10. Veblen, O., Institute for Advanced Study
11. Hopf, H., Federal Technical School, Switzerland
12. Newman, M. H. A., University of Manchester, England
13. Hodge, W. V. D., Cambridge, England
14. Dirac, P. A. M., Cambridge University, England
15. Hua, L. K., Tsing Hua University, China
16. Tukey, J. W., Princeton
17. Harrold, O. G., Princeton
18. Mayer, W., Institute for Advanced Study
19. Mautner, F. I., Institute for Advanced Study
20. Gödel, K., Institute for Advanced Study
21. Levinson, N., Massachusetts Institute of Technology
22. Cohen, I. S., University of Pennsylvania
23. Seidenberg, A., University of California
24. Kline, J. R., University of Pennsylvania
25. Eilenberg, S., Indiana University
26. Fox, R. H., Princeton
27. Wiener, N., Massachusetts Institute of Technology
28. Rademacher, H., University of Pennsylvania
29. Salem, R., Massachusetts Institute of Technology
30. Tarski, A., University of California
31. Bargmann, V., Princeton
32. Jacobson, N., The Johns Hopkins University
33. Kac, M., Cornell University
34. Stone, M. H., University of Chicago
35. Von Neumann, J., Institute for Advanced Study
36. Hedlund, G. A., University of Virginia
37. Zariski, O., University of Illinois
38. Whyburn, G. T., University of Virginia
39. McShane, E. J., University of Virginia
40. Quine, W. V., Harvard
41. Wilder, R. L., University of Michigan
42. Kaplansky, I., Institute for Advanced Study
43. Bochner, S., Princeton
44. Leibler, R. A., Institute for Advanced Study
45. Hildebrandt, T. H., University of Michigan
46. Evans, G. C., University of California
47. Widder, D. V., Harvard
48. Hotelling, H., University of North Carolina
49. Peck, L. G., Institute for Advanced Study
50. Synge, J. L., Carnegie Institute of Technology
51. Rosser, J. B., Cornell
52. Murnaghan, F. D., The Johns Hopkins University
53. Mac Lane, S., Harvard
54. Cairns, S. S., Syracuse University
55. Brauer, R., University of Toronto, Canada
56. Schoenberg, I. J., University of Pennsylvania
57. Shiffman, M., New York University
58. Milgram, A. N., Institute for Advanced Study
59. Walker, R. J., Cornell
60. Hurewicz, W., Massachusetts Institute of Technology
61. McKinsey, J. C. C., Oklahoma Agricultural and Mechanical
62. Church, A., Princeton
63. Robertson, H. D., Princeton
64. Bullitt, W. M., Bullitt and Middleton, Louisville, Ky.
65. Hille, E., Yale University
66. Albert, A. A., University of Chicago
67. Rado, T., The Ohio State University
68. Whitney, H., Harvard
69. Ahlfors, L. V., Harvard
70. Thomas, T. Y., Indiana University
71. Crosby, D. R., Princeton
72. Weyl, H., Institute for Advanced Study
73. Walsh, J. L., Harvard
74. Dunford, N., Yale
75. Spencer, D. C., Stanford University
76. Montgomery, D., Yale
77. Birkhoff, G., Harvard
78. Kleene, S. C., University of Wisconsin
79. Smith, P. A., Columbia University
80. Youngs, J. W. T., Indiana University
81. Steenrod, N. E., University of Michigan
82. Wilks, S. S., Princeton
83. Boas, R. P., Mathematical Reviews, Brown University
84. Doob, J. L., University of Illinois
85. Feller, W., Cornell University
86. Zygmund, A., University of Pennsylvania
87. Artin, E., Princeton
88. Bohnenblust, H. F., California Institute of Technology
89. Allendoerfer, C. B., Haverford College
90. Robinson, R. M., Princeton
91. Bellman, R., Princeton
92. Begle, E. G., Yale
93. Tucker, A. W., Princeton

Key to photograph of conference participants

Introductory note to *1946*

In 1946, the year of its bicentennial, Princeton University organized several conferences. The conference on problems of mathematics took place on 17–19 December 1946. Its more than seventy participants included most of the leading American mathematicians and several from abroad. Its sessions included discussions of practically the whole range of pure mathematics. The university printed a pamphlet containing a summary of the discussion in each of the sessions (*Princeton University 1947*). Alfred Tarski was listed as "discussion leader" in the session on mathematical logic, held on 17 December, and evidently gave the principal invited talk. Gödel's remarks took their point of departure from Tarski's talk, and may have been invited or planned as a comment.

No proceedings of the conference were published, but a larger volume entitled *Problems of mathematics* was planned at the time. Gödel sent his paper to J. C. C. McKinsey early in 1947 for this purpose. Correspondence with J. W. Tukey indicates that a chapter was planned on the Hilbert problems, for which Gödel was asked to write on the first two (the continuum problem and the problem of the consistency of analysis). Gödel agreed to do the first but declined to do the second, on the ground that it ought to be done by someone more sympathetic to Hilbert's views.

Although it could not have been unknown, the paper *Gödel 1946* seems to have had little circulation before its publication in *Davis 1965*.[a] Its main new technical idea, ordinal definability, was rediscovered independently by several others (see below). The reprint *Gödel 1968* contained several changes in the text. Correspondence between Gödel and Mario Casolini, evidently an editor with the publisher of *1968*, shows that the principal changes were proposed by Gödel.

Gödel's paper consists largely of an exploration of possible absolute notions of demonstrability and definability, which would not have to be relative to a particular formal system or formalized language. From the summary of the discussion (*Princeton University 1947*, pages 10–12), evidently much of the session was concerned with decision problems; Gödel thus began by noting that, with the concept of general recursiveness or Turing computability, "one has for the first time succeeded in giving an absolute definition of an interesting epistemological notion, i.e., one not

[a]Davis has informed me that he did not know of the existence of this paper until S. C. Kleene suggested its inclusion in *The undecidable* and supplied a copy. Kleene has indicated that the paper was published there with Gödel's approval.

depending on the formalism chosen" (*Gödel 1946*, page 1). Gödel refers here not primarily to the equivalence of different formulations such as Turing computability, λ-definability and Herbrand–Gödel general recursiveness, but to the absence of the sort of relativity to a given language that leads to stratification of the notion, such as (in the case of definability in a formalized language) into definability in languages of greater and greater expressive power. Such stratification is driven by diagonal arguments. But, since a function enumerating the recursive functions is not recursive and there is no reason to think it computable, the diagonal function it gives rise to is simply non-recursive, rather than "recursive at the next level". One can of course effectively enumerate computing procedures (partial recursive functions), but then the diagonal procedure simply leads to partial recursive functions that must be undefined for certain arguments (and to the undecidability of the question whether an arbitrary partial recursive function is defined for a given argument).

Gödel is thus encouraged to search for absolute notions of demonstrability and definability. His remarks on the former notion (pages 1–2) are brief. Reflection on a formalism that makes the notion of provability precise "gives rise to new axioms which are exactly as evident and justified as those with which you started". (That the new axioms are *exactly* as evident might be questioned, even in the case of the weakest, such as the statement of the consistency of the formalism.) The process of transfinite iteration of such extension, which he then mentions, had already been studied in *Turing 1939*; completeness results of the sort envisaged in the remark that all steps of such an extension process "could be described and collected together in some non-constructive way" were proved in *Feferman 1962*, but such results are essentially arithmetic in character. In set theory, Gödel suggests that absolute provability would have to incorporate a notion of proof using stronger and stronger axioms of infinity. The notion of an "axiom of infinity" could not be given a "combinational and decidable" characterization, but an axiom of infinity might be characterized as a sentence of a certain formal structure that is *true*; then by absolute demonstrability one might mean proof with the help of axioms of infinity in that sense. But evidently Gödel thought this suggestion very speculative. It was perhaps bound up with the hope, so far disappointed, that suitable axioms of infinity would suffice to decide the continuum problem.

Gödel now turns to mathematical definability, where he says he "can give somewhat more definite suggestions". A hierarchy of concepts of definability is forced on us by a "finitistic concept of language", but to collect into one notion all the stages of a hierarchy of this kind requires "as many primitive terms as you wish to consider steps in this hierarchy of languages, i.e., as many as there are ordinal numbers" (page 2). It is noteworthy that he considers immediately a hierarchy indexed by *arbitrary* ordinals in the set-theoretic sense. He is led immediately to the concept of ordinal definability (pages 2–3):

> The simplest way of doing it is to take the ordinals themselves as primitive terms. So one is led to the concept of definability in terms of ordinals, i.e., definability by expressions containing names of ordinal numbers and logical constants, including quantification referring to sets.

Gödel observes that one obtains no new ordinal-definable sets by extending the language by a truth predicate, presumably for formulas containing names of arbitrary ordinals. This implies immediately that "x is an ordinal-definable set" is expressible in the language of ZF, an assertion that has usually been proved by means of Levy's reflection principle.[b] An extension of that argument proves Gödel's assertion about the ordinal-definable sets (OD).[c] It is reasonable to conjecture that Gödel knew the reflection principle at the time. Otherwise he would have had to have a quite different argument in mind; moreover, an argument like the proof of the principle figures in the proof in *Gödel 1940* that $V = L$ (where V is the universe of all sets and L is the class of all constructible sets) implies the generalized continuum hypothesis.

Gödel remarks that his notion of constructible set is also a kind of definability in terms of ordinals (see *Gödel 1939a*). Since it admits quantification only over constructible sets and not over sets in general, however, "you can actually define sets, and even sets of integers, for which you cannot prove that they are constructible" (*1946*, page 3). For this reason he considers constructibility not satisfactory as a notion of definability. Gödel's language is puzzling, since he seems to be saying that, for some definition of a set of integers, it is unprovable that that set is

[b] *Myhill and Scott 1971*, p. 272; *Krivine 1968* or *1971*, Chapter 6.

[c] Let $Sat(x, y)$ mean: x is a formula of the language of set theory, y is an assignment of objects to the free variables of x, and y satisfies x. Now let $A(x)$ be a formula of the language of set theory augmented by *Sat*, with ordinal parameters, that is uniquely true of the set x_0. Then, by the extended reflection principle, we can find an ordinal β such that $x_0 \in V_\beta$, the ordinal parameters are less than β and $\langle V_\beta, \epsilon\rangle$ is an elementary substructure of the universe. In other words, for all formulas u of the language of set theory and all assignments y of objects in V_β,

$$(1) \qquad Sat(u, y) \leftrightarrow Sat_0(u, V_\beta, y),$$

where $Sat_0(x, z, y)$ is the satisfaction predicate for formulas with quantifiers interpreted to range over z, and moreover V_β reflects the formula $A(x)$, that is, for any $x \in V_\beta$,

$$(2) \qquad A(x) \leftrightarrow A^{V_\beta}(x).$$

But now (1) implies that in $A^{V_\beta}(x)$ we may replace any subformula $Sat(u, v)$ by $Sat_0(u, V_\beta, v)$, which is a formula of ZF. It follows by (2) that x_0 is definable in ZF from the given ordinal parameters and β, that is, $x_0 \in OD$.

constructible. How would he have known this then? The first published proof of the consistency of the existence of non-constructible sets of integers is in *Cohen 1963*.[d]

The remainder of the paper deals with the claim that the notion of ordinal definability is a satisfactory absolute notion and with the mathematical interest of this notion. Reversing Gödel's order, we note that he conjectures that the ordinal-definable sets satisfy the axioms of set theory "and so will lead to another and probably simpler proof of the consistency of the axiom of choice" (page 4). He does not claim to be in possession of such a proof, which has since been given by others by means of the observation that the *hereditarily* ordinal-definable sets (HOD) are an inner model satisfying the axiom of choice. Since the proof of this is rather straightforward (see *Myhill and Scott 1971*), Gödel's caution may mean simply that he had not worked it out in detail. Gödel's second and closing remark is that, although it can be proved (in ZF) that there is a bound on the ordinals needed to obtain ordinal-definable sets of integers, he doubts that one can prove that the bound is ω_1, as in the case of the constructible sets. Therefore the proof of the consistency of the axiom of choice will not extend to the continuum hypothesis as did that by the constructible sets. It was subsequently proved in *McAloon 1966* that $V = HOD$ is consistent with the negation of the continuum hypothesis, thus verifying this conjecture of Gödel's.[e] Gödel may have thought this a virtue of the notion, since $V = HOD$ might then be compatible with the sort of axiom Gödel speculates about in *1947* that would refute the continuum hypothesis.[f] He offers no direct comment on the question of the *truth* of $V = OD$.[g]

Gödel's philosophical remarks are prompted by the obvious objection that admitting all ordinals as primitive terms makes his notion no longer

[d]It was later shown by Solovay (*1967*) that strong axioms of infinity, such as the existence of a measurable or Ramsey cardinal, imply the existence of even analytically definable sets (of integers) that are not constructible.

[e]Gödel wished to add to *1968* a note stating this result, but it was received too late by the publisher. The note is included in the present volume.

[f]In default of the axioms, there cannot be definite results on this question. However, $V = OD$ has been shown consistent with strong axioms of infinity incompatible with $V = L$. Let M be the proposition that there exists a measurable cardinal. *McAloon 1966* shows that, if $\mathrm{ZF} + M$ is consistent, then $\mathrm{ZF} + M + V = OD$ is consistent. (This follows from Corollary 6.5 of *Kunen 1970*.) It is extended from measurable to supercompact cardinals in *Menas 1973*.

Yet it seems doubtful that Gödel's primary interest in the notion was as a means of proving the consistency of the axiom of choice in settings where $V = L$ fails. Wang reports that in 1941 Gödel had another general method of proving the consistency of the axiom of choice (*Wang 1981*, p. 657).

[g]Note that, since $V_\alpha \in OD$ and HOD is transitive, if $HOD = OD$ then $V = HOD = OD$. Thus $V = OD$, $HOD = OD$, and $V = HOD$ are all equivalent.

a notion of *definability*. He finds it plausible that "all things conceivable by us are denumerable". Does this mean that for any x, if x is conceivable by us, then x is denumerable, or that there are only denumerably many things conceivable by us? The latter reading seems more likely. Gödel remarks that, because of the paradox of the least indefinable ordinal, a notion of mathematical definability that makes the notion itself mathematically definable will have to have all ordinals definable. But what follows is that a notion satisfying the "postulate of denumerability" must involve some "extramathematical element concerning the psychology of the being which deals with mathematics" (page 4). It seems to me that the point is not just that to characterize such a notion will require some extramathematical vocabulary, for what will rule out a least ordinal not definable with the help of the extra vocabulary? Apparently the extra vocabulary must have the property that "definable with the help of the extra vocabulary" is no longer definable with the help of the extra vocabulary. Gödel finally argues that ordinal definability at least captures the notion of "being formed according to a law" as opposed to "being formed by a random choice of the elements" (page 4). In particular, there is not a random element in the ordinals themselves.

I would note further that, since the notion of definability contains a modal element, the question whether ordinal definability is a genuine notion of definability depends on the underlying modal notion, the "can" in "can be defined". Admitting all ordinals as definable might be viewed as an extreme extension of the notion of abstract mathematical possibility that arises in other contexts in the foundations of mathematics, such as that of computable function, where complete abstraction is made both from the limitations of "hardware" and from feasibility in terms of the time required for a computation. It should be kept in mind that an ordinal definition of a set requires only finitely many ordinals (which can be reduced to one). To deny that an ordinal-definable set is "really" definable implies the existence of ordinals that are not really definable, no matter how our means of such definition might be extended. It is hard to see how a case could be made out for this so long as one stays on the abstract mathematical plane and does not introduce notions concerning "the psychology of the being which deals with mathematics", at least in a broad sense.[h]

It appears that subsequent work on ordinal definability was done almost entirely independently of Gödel's. In 1952, Post rediscovered the

[h]In unpublished work, Allen Hazen argues that a physicalist theory of the mind implies that there is an absolute notion of definability satisfying the condition of denumerability, or at least some other cardinality restriction that makes the paradox of the least indefinable ordinal a genuine problem.

notion and proved some of its principal properties (see *Myhill and Scott 1971*, page 278), but his abstract *1953* was evidently too cryptic to be understood, and Post died soon afterward. The idea was also implicit in *Takeuti 1961*. The concept became more widely known through the work of Myhill and Scott in the early 1960s. They showed the definability of "x is an ordinal-definable set", carried out the relative consistency proof for the axiom of choice, and also proved the equivalence of ordinal definability and constructibility if both are phrased in terms of definability in second-order logic. Further work was stimulated by Cohen's discovery of the method of forcing. Since obviously $L \subseteq HOD \subseteq OD \subseteq V$, and $HOD = OD$ implies $V = OD$ (see footnote g), the question naturally arises as to what can consistently be assumed about the equality and inequality of L, HOD, and V. Let us define $A \subset B$ to be $A \subseteq B$ and $A \neq B$. Levy (*1965*) proved that Cohen's model showing the independence of $V = L$ from ZFC $+ GCH$ satisfies $L = HOD$, thus establishing the consistency of $L = HOD \subset V$. McAloon (*1966*) showed the consistency of $L \subset HOD = V$, both with GCH and with CH false (see his *1971*). In *1966* he also established the consistency of $L \subset HOD \subset V$.

Classes related to HOD have been used in model constructions for other purposes. A noteworthy example occurs in *Solovay 1970*. The main result of his paper is the consistency relative to ZF $+ DC$ (the axiom of dependent choices) $+ I$ ('there exists a strongly inaccessible cardinal') of a list of propositions including 'every set of reals is Lebesgue measurable'. One constructs a forcing extension N of a ground model M of ZFC $+ I$. With the help of the fact that, in N, a strongly inaccessible cardinal of M is collapsed to $\aleph_1$, it is shown that, although the axiom of choice still holds in N, all sets of reals definable from a *countable sequence* of ordinals are Lebesgue measurable (Theorem 2). Within N, one then constructs the submodel N_1 of sets hereditarily definable from countable sequences of ordinals; in N_1 all sets of reals are Lebesgue measurable (*1970*, page 52).

The exploration in *1946* of the notions of absolute demonstrability and absolute definability, and the development in the latter context of the notion of ordinal definability, are instances of the application to concrete problems of Gödel's realistic point of view.[i] But it should be noted that Gödel qualifies the "absoluteness" of the notions he considers: they are "not absolute in the strictest sense, but only with respect to a certain system of things, namely the sets as described in axiomatic set theory" (*1946*, page 4). The question whether such notions can be treated in a "completely absolute way" is left open.

[i]See my introductory note to *Gödel 1944* in this volume.

The pagination in the text follows the original typescript, whose pagination differs somewhat from that of *Davis 1965*.

Charles Parsons[j]

[j]I am indebted to John P. Burgess, Martin Davis, John W. Dawson, Jr., Solomon Feferman, Kenneth McAloon, Dana Scott and Hao Wang for information, assistance and suggestions.

Remarks before the Princeton bicentennial conference on problems in mathematics (*1946*)

Tarski has stressed in his lecture (and I think justly) the great importance of the concept of general recursiveness (or Turing's computability). It seems to me that this importance is largely due to the fact that with this concept one has for the first time succeeded in giving an absolute definition of an interesting epistemological notion, i.e., one not depending on the formalism chosen.[1] In all other cases treated previously, such as demonstrability or definability, one has been able to define them only relative to a given language, and for each individual language it is clear that the one thus obtained is not the one looked for. For the concept of computability, however, although it is merely a special kind of demonstrability or decidability, the situation is different. By a kind of miracle it is not necessary to distinguish orders, and the diagonal procedure does not lead outside the defined notion. This, I think, should encourage one to expect the same thing to be possible also in other cases (such as demonstrability or definability). It is true that for these other cases there exist certain negative results, such as the incompleteness of every formalism or the paradox of Richard. But closer examination shows that these results do not make a

[1][*Footnote added in 1965*: To be more precise: a function of integers is computable in any formal system containing arithmetic if and only if it is computable in arithmetic, where a function f is called computable in S if there is in S a computable term representing f.]

definition of the absolute notions concerned impossible under all circumstances, but only exclude certain ways of defining them, or, at least, that certain very closely related concepts may be definable in an absolute sense.

Let us consider, e.g., the concept of demonstrability. It is well known that, in whichever way you make it precise by means of a formalism, the contemplation of this very formalism gives rise to new axioms which are exactly as evident and justified as those with which you started, and that this process of extension can be iterated into the transfinite. So there cannot exist any formalism which would embrace all these steps; but this does not exclude that all these steps (or at least all of them which give something new for the domain | of propositions in which you are interested) could be described and collected together in some non-constructive way. In set theory, e.g., the successive extensions can most conveniently be represented by stronger and stronger axioms of infinity. It is certainly impossible to give a combinational and decidable characterization of what an axiom of infinity is; but there might exist, e.g., a characterization of the following sort: An axiom of infinity is a proposition which has a certain (decidable) formal structure and which in addition is true. Such a concept of demonstrability might have the required closure property, i.e., the following could be true: Any proof for a set-theoretic theorem in the next higher system above set theory (i.e., any proof involving the concept of truth which I just used) is replaceable by a proof from such an axiom of infinity. It is not impossible that for such a concept of demonstrability some completeness theorem would hold which would say that every proposition expressible in set theory is decidable from the present axioms plus some true assertion about the largeness of the universe of all sets. 2

Let me consider a second example where I can give somewhat more definite suggestions, namely the concept of definability (or, to be more exact, of mathematical definability). Here also you have, corresponding to the transfinite hierarchy of formal systems, a transfinite hierarchy of concepts of definability. Again it is not possible to collect together all these languages in one, as long as you have a finitistic concept of language, i.e., as long as you require that a language must have a finite number of primitive terms. But, if you drop this condition, it does become possible (at least as far as it is necessary for the purpose), namely, by means of a language which has as many primitive terms as you wish to consider steps in this hierarchy of languages, i.e., as many as there are ordinal numbers. The simplest way of doing it is to take the ordinals themselves as primitive terms. So one is led to the concept of definability in terms of ordinals, | i.e., definability by expressions containing names of ordinal numbers and logical constants, including quantification referring to sets. This concept should, I think, be investigated. It can be proved that it has the required closure property: By introducing the notion of truth for this whole transfinite language, i.e., by going over to the next language, you will obtain no new 3 definable sets (although you will obtain new definable properties of sets).

The concept of constructible set I used in the consistency proof for the continuum hypothesis can be obtained in a very similar way, i.e., as a kind of definability in terms of ordinal numbers; but, comparing constructibility with the concept of definability just outlined, you will find that not all logical means of definition are admitted in the definition of constructible sets. Namely, quantification is admitted only with respect to constructible sets and not with respect to sets in general. This has the consequence that you can actually define sets, and even sets of integers, for which you cannot prove that they are constructible (although this can of course be consistently assumed). For this reason, I think constructibility cannot be considered as a satisfactory formulation of definability.

But now, coming back to the definition of definability I suggested, it might be objected that the introduction of all ordinals as primitive terms is too cheap a way out of the difficulty, and that the concept thus obtained completely fails to agree with the intuitive concept we set out to make precise, because there exist undenumerably many sets definable in this sense. There is certainly some justification in this objection. For it has some plausibility that all things conceivable by us are denumerable, even if you disregard the question of expressibility in some language. But, on the other hand, there is much to be said in favor of the concept under consideration; namely, above all it is clear that, if the concept of mathematical definability is to be itself mathematically definable, it must necessarily be so that all ordinal numbers are definable, because otherwise you could define the first ordinal number not definable, and | would thus obtain a contradiction. I think this does not mean that a concept of definability satisfying the postulate of denumerability is impossible, but only that it would involve some extramathematical element concerning the psychology of the being who deals with mathematics.

4

But, irrespective of what the answer to this question may be, I would think that "definability in terms of ordinals", even if it is not an adequate formulation for "comprehensibility by our mind", is at least an adequate formulation in an absolute sense for a closely related property of sets, namely, the property of "being formed according to a law" as opposed to "being formed by a random choice of the elements". For, in the ordinals there is certainly no element of randomness, and hence neither in sets defined in terms of them. This is particularly clear if you consider von Neumann's definition of ordinals, because it is not based on any well-ordering relations of sets, which may very well involve some random element.

You may have noticed that, in both examples I gave, the concepts arrived at or envisaged were not absolute in the strictest sense, but only with respect to a certain system of things, namely the sets as conceived in axiomatic set theory; i.e., although there exist proofs and definitions not falling under these concepts, these definitions and proofs give, or are to give, nothing new within the domain of sets and of propositions expressible in terms of "set", "ϵ" and the logical constants. The question whether the two epistemological concepts considered, or any others, can be treated in a completely absolute way is of an entirely different nature.

In conclusion I would like to say that, irrespective of whether the concept of definability suggested in this lecture corresponds to certain intuitive notions, it has some intrinsic mathematical interest; in particular, there are two questions arising in connection with it: (1) Whether the sets definable in this sense satisfy the axioms of set theory. I think this question is to be answered in the affirmative, and so will lead to another, and probably simpler, proof for the consistency of the axiom of choice. (2) It follows from the axiom of replacement that the ordinals necessary to define all sets of integers which can at all be defined in this way will have an upper limit. I doubt that it will be possible to prove that this upper limit is ω_1, as in the case of the constructible sets.[2]

[2][*Footnote added on 26 June 1968*: I have recently been informed that this conjecture has been verified by Kenneth McAloon in a dissertation at the University of California at Berkeley: to be more precise, that Dr. McAloon, using Cohen's method, has proved the consistency (with the Zermelo–Fraenkel axioms of set theory) of the assumption that all sets are 'ordinal definable' and that $2^{\aleph_0}$ is much greater than $\aleph_1$.]

Introductory note to *1947* and *1964*

1. Introduction

Cantor's continuum problem served as one of the principal and periodic foci for Gödel's research from 1935 until his death more than four decades later. His article *1947* (substantially revised and expanded to become *1964*) originated from a request, made in 1945 by the editor of the *American mathematical monthly*, for a paper on the continuum problem. The result was an expository article written in the style for which the *Monthly* is well known, but having a flavor that reflected Gödel's distinctive blend of mathematical and philosophical interests. Although *1947* contains no new technical results, it gives considerable insight into his philosophical views on set theory and on what would and would not, in his opinion, constitute a solution to the continuum problem. In one sense, *1947* can be regarded as a continuation, and as a variation in a different key, of his reflections in *1944* on Russell and mathematical logic. Like *1944*, the article *1947* originated from a request for a contribution by Gödel, and included both technical hints for possible future research in mathematics and cogent philosophical arguments in favor of Platonism. But *1947*, unlike *1944*, was expository (indeed, the only expository article that Gödel ever published) and concerned a specific mathematical problem rather than a philosopher's contribution to logic.

This introductory note has seven sections, which serve different purposes. Section 2 places *1947* in a historical context by tracing the continuum problem from its origins to Gödel's attempts (circa 1938–1942) to establish the independence of the continuum hypothesis. Section 3 recounts the circumstances which led Gödel to write *1947*. The content of *1947* is analyzed in Section 4, while Section 5 indicates how Gödel's perspective changed in the revised version *1964* (and in his 1966 plans for a third version of the paper). Section 6 discusses the effect of recent mathematical developments on Gödel's claims in *1947* and *1964*. Finally, Section 7 concerns his two unpublished articles on the continuum hypothesis, both written about 1970.

2. Historical background to the continuum problem, including Gödel's work before *1947*

The continuum problem, which Cantor first posed in *1878*, grew out of research that he began in 1873. At that time, in a letter to Dedekind,

Cantor posed the question whether the set $\mathbb{R}$ of real numbers can be put in one-to-one correspondence with the set $\mathbb{N}$ of natural numbers. Although Dedekind at first doubted the importance of this question, he was pleased when Cantor discovered a proof that such a correspondence cannot exist. In January 1874 Cantor posed a further question to Dedekind: Can a line segment be put in one-to-one correspondence with a square and its interior? Three years passed before Cantor succeeded in showing that there exists such a correspondence between a line segment and n-dimensional space for any n.[a] At the end of the article (*1878*) detailing this proof, Cantor stated that every uncountable set of real numbers can be put in one-to-one correspondence with the set of all real numbers, i.e., that there is no cardinal number strictly between that of $\mathbb{N}$ and that of $\mathbb{R}$. This proposition was the original form of the continuum hypothesis. Since there is no standard terminology for this form, we shall call it the *weak continuum hypothesis.*

When in *1883* Cantor developed the notion of well-ordering and asserted that every set can be well-ordered, he gave a second and more elegant form to this hypothesis: $\mathbb{R}$ has the same power as the set of countable ordinals. In his aleph notation of *1895* this can be stated as $2^{\aleph_0} = \aleph_1$, the form in which the continuum hypothesis (*CH*) is now known. (It is easily seen that *CH* is equivalent to the conjunction of the weak continuum hypothesis and the proposition that $\mathbb{R}$ can be well-ordered.) Cantor himself never used the term "continuum hypothesis"; instead, in his 1882 correspondence with Dedekind, he referred to the weak continuum hypothesis as the "two-class theorem".

In *1883* Cantor began to generalize *CH*, asserting that the set of all real functions has the third infinite power; in his later notation, this stated that $2^{\aleph_1} = \aleph_2$. He never discussed any more general form of *CH*, perhaps because he saw no use for such a generalization. The generalized continuum hypothesis (*GCH*), which states that $2^{\aleph_\alpha} = \aleph_{\alpha+1}$ for all ordinals α, was first formulated by Hausdorff (*1908*, pages 487, 494) and was given this name by Tarski (*1925*).

Despite very intense research, especially during 1884, Cantor never succeeded in demonstrating *CH*. However, he obtained a special case of the weak continuum hypothesis: Every uncountable *closed* subset of $\mathbb{R}$ has the power of $\mathbb{R}$ (*1884*). For a while that year, during August and again during October, he believed that he had proved *CH*, and then, for a brief period in November, that he had refuted *CH* (*Moore 1982*, pages 43–44).

In August 1904 a Hungarian mathematician, J. König, also claimed to have disproved *CH*. This occurred in a lecture he gave at the Inter-

[a]See *Noether and Cavaillès 1937*, pp. 12–13, 20–21, 25.

national Congress of Mathematicians at Heidelberg. However, the next day E. Zermelo found the gap in König's argument. When revised for publication (*König 1905*), König's result was that the power of $\mathbb{R}$ cannot equal $\aleph_{\alpha+\omega}$ for any ordinal α. In the light of Hausdorff's 1906–1908 researches on cofinality, the result was extended to the following: $2^{\aleph_0}$ cannot equal $\aleph_\beta$ for any β of cofinality ω. In *1947* Gödel observed that nothing beyond this was known about the cardinality of $\mathbb{R}$.

As F. Bernstein noted (*1901*, page 14), one line of research on the continuum problem consisted in trying to extend, to larger and larger classes of subsets of $\mathbb{R}$, Cantor's result that the weak continuum hypothesis holds for the closed subsets of $\mathbb{R}$. The hierarchy soon used for this purpose was that of the Borel sets, introduced by E. Borel (*1898*) and first extended to transfinite levels by H. Lebesgue (*1905*). In *1903* W. H. Young strengthened Cantor's result by showing that every uncountable G_δ subset of $\mathbb{R}$ has the power of $\mathbb{R}$. A decade later Hausdorff succeeded in extending the result further, first to the $G_{\delta\sigma\delta}$ sets (*1914a*) and then to the entire Borel hierarchy (*1916*).

For the next two decades, almost all progress on *CH* had a close connection with N. Luzin and his students (such as P. S. Aleksandrov and M. Suslin), who together made up the Moscow school of function theorists. The school's first result occurred when Aleksandrov (*1916*) obtained the above-mentioned theorem on the Borel hierarchy at the same time that Hausdorff did. In 1917 Luzin and Suslin extended the Borel hierarchy by introducing the analytic sets, the first level of what later became the projective hierarchy. Suslin established that the weak continuum hypothesis holds for the analytic sets, now called the $\mathbf{\Sigma}^1_1$ sets, since every uncountable analytic set has a perfect subset.[b] Yet, as Gödel observed in *1947* (page 517), progress stopped there; for it had not been shown that the weak continuum hypothesis holds for every $\mathbf{\Pi}^1_1$ set but only that an uncountable $\mathbf{\Pi}^1_1$ set has either the cardinality $\aleph_1$ or that of $\mathbb{R}$—a result due to K. Kuratowski (*1933*, page 246).

A second approach to the continuum problem was begun by Luzin (*1914*) and pursued vigorously in Poland by his collaborator W. Sierpiński. In this approach, various propositions were shown to be consequences of *CH*. By assuming *CH* as a hypothesis, set theorists gained knowledge about its strength and were able to settle various open problems. Sierpiński, beginning in *1919*, was especially concerned to find interesting propositions equivalent to *CH*. He summarized his results in a book, *Hypothèse du continu* (*1934*), the source for the "paradoxical" consequences of *CH* that Gödel cited in *1947*.

[b] *Luzin 1917*. For discussion of the projective hierarchy, as well as the definition of $\mathbf{\Sigma}^1_n$ and $\mathbf{\Pi}^1_n$ sets, see p. 13 above of the introductory note to *1938*.

In 1923 D. Hilbert claimed that his recently developed proof theory could not only provide a foundation for mathematics but could even settle classical unsolved problems of set theory such as the continuum problem (*1923*, page 151). Three years later he published his attempt to sketch a proof, based on definability considerations, of what he called the "continuum theorem" (*1926*). This attempted proof of *CH* met with widespread skepticism, in particular from Fraenkel (*1928*) and from Luzin (*1929*). In 1935 Luzin returned to this question, arguing that there was not in fact one continuum hypothesis but rather several continuum hypotheses; he dubbed as the "second continuum hypothesis" the following proposition contradicting *CH*:

$$2^{\aleph_0} = 2^{\aleph_1}.$$

Finally, he argued that the second continuum hypothesis accorded with a proposition (contradicting *CH*) of whose truth he felt certain: Every subset of $\mathbb{R}$ having power $\aleph_1$ is a $\mathbf{\Pi}^1_1$ set (*1935*, pages 129–131). Gödel referred in passing to these matters (*1947*, page 523) while mentioning that Luzin, like Gödel himself, believed *CH* to be false.

In the absence of a proof or refutation of *CH*, mathematicians could try to establish its undecidability on the basis of the accepted axioms of set theory. As early as 1923, T. Skolem conjectured that *CH* cannot be settled by Zermelo's 1908 axiom system (*Skolem 1923a*, page 229). But, when Skolem wrote, the understanding of models of set theory was still very rudimentary. Luzin hoped that Hilbert's proof theory would supply a consistency proof for the "second" continuum hypothesis as well as for *CH* (*Luzin 1935*, pages 129–131).

During the 1920s it was also uncertain whether models of set theory should be studied within second-order logic, as did Fraenkel (*1922a*) and Zermelo (*1929, 1930*), or within first-order logic, as Skolem proposed (*1923a, 1930*). In *1930* Zermelo showed that all second-order models of Zermelo–Fraenkel set theory (ZF) consist of the αth stage of the cumulative type hierarchy, where α is a strongly inaccessible ordinal. In an unpublished report of about 1930 to the Emergency Society of German Science, Zermelo pointed out that *CH* is either true in all of these models or false in all of them, so that in either case *CH* is decided in second-order ZF.[c] This result contrasts with the later discoveries of

[c]This report is printed in *Moore 1980*, pp. 130–134, and the observation on *CH* can be found on p. 134. Kreisel (*1967a*, pp. 99–100) also emphasized this point, though unaware that Zermelo had formulated it almost four decades earlier; however, L. Kalmár (*1967*, p. 104) and A. Mostowski (*1967a*, p. 107) reacted negatively to Kreisel's observation, and the second-order version of *CH* has been little studied.

Gödel and P. J. Cohen that *CH* is undecided in the first-order version of ZF.

About 1935 Gödel realized that if Zermelo's cumulative hierarchy were restricted at each level to the sets first-order definable from those obtained at previous levels, then one would have a class model of first-order ZF in which various important propositions held. Originally, in 1935, he proved only that the axiom of choice is such a proposition, but by 1937 he had shown that *GCH* holds in the model as well. In *1938* he was inclined to accept the axiom of constructibility as true, referring to it as "a natural completion of the axioms of set theory" (page 557), and hence to believe that the generalized continuum hypothesis is also true. Yet Gödel refrained, for more than a year, from publishing an announcement of these relative consistency results. A clue to his silence can be found in his letter, written in December 1937 to Karl Menger, which reveals Gödel's hopes for an even stronger result about *CH*:

> I continued my work on the continuum problem last summer, and I finally succeeded in proving the consistency of the continuum hypothesis (even in the generalized form $2^{\aleph_\alpha} = \aleph_{\alpha+1}$) with respect to general set theory. But I ask you, for the time being, please not to tell anyone about this. So far, except for you, I have communicated this result only to von Neumann .... Right now I am also trying to prove the independence of the continuum hypothesis, but do not yet know whether I will succeed with it ....

Unfortunately, Gödel did not succeed in proving the independence of *CH*, despite repeated attempts.

On the other hand, Gödel's efforts to show the independence of the axiom of choice, and consequently of the axiom of constructibility as well, were more fruitful. When Cohen received the Fields Medal for establishing the independence of *CH*, A. Church pointed out, in his speech awarding the medal (*1968*, page 17), that

> Gödel ... in 1942 found a proof of the independence of the axiom of constructibility in ⟦finite⟧ type theory. According to his own statement (in a private communication), he believed that this could be extended to an independence proof of the axiom of choice; but due to a shifting of his interests toward philosophy, he soon afterward ceased to work in this area, without having settled its main problems. The partial result mentioned was never worked out in full detail or put into form for publication.

Gödel also commented on his independence results in a letter of 1967 to W. Rautenberg, who had written to Gödel inquiring about Mostowski's claim that Gödel, about 1940, had obtained most of Cohen's independence results. In his reply (written in German and translated here), Gödel confirmed what Church had stated:

> In reply to your inquiry I would like to refer to the presentation of the facts that Professor Alonzo Church gave in his lecture at the last International Congress of Mathematicians.
>
> Mostowski's assertion is incorrect insofar as I was merely in possession of certain partial results, namely, of proofs for the independence of the axiom of constructibility and of the axiom of choice in type theory. Because of my highly incomplete records from that time (i.e., 1942) I can only reconstruct the first of these two proofs without difficulty. My method had a very close connection with that recently developed by Dana Scott ⟦Boolean-valued models⟧ and had less connection with Cohen's method.
>
> I never obtained a proof for the independence of the continuum hypothesis from the axiom of choice, and I found it very doubtful that the method that I used would lead to such a result.

Thus there can be no doubt that Gödel believed that he had obtained some significant independence results, but not for *CH*.

By the time that Gödel composed *1947* he had become convinced, contrary to the views he expressed in *1938*, that *CH* (and hence the axiom of constructibility as well) was false.

3. The origins of *Gödel 1947*

Gödel undertook to write the article *1947* at the request of Lester R. Ford, the editor of the *American mathematical monthly*. "For some time we have been running a series of papers ...", Ford wrote Gödel on 30 November 1945,

> which we call the "What Is?" series. In these papers the authors have presented some small aspect of higher mathematics in as simple, elementary and popular a way as they possibly can. We have had papers by both Birkhoffs, Morse, Kline, Wilder and several others.
>
> I am writing this to ask if you would like to prepare such a paper. The subject would be of your own choosing, but I had thought of "What is the problem of ⟦the⟧ continuum?"

When Gödel did not respond, Ford wrote again on 31 January 1946. On 14 February, Gödel, who had not received the earlier letter, expressed his willingness to consider the matter, adding that "in any case I could not write the paper immediately, because I am unfortunately very busy with other things at present." A week later, Ford replied: "Let me know as promptly as you can whether you can write this paper. I ought to have it by the month of July. It will not be a long paper and its writing ought not to take a great deal of time."

Ford did not realize that, when composing an article, Gödel was an extreme perfectionist. Another year passed before Gödel completed the paper that, in March 1946, he agreed to write. On 13 August 1946 Ford inquired about the paper, since he wished to print it before his editorship ended in December. Gödel answered on 31 August: "The paper about the continuum problem ... was finished and typewritten a few weeks ago, but on rereading it, I found some insertions desirable, which I have now about completed." Once again, this was not to be.

Finally, on 29 May 1947, Gödel sent the paper to the new editor, C. V. Newsom. In his covering letter, Gödel mentioned that he had "inserted a great number of footnotes whose order does not completely agree with the order in which they occur in the text." He suggested that the new footnotes be printed after the text of the article. Unfortunately, as Gödel learned when he saw the article in print, the footnotes had been renumbered in page proof without changing the internal references to them.[d] He had received no page proofs, having returned his galley proofs at the last moment. Newsom apologized for the errors, which occurred when the compositor tried to make sense of the footnotes, and added by way of compensation: "Your paper has brought many compliments; it is by far the best article in volume 54."

4. How Gödel viewed the continuum problem in *1947*

Gödel's essay *1947* consists of four sections: (1) a discussion of the notion of cardinal number, (2) a survey of the known results about the power $2^{\aleph_0}$ of the continuum $\mathbb{R}$, (3) a philosophical analysis of set theory, and (4) a proposal for solving the continuum problem.

In Section 1, Gödel stressed that Cantor's notion of cardinal number is unique, provided one accepts the minimal requirement that if two sets have the same cardinal number, then there exists a one-to-one correspondence between them. Here Gödel did not discuss how the notion of cardinal number might be defined, contenting himself with the definition

[d] These errors, which Gödel noted in volume 55 of the *Monthly*, are corrected in the text of *1947* printed in the present volume.

of equality between cardinal numbers. In this context he introduced the continuum problem as the question of how many points there are on a Euclidean straight line (or equivalently, how many sets of integers exist). This problem would lack meaning, he observed, if there were not a "natural" representation for the infinite cardinal numbers. But since the alephs $\aleph_\alpha$ provide such a representation and since, by the axiom of choice, the cardinal number of every set is an aleph, it follows that the continuum problem is meaningful. In footnote 2 he defended such uses of the axiom of choice by arguing, on the one hand, that this axiom is consistent relative to the usual axioms for set theory (as shown in his *1940*); on the other hand, he asserted that the axiom of choice is quite as self-evident as the usual axioms for the notion of arbitrary set and is even provable for "sets in the sense of extensions of definable properties" (that is, for the constructible sets, as well as for the ordinal-definable sets of his *1946*).

In Section 2, Gödel reformulated the continuum problem as the question:

Which $\aleph_\alpha$ is the cardinal number of $\mathbb{R}$?

He noted that Cantor had conjectured *CH* as an answer. But he did not mention that Cantor not only conjectured the truth of *CH* but also, on numerous occasions, claimed in print to have proved *CH*. (In fact, many mathematicians took *CH* as true during the 1880s and 1890s.) Nor did Gödel distinguish between *CH* and the weak continuum hypothesis, regarding them as equivalent since he assumed the axiom of choice. Later researchers, however, would find it necessary to distinguish carefully between *CH* and the weak continuum hypothesis when they attempted to solve the continuum problem (especially when the axiom of determinacy was involved; cf. Section 6 below).

Gödel stressed how little was known about the power $2^{\aleph_0}$ of $\mathbb{R}$, despite the many years that had passed since Cantor formulated *CH*. Indeed, Gödel remarked that only two facts were known: (a) $2^{\aleph_0}$ does not have cofinality ω and (b) the weak continuum hypothesis holds for the Σ^1_1 sets (the analytic sets), which, however, are only a tiny fraction of all the subsets of $\mathbb{R}$. In particular, he added, it was not known whether:

(i) There is some given aleph that is an upper bound for $2^{\aleph_0}$,

(ii) $2^{\aleph_0}$ is accessible or is weakly inaccessible,

(iii) $2^{\aleph_0}$ is singular or regular,

or

(iv) $2^{\aleph_0}$ has any restrictions on its cofinality other than König's result that its cofinality is uncountable.

What was known, he continued, was merely a large number of propositions that follow from *CH* as well as several propositions that are equivalent to it.[e]

Gödel observed that our ignorance about the power of the continuum was part of a greater ignorance about infinite cardinal products. In particular, the power of the continuum, $2^{\aleph_0}$, is the simplest non-trivial cardinal product, namely, the product of $\aleph_0$ copies of 2. He added that it was not even known whether

(v) there is some given cardinal that is an upper bound for some infinite product of cardinals greater than 1.

All that was known were certain lower bounds on infinite products, such as Cantor's theorem that the product of $\aleph_0$ copies of 2 is greater than $\aleph_0$ and the Zermelo–König theorem that if $\mathfrak{m}_\alpha < \mathfrak{n}_\alpha$ for all α in some given set I, then

$$\sum_{\alpha\epsilon I} \mathfrak{m}_\alpha < \prod_{\alpha\epsilon I} \mathfrak{n}_\alpha.$$

Thus it was not even known whether the product of $\aleph_0$ copies of 2 is less than the product of $\aleph_1$ copies of 2, that is, whether

$$2^{\aleph_0} < 2^{\aleph_1}.$$

In Section 3 Gödel argued that this lack of knowledge was not due entirely to a failure to find the appropriate proofs, but stemmed from the fact that the concept of set required "a more profound ⟦conceptual⟧ analysis ... than mathematics is accustomed to give" (page 518). He began his philosophical analysis of this concept by rejecting intuitionism, because it is destructive of set theory, and by laying aside the semi-intuitionistic viewpoints of Poincaré and Weyl for the same reason. Instead, he insisted that axiomatic set theory provides the proper foundation for Cantorian set theory. Protecting himself against the objection that the paradoxes threaten set theory, he asserted that no paradox has ever emerged for the iterated notion of "set of" (the cumulative type hierarchy V_α).[f] Here Gödel permitted a set of urelements (the integers, for example) as the basis from which the cumulative hierarchy is built up; incidentally, this corroborates the view that he adopted the cumulative hierarchy from *Zermelo 1930.* Finally, Gödel insisted that the continuum problem—if formulated in a combinatorial way as the question whether *CH* can be deduced from the axioms of set theory—retains

[e]What is now known about (i)–(iv) is discussed in Section 6 below.

[f]The cumulative hierarchy V_α is also called $R(\alpha)$. On this hierarchy, see p. 4 above of the introductory note to *1938*.

a meaning, independently of one's philosophical standpoint, even for the most extreme intuitionist.

If the usual axioms of set theory are consistent, Gödel remarked, then *CH* is either provable, disprovable, or undecidable. After noting that his *1940* ruled out the second possibility, he asserted that the third one is probably correct. To attempt to establish that *CH* is undecidable, he insisted, was the most promising way of attacking the problem.[g]

What is especially important, however, is this: Although Gödel argued that *CH* is almost certainly independent from ZF (as formulated in first-order logic), he insisted strongly that a proof of its independence would *not* solve the continuum problem. Indeed, he emphasized, as Zermelo (*1930*) had done, that "the axioms of set theory by no means form a system closed in itself, but, quite on the contrary, the very concept of set on which they are based ⟦the cumulative hierarchy⟧ suggests their extension by new axioms which assert the existence of still further iterations of the operation 'set of' " (*1947*, page 520). Consequently, he urged mathematicians to search for new large cardinal axioms which would, he hoped, decide *CH*. He added, with his incompleteness theorems in mind, that such axioms would settle questions about Diophantine equations undecidable by the usual axioms.

Here Gödel's strongly held Platonism was visible, as it had been in *1944* and as it would be even more strongly in *1964*. If the undecidability of Cantor's conjecture *CH* were established, he stressed, this would not settle the continuum problem—for essentially philosophical reasons. In fact, he wrote (*1947*, page 520),

> only someone who (like the intuitionist) denies that the concepts and axioms of classical set theory have any meaning (or any well-defined meaning) could be satisfied with such a solution, not someone who believes them to describe some well-determined reality. For in this reality Cantor's conjecture must be either true or false, and its undecidability from the axioms as known today can only mean that these axioms do not contain a complete description of this reality.

After granting that all large cardinal axioms known at the time failed to settle *CH*, since all of them were consistent with the axiom of constructibility, Gödel made an eloquent plea for new axioms (*1947*, page 521):

[g]On the other hand, Gödel did not mention that in *1923a* Skolem had also argued for the independence of *CH*, nor that he himself had worked intensively at establishing its independence during 1942 (as his *Arbeitshefte* attest).

> Even disregarding the intrinsic necessity of some new axiom, and even in case it had no intrinsic necessity at all, a decision about its truth is possible also ... inductively by studying ... its fruitfulness in consequences and in particular in ... consequences demonstrable without the new axiom, whose proofs by means of the new axiom, however, are considerably simpler and easier to discover, and make it possible to condense into one proof many different proofs .... There might exist axioms so abundant in their verifiable consequences, shedding so much light upon a whole discipline, and furnishing such powerful methods for solving given problems (and even solving them, as far as that is possible, in a constructivistic way) that quite irrespective of their intrinsic necessity they would have to be assumed at least in the same sense as any well-established physical theory.

This allusion to physics illustrates his view (already stated in *1944*, page 137) that the assumption of an underlying reality is as "necessary to obtain a satisfactory theory of mathematics" as the assumption of the reality of physical objects is "necessary for a satisfactory theory of our sense perceptions".

In Section 4, Gödel returned to his conjecture that *CH* is not decided by the usual axioms for set theory, arguing that there were at least two reasons for expecting such undecidability. The first was that there exist two quite different classes satisfying the usual axioms: the class of constructible sets and the class of "sets in the sense of arbitrary multitudes" (page 521). Thus he believed that one could not expect *CH* to be settled if one did not specify axiomatically which of these two classes was being considered. (He did not mention here, perhaps for philosophical reasons, a third such class, namely the class of ordinal-definable sets, to which he alluded in footnote 26.[h]) Half of his conjecture about undecidability had already been verified, namely the relative consistency of *CH* with the usual axioms, since *CH* is true in the class of constructible sets.

Gödel then made the important suggestion that "from an axiom in some sense directly opposite to this ⟦axiom of constructibility⟧ the negation of Cantor's conjecture ⟦*CH*⟧ could perhaps be derived" (page 522). The difficulty, of course, with Gödel's suggestion resides in the phrase "directly opposite", since he himself rightly believed that the mere nega-

[h]At first glance it might appear that in footnote 20 he conflated the class of ordinal-definable sets, introduced in *1946*, with the class of constructible sets. However, by comparing footnote 20 with footnote 26, one sees that in the earlier footnote he had in mind the constructible sets and, in the latter, the ordinal-definable sets. Likewise, in footnote 21 of *1964* he meant the constructible sets rather than the ordinal-definable ones.

tion of the axiom of constructibility would not suffice for this purpose (see Section 5). Yet insofar as the axiom of constructibility is a minimality axiom (expressing that the power set of a set, and hence the universe, is as small as possible), he may have had in mind here some kind of maximality axiom, as he certainly did in *1964* (see pages 167–168 below).

Gödel's second reason for expecting the independence of *CH* was that *CH* has certain "paradoxical" consequences which he found unlikely to be true—in particular, the existence of certain very thin subsets of $\mathbb{R}$ that have the power $2^{\aleph_0}$. The first effect of *CH* was to ensure that some kinds of thin subsets of $\mathbb{R}$, proved in ZFC to have instances that are uncountable, can actually have the power $2^{\aleph_0}$. Examples of such sets are

(1) sets of first category on every perfect subset of $\mathbb{R}$,

and

(2) sets carried into a set of measure zero by every continuous one-to-one mapping of $\mathbb{R}$ onto itself.

The second effect of *CH* was to imply that certain kinds of thin subsets of $\mathbb{R}$ can have the power $2^{\aleph_0}$ even though, in ZFC, no instances of these kinds are known that are uncountable. Here he gave as an example the sets of absolute measure zero (by definition, such a set is coverable by a given sequence of intervals of arbitrarily small positive lengths). He then gave several other examples, such as a subset of $\mathbb{R}$ including no uncountable set of measure zero.[i]

Gödel attempted to protect himself against the rejoinder that many kinds of point-sets obtained without *CH* (such as a Peano curve) are highly counterintuitive. In these cases, he argued, the implausibility of the point-sets was due to "a lack of agreement between our intuitive geometrical concepts and the set-theoretical ones occurring in the theorems" (page 524).

Nevertheless, there appears to be little evidence that analysts and set theorists now regard as "paradoxical" the kinds of thin sets cited by Gödel. For example, P. J. Cohen, when asked his opinion of these thin sets of power $2^{\aleph_0}$, was not troubled by them.[j] Likewise, in an article surveying recent work on *CH*, D. A. Martin responded negatively to Gödel's claim: "While Gödel's intuitions should never be taken lightly, it is very hard to see that the situation ⟦with *CH*⟧ *is* different from that of Peano curves, and it is even hard for some of us to see why the examples Gödel cites are implausible at all" (*1976*, page 87).

In the conclusion to his article, Gödel insisted that "it is very suspi-

[i]This particular example, however, was dropped in his *1964* version of the article.

[j]Personal communication from P. J. Cohen, April 1984.

cious that, as against the numerous plausible propositions which imply the negation of the continuum hypothesis, not one plausible proposition is known which would imply the continuum hypothesis" (*1947*, page 524). What are these "numerous plausible propositions"? We cannot be certain, since Gödel did not mention even one of them explicitly. Perhaps he simply intended such propositions to be the negations of those that he had called "paradoxical". In any case, here he was uncharacteristically incautious in his assertion. In 1970 he himself would find a proposition, which he then regarded as quite plausible, that implies *CH* (see Section 7).

5. Gödel's altered perspective in *1964*

The article *1964* resulted from a request, made to Gödel by P. Benacerraf and H. Putnam, for permission to reprint both of the essays *1944* and *1947* in their forthcoming source book *Philosophy of mathematics: Selected readings.* At first, Gödel hesitated to grant permission, fearing that the introduction to their book would subject his article to positivistic attacks. He asked Benacerraf, in conversation, for what amounted to editorial control of the editor's introduction to the source book. As an alternative, since such control could not be granted, Benacerraf assured Gödel that he would be shown the introduction and, furthermore, that the editors did not intend it to make a major philosophical statement but rather to outline the issues. Thus placated, Gödel gave permission to reprint his two essays, and began extensively revising *1947.* Benacerraf met with Gödel a number of times to go over the revisions, since Gödel felt that he did not know English "well enough". Yet Benacerraf knew no one with a more subtle grasp of the various ways in which an English text could be interpreted. While considering the proposed changes, Gödel repeatedly pointed out to Benacerraf various of their unwanted consequences.[k]

Whereas Gödel made no substantive modifications in reprinting *1944*, merely adding an initial footnote, he introduced more than one hundred separate alterations in *1947* in the course of preparing *1964*. Most of these changes were stylistic and reflected his increasing acquaintance with the nuances of the English language. In particular, a number of long and rather Germanic sentences were divided into shorter and more idiomatic ones.

[k]Personal communications from P. Benacerraf, July 1982 and March 1986.

Nevertheless, a substantial number of his changes were more than stylistic. A minor example is his reference in *1947* to a "natural" representation of the infinite cardinal numbers (the alephs), replaced in *1964* with a reference to a "systematic" representation. Far more surprising is his omission in *1964* of all reference to the ordinal-definable sets, which in *1947* he had discussed on page 522 and in footnote 26. It is uncertain what prompted him to omit this notion of set that he had introduced in his *1946*.

In Gödel's *Nachlass* there exist two drafts of his *1964*, each an offprint of *1947* with alterations written on it. The second of these contains a revision, not incorporated into *1964*, that credits Zermelo (*1930*) with "substantially the same solution of the paradoxes" as is embodied in the cumulative type hierarchy, which Gödel designates by his notion "set of". Again, it is unknown why he intended to credit Zermelo and then decided not to do so.

One particularly important addition occurred in footnote 20 of *1964*, where large cardinal axioms were discussed. Here he remarked that D. Scott (*1961*) had proved that the existence of a measurable cardinal contradicts the axiom of constructibility—in contrast to earlier large cardinal axioms, such as those of Mahlo (*1911, 1913*), which are consistent with that axiom. Consequently, he continued, the relative consistency proof for *CH* by means of the class of constructible sets fails if one assumes that there is a measurable cardinal. (In *1971a*, however, J. Silver established that *GCH* holds in the class of sets constructible from a countably additive measure on the least measurable cardinal. In *1967*, Levy and Solovay had already shown, by means of forcing, that *CH* is relatively consistent with a measurable cardinal; see footnote p below.) Gödel then added that it was not yet certain whether "the general concept of set" implies the existence of a measurable cardinal in the same way as it implies Mahlo's axioms. By contrast with this uncertainty, in Gödel's unpublished revision of September 1966 he argued for the existence of a measurable cardinal since this follows "from the existence of generalizations of Stone's representation theorem to Boolean algebras with operations on infinitely many elements" (page 261 below).[1]

Another noteworthy addition occurred in footnote 23 of *1964*. Whereas in the *1947* version of this footnote, Gödel had argued that *CH* might be decided by means of some axiom diametrically opposite to the axiom of constructibility, in *1964* he spelled out what he meant:

[1]See also Gödel's oral comments about measurable cardinals to Solovay on p. 19 above.

> I am thinking of an axiom which (similar to Hilbert's completeness axiom in geometry) would state some maximum property of the system of all sets, whereas axiom A ⟦the axiom of constructibility⟧ states a minimum property. Note that only a maximum property would seem to harmonize with the concept of set explained in footnote 14 ⟦arbitrary sets of the cumulative type hierarchy⟧.

Hilbert's axiom of completeness (*1902*), which belongs to second-order logic, had characterized Euclidean geometry (and, analogously, the real numbers) as the maximal structure satisfying his other axioms. What Gödel proposed for set theory was vague but suggestive; in particular, the various large cardinal axioms can be regarded as steps in the direction of maximality. His meaning is made more definite by a letter he wrote to S. Ulam (quoted in *Ulam 1958*, page 13) apropos of von Neumann's axiom (*1925*) that a class S is a proper class if and only if S is equipotent with the class V of all sets:

> The great interest which this axiom has lies in the fact that it is a maximum principle, somewhat similar to Hilbert's axiom of completeness in geometry. For, roughly speaking, it says that any set which does not, in a certain well-defined way, imply an inconsistency exists. Its being a maximum principle also explains the fact that this axiom implies the axiom of choice. I believe that the basic problems of abstract set theory, such as Cantor's continuum problem, will be solved satisfactorily only with the help of stronger axioms of *this* kind, which in a sense are opposite or complementary to the constructivistic interpretation of mathematics.

More recent attempts to formulate such a maximum principle have not been completely successful. J. Friedman (*1971*) proposed one such proposition, called the generalized maximization principle, and showed it to be equivalent to *GCH*; thus far it has attracted little attention. Recently, S. Shelah's strong version of his proper forcing axiom, PFA+ (by which, in *1982*, he generalized Martin's axiom in the direction of maximality), and the principle dubbed "Martin's maximum" by M. Foreman, M. Magidor and Shelah have each been shown (by them in *198?*, and independently by S. Todorcevic) to imply that $2^{\aleph_0} = \aleph_2$; more recently, Todorcevic has announced a proof that $2^{\aleph_0} = \aleph_2$ already follows from PFA. At present, there is no consensus among set theorists as to the truth of these hypotheses. Nor does the author wish to conjecture what Gödel would have thought of them.

By far the most substantial alteration in *Gödel 1964* was the addition of a long supplement, together with a brief postscript noting that Cohen (*1963, 1964*) had just established the independence of *CH* and thereby had verified Gödel's *1947* claim that *CH* would not be settled by the usual axioms for set theory. The supplement consists of a discussion of new results that Gödel considered important, along with an extended philosophical defense of his Platonist position on *CH*.

Ostensibly, this defense was stimulated by A. Errera's article *1952*, claiming that if *CH* is not decided by the usual axioms for set theory, then the question whether *CH* is true will lose its meaning, just as happened to the parallel postulate when non-Euclidean geometry was proved consistent. Gödel insisted that, on the contrary, "the situation in set theory is very different from that in geometry, both from the mathematical and from the epistemological point of view" (*1964*, page 270). Here he stressed the asymmetry between assuming that there is, and assuming that there is not, a strongly inaccessible cardinal. The former assumption was fruitful in the sense of having consequences for number theory, while the latter was not. Likewise, he continued, *CH* "can be shown to be sterile for number theory ..., whereas for some other assumption about the power of the continuum this perhaps is not so" (page 271). This "sterility", for first-order number theory, was due to the fact that $\mathbb{N}$ is absolute for L, the class of all constructible sets. (In his revisions of 1966–1967, discussed below, he here replaced *CH* by *GCH*, and "power of the continuum" by "power of $2^{\aleph_\alpha}$".)

By using later results, we can say more. In *1969* R. A. Platek established that if a sentence of second-order number theory is provable from *CH*, then it is already provable from the usual axioms of set theory along with the axiom of choice; moreover, he showed that the same holds for any Π^2_1 sentence of third-order number theory.[m] (No further extension was possible, since *CH* itself is a Σ^2_1 sentence.) By 1965 Solovay had independently found Platek's result on *CH*, and in addition had discovered a corresponding result for not-*CH*: If a Π^1_3 sentence of second-order number theory is provable from not-*CH*, then it is already provable from ZF and the axiom of choice.[n] In this sense, then, both *CH* and not-*CH* are sterile for number theory.

The Platonist views put forward by Gödel in *1947* were strengthened in *1964*, not only in the supplement but in the text as well, where he described himself as "someone who considers mathematical objects to exist independently of our constructions" (page 262). Nevertheless, his Platonism was most visible in the supplement, where on page 271 he pursued at some length the analogy between mathematics and physical theories that he had already broached in *1947*:

[m]S. Kripke and J. Silver had each independently arrived at the same result (*Platek 1969*, p. 219).

[n]Personal communication from R. M. Solovay, 27 October 1984.

> Despite their remoteness from sense experience, we do have something like a perception also of the objects of set theory, as is seen from the fact that the axioms force themselves upon us as being true. I don't see any reason why we should have less confidence in this kind of perception, i.e., in mathematical intuition, than in sense perception, which induces us to build up physical theories and to expect that future sense perceptions will agree with them, and, moreover, to believe that a question not decidable now has meaning and may be decided in the future.

In September 1966, Gödel wrote an addendum called "Changes to be made in 3rd edition", anticipating that *1964* would be reprinted.[o] Already in the postscript to *1964*, which was added when *1964* was almost in press, Gödel had mentioned Cohen's 1963 proof of the independence of *CH*. But in the 1966 addendum Gödel expressed himself more strongly: "Cohen's work ... is the greatest advance in the foundations of set theory since its axiomatization". He added that Cohen's forcing "has been used to settle several other important independence questions"; yet he mentioned only one result, namely, that all known large cardinal axioms "are not sufficient to answer the question of the truth or falsehood of Cantor's continuum hypothesis" (page 270 below). Although he did not give a reference, he was almost certainly referring to the result of Levy and Solovay that, for all known large cardinals κ (and in particular for measurable cardinals), if there is a model of set theory containing κ, then there is a model containing κ in which *CH* is true and another model containing κ in which *CH* is false.[p]

6. Later research affecting *1947* and *1964*

There were two major developments that affected Gödel's program, as proposed in *1947* and *1964*, for settling *CH*. The first of these was research on large cardinals, and the second consisted of new independence results obtained by Cohen's method of forcing. In fact, there has been an extremely fruitful interaction, which still continues, between these two lines of development.

[o]These changes have been incorporated into the text of *1964* in the present volume, where they are printed in square brackets. Gödel made additional changes in a manuscript of October 1967. The textual notes record the exact changes to *1964* made in 1966 and 1967. On the other hand, the reprinting of *1964* in *Benacerraf and Putnam 1983* does not include these alterations and additions.

[p]This result, announced in *Levy 1964* and independently in *Solovay 1965a*, was proved in detail in *Levy and Solovay 1967*.

The question of the relationship of *CH* to large cardinal axioms, and to new axioms such as the axiom of determinacy (*AD*), has turned out to be unexpectedly complicated. Large cardinal axioms are now known to affect the class of sets for which the weak continuum hypothesis is true. In particular, Solovay showed (*1969*) that if there exists a measurable cardinal, then the weak continuum hypothesis is true for Σ^1_2 sets. Moreover, *AD*, which may be regarded as a kind of large cardinal axiom, implies that the weak continuum hypothesis holds for every subset of $\mathbb{R}$. Unfortunately, *AD* contradicts *CH*, since it implies that the real numbers cannot be well-ordered (*Mycielski 1964*, page 209), and so was surely unacceptable to Gödel as a solution to the continuum problem. On the other hand, the axiom of projective determinacy (that is, *AD* restricted to the projective sets) is also a kind of large cardinal axiom and has recently been shown to be consistent with the axiom of choice, provided a sufficiently large cardinal exists. Indeed, D.A. Martin and J.R. Steel (*198?*) have recently established, among other things, that if there is a supercompact cardinal (or, what is weaker, infinitely many Woodin cardinals), then projective determinacy is true and hence the weak continuum hypothesis is true for all projective sets.[q]

The second line of development, independence proofs, profoundly affected Gödel's program. In 1963 Cohen established not only that *CH* is independent but also that $2^{\aleph_0}$ can be arbitrarily large among the alephs. Feferman then showed that it is consistent with ZF to have $2^{\aleph_0} = 2^{\aleph_1}$, Luzin's second continuum hypothesis (*Cohen 1964*, page 110). From Cohen's work it followed, in regard to (i)–(iv) on page 161 above, that $2^{\aleph_0}$ is not bounded above by any given aleph and can be either accessible or weakly inaccessible, singular or regular; moreover, there are no restrictions on the cofinality of $2^{\aleph_0}$ other than König's theorem. Solovay independently determined the α for which $2^{\aleph_0} = \aleph_\alpha$ is consistent, namely all $\aleph_\alpha$ of uncountable cofinality (*1965*). Thus it was shown that our ignorance regarding (i)–(iv) is inevitable if we assume only the usual first-order axioms of set theory. (In *1964*, Gödel was inclined to believe that $2^{\aleph_0}$ is rather large, and favored the proposition that $2^{\aleph_0}$ is the first weakly inaccessible cardinal (*1964*, page 270).)

Shortly after Cohen announced his results in 1963, research on the continuum problem turned to establishing what are the possibilities for the continuum function $F(\aleph_\alpha) = 2^{\aleph_\alpha}$, defined on all ordinals. The first major breakthrough was Easton's theorem (*1964*, *1970*) that the continuum function F can, on regular cardinals, be any nondecreasing function

[q]By combining this result with earlier work of Woodin, one obtains from a supercompact cardinal the existence of a transitive class model of ZF + *AD* + *DC* containing all real numbers and all ordinals.

for which the cofinality of $F(\aleph_\alpha)$ is greater than $\aleph_\alpha$. For a decade there was a consensus among set theorists that something analogous to Easton's result would also be shown for singular cardinals. Both Easton and Solovay, among others, attempted to solve what came to be called the singular cardinals problem.

Consequently, set theorists were quite surprised in 1974 when Silver established that if *GCH* holds below a singular cardinal κ of uncountable cofinality, then it holds at κ as well (*Silver 1975*). Even this result, however, by no means settled the singular cardinals problem—provided that this problem is taken as asking for all the laws about cardinal exponentiation relative to singular cardinals. A first step occurred when Bukovský (*1965*) proved that cardinal exponentiation is determined by the so-called gimel function $\aleph_\alpha^{\mathrm{cf}(\aleph_\alpha)}$—a result that Gödel had stated but not proved in *1947* (page 517).

One important spinoff of Silver's result was Jensen's covering theorem (*Devlin and Jensen 1975*), which states that if the large cardinal axiom asserting the existence of $0^\#$ is false, then the singular cardinals hypothesis is true.[r] This hypothesis asserts that the continuum function $F(\aleph_\alpha) = 2^{\aleph_\alpha}$ is determined by its behavior at regular $\aleph_\alpha$. Thus, although known large cardinal axioms did not settle *CH*, the negation of a large cardinal axiom settled the behavior of the continuum function F at singular cardinals.

Silver's result was extended by Galvin and Hajnal (*1975*) for the case where κ is a singular strong limit cardinal of uncountable cofinality. For such a κ, they found an upper bound on 2^κ in terms of the behavior of 2^λ for a stationary set of $\lambda < \kappa$. Somewhat earlier, in *1974*, Solovay proved that if κ is strongly compact, then there is a proper class of cardinals for which *GCH* holds, namely, the class of singular strong limit cardinals greater than κ.

Magidor (*1977*) established that Silver's assumption of uncountable cofinality is necessary. In particular, Magidor showed, using a very large cardinal, that if *GCH* holds below $\aleph_\omega$, then it may happen that $2^{\aleph_\omega} = \aleph_{\omega+2}$. Shelah (*1982*) obtained a bound on $2^{\aleph_\omega}$ under the assumption that *GCH* holds below $\aleph_\omega$. Furthermore, Shelah discovered an analogue of the Galvin–Hajnal result for singular cardinals of countable cofinality. Finally, using a large cardinal assumption, Foreman and Woodin found a model of ZFC in which *GCH* fails everywhere; Woodin later improved this to $2^{\aleph_\alpha} = \aleph_{\alpha+2}$ for all α. (It is known, thanks to an earlier result of L. Patai, that if, for all α and for a fixed β, $2^{\aleph_\alpha} = \aleph_{\alpha+\beta}$, then β is finite; see *Jech 1978*, pages 48 and 580.)

[r]Concerning $0^\#$, see p. 21 above of the introductory note to *1938*.

Recently Foreman (*1986*) has proposed the axiom of resemblance, which he regards as a generalization of large cardinal axioms, and has announced that it implies both *GCH* and the axiom of projective determinacy. (He has shown in *1986* that, from *CH* and the axiom of resemblance, *GCH* follows.) For Gödel, however, the fact that the axiom of resemblance implies *GCH* would probably have disqualified it as settling the continuum problem.[s]

7. Gödel's unpublished papers on *CH*

After his proposal for using large cardinal axioms to decide *CH* did not succeed, Gödel introduced other axioms that he hoped would decide it. In January 1964, before he knew that such axioms, and in particular the existence of a measurable cardinal, did not settle *CH*, he wrote to Cohen about a related question:

> Once the continuum hypothesis is dropped, the key problem concerning the structure of the continuum, in my opinion, is the question of whether there exists a set of sequences of integers of power $\aleph_1$ which, for any given sequence of integers, contains one majorizing it from a certain point on .... I always suspected that, in contrast to the continuum hypothesis, this proposition is correct and perhaps even demonstrable from the axioms of set theory.

Six years later, Gödel postulated the existence of such a set of sequences as one of his axioms, now called Gödel's square axioms, which were intended to resolve the continuum problem.

The square axioms are an axiom schema stating that, for each natural number n, there exists a scale, of type ω_{n+1}, of functions from ω_n to ω_n.[t] Perhaps Gödel was led to formulate the square axioms by reading *Borel 1898*, which he cites. On page 116, Borel claimed that there exists a scale for the case $n = 0$ for all "effectively defined" functions, though he did not give a proof of his claim.

Gödel introduced these axioms in his final contribution to solving the continuum problem, a short paper written in 1970 and entitled "Some considerations leading to the probable conclusion that the true power of the continuum is $\aleph_2$", which he intended to publish in the *Proceedings*

[s]A recent argument that $2^{\aleph_0} \geq \aleph_\omega$ can be found in *Freiling 1986*.

[t]In other words, let F be the set of functions from ω_n to ω_n; then F has a subset S of power $\aleph_{n+1}$ such that for any function f in F there is some function g in S such that for some α and for all $\beta > \alpha$, $f(\beta) < g(\beta)$.

of the National Academy of Sciences.[u] In this paper he proposed four axioms (or axiom schemas), of which the square axioms were the first. The second axiom asserted that there are exactly $\aleph_n$ initial segments of the scale given by the square axioms. The third axiom was that there exists a maximal scale of functions from $\mathbb{N}$ to $\mathbb{R}$ such that "every ascending or descending sequence has cofinality ω". The fourth and final axiom consisted of the Hausdorff continuity axiom for the scale given by Gödel's third axiom. (Axiom 4 implies that $2^{\aleph_0} = 2^{\aleph_1}$.)

Gödel mailed his paper to Tarski, who then asked Solovay to examine its correctness. D. A. Martin, to whom Solovay had sent a copy of the paper, found that a result in it contradicted a theorem of Solovay's. In particular, Martin observed, since Solovay had shown that the square axioms do not put an upper bound on the size of $2^{\aleph_0}$, Gödel had to be mistaken in his claim that these axioms yield the result that $2^{\aleph_0}$ is bounded by $\aleph_2$.[v] On 19 May 1970 Tarski returned the paper to Gödel, adding in his covering letter that "you will certainly hear still in this matter either from me or from somebody else in Berkeley."

The whole matter was tinged with irony. For by 1965, having become convinced of the proposition that the square axioms do put an upper bound on $2^{\aleph_0}$, Gödel discussed this proposition with Solovay at the Institute for Advanced Study. At Gödel's request Solovay looked into the matter and found that there are models of set theory satisfying the square axioms but having $2^{\aleph_0}$ arbitrarily large. Gödel remained unconvinced, despite K. Prikry's assurances that Solovay was correct.[w] Solovay's result had to be rediscovered independently by E. Ellentuck (about 1973) before Gödel came to accept it.[x]

In 1970, not long after receiving Tarski's letter, Gödel drafted a second version of his paper on *CH*, entitled "A proof of Cantor's continuum hypothesis from a highly plausible axiom about orders of growth". His title represented a sudden and unexpected shift in his longstanding rejection of *CH*. This change in attitude appears to have been due to his belief that the square "axioms for $\aleph_n$ (or even any regular ordinal) are highly plausible, much more so than the continuum hypothesis." Indeed, he claimed that *CH* follows from the square axiom for $\aleph_1$ (that is, for $n = 1$). In conclusion, he wrote:

[u]The various versions of this paper are being considered for inclusion in Volume III of these *Collected works*.

[v]Personal communication from D. A. Martin and R. M. Solovay, 13 February 1984.

[w]Personal communication from R. M. Solovay, 4 April 1984.

[x]Ellentuck only learned of Solovay's priority for this result after finding it himself; see Ellentuck's note, dated February 1973, in Gödel's *Nachlass*.

> It seems to me this argument gives *much* more likelihood to the truth of Cantor's continuum hypothesis than any counterargument set up to now gave to its falsehood, and it has at any rate the virtue of deriving the power of the set of *all* functions $\omega \to \omega$ from that of certain *very* special sets of these functions. Of course the argument can be applied to higher cases of the generalized continuum hypothesis (in particular to all $\aleph_n$). It is, however, questionable whether the whole generalized continuum hypothesis follows.

At the top of this second version Gödel had written "nur für mich geschrieben" ("written only for myself"). It is unclear who, if anyone, saw this version before Gödel's death.

A third version of the paper (so Gödel described it) was a draft of a letter to Tarski, apparently never sent, that survives in Gödel's *Nachlass.* This letter is much closer in spirit to the first version of the paper than to the second. In the letter Gödel stated that he had written the first version hurriedly right after an illness for which he had been taking medication. What he had proved, he now believed, "is a nice equivalence result for the generalized continuum hypothesis ... ⟦showing that it⟧ follows from certain very special and weak cases of it." Gödel concluded the letter with some speculations:

> My conviction that $2^{\aleph_0} = \aleph_2$ of course has been somewhat shaken. But it still seems plausible to me. One of my reasons is that I don't believe in any kind of irrationality such as, e.g., random sequences in any absolute sense. Perhaps $2^{\aleph_0} = \aleph_2$ does follow from my axioms 1–4, but unfortunately Axiom 4 is rather doubtful, while axioms 1–3 seem *extremely likely* to me.

Yet he conceded that Axioms 1–3 do not imply $2^{\aleph_0} \leq \aleph_2$.[y]

Thus ended Gödel's last attempt to settle the continuum problem, which he had analyzed so brilliantly in *1947* and *1964*.[z]

Gregory H. Moore

[y]Before 1973, Gödel's square axioms were studied by G. Takeuti, who established that the existence of a scale from ω_1 to ω_0 implies *CH*. These axioms are also investigated in *Ellentuck 1975*, *Takeuti 1978* and *P. E. Cohen 1979*.

[z]I would like to thank S. Feferman for many substantive suggestions, and J. Dawson for many stylistic ones, to an earlier draft of this introductory note; Dawson has also been of considerable assistance on archival matters. I am especially grateful to R. M. Solovay for his many useful suggestions regarding Section 6.

What is Cantor's continuum problem? (*1947*)

1. The concept of cardinal number

Cantor's continuum problem is simply the question: How many points are there on a straight line in Euclidean space? In other terms, the question is: How many different sets of integers do there exist?

This question, of course, could arise only after the concept of "number" had been extended to infinite sets; hence it might be doubted if this extension can be effected in a uniquely determined manner and if, therefore, the statement of the problem in the simple terms used above is justified. Closer examination, however, shows that Cantor's definition of infinite numbers really has this character of uniqueness, and that in a very striking manner. For whatever "number" as applied to infinite sets may mean, we certainly want it to have the property that the number of objects belonging to some class does not change if, leaving the objects the same, one changes in any way whatsoever their properties or mutual relations (e.g., their colors or their distribution in space). From this, however, it follows at once that two sets (at least two sets of changeable objects of the space-time world) will have the same cardinal number if their elements can be brought into a one-to-one correspondence, which is Cantor's definition of equality between numbers. For if there exists such a correspondence for two sets A and B it is possible (at least theoretically) to change the properties and relations of each element of A into those of the corresponding element of B, whereby A is transformed into a set completely indistinguishable from B, hence of the same cardinal number. For example, assuming a square and a line segment both completely filled with mass points (so that at each point of them exactly one mass point is situated), it follows, owing to the demonstrable fact that there exists a one-to-one correspondence between the points of a square and of a line segment, and, therefore, also between the corresponding mass points, that the mass points of the square can be so rearranged as exactly to fill out the line segment, and vice versa. Such considerations, it is true, apply directly only to physical objects, but a definition of the concept of "number" which would depend on the kind of objects that are numbered could hardly be considered as satisfactory.

So there is hardly any choice left but to accept Cantor's definition of equality between numbers, which can easily be extended to a definition of "greater" and "less" for infinite numbers by stipulating that the cardinal number M of a set A is to be called less than the cardinal number N of a set B if M is different from N but equal to the cardinal number of some subset of B. On the basis of these definitions it becomes possible to prove that there exist infinitely many different infinite cardinal numbers or "powers", and that, in particular, the number of subsets of a set is always greater than the number of its elements; furthermore, it becomes possible to extend (again without any arbitrariness) the arithmetical operations to infinite numbers (including sums and products with any infinite | number of terms or factors) and to prove practically all ordinary rules of computation. 516

But, even after that, the problem to determine the cardinal number of an individual set, such as the linear continuum, would not be well-defined if there did not exist some "natural" representation of the infinite cardinal numbers, comparable to the decimal or some other systematic denotation of the integers. This systematic representation, however, does exist, owing to the theorem that for each cardinal number and each set of cardinal numbers[1] there exists exactly one cardinal number immediately succeeding in magnitude and that the cardinal number of every set occurs in the series thus obtained.[2] This theorem makes it possible to denote the cardinal number immediately succeeding the set of finite numbers by $\aleph_0$ (which is the power of the "denumerably infinite" sets), the next one by $\aleph_1$, etc.; the one immediately succeeding all $\aleph_i$ (where i is an integer), by $\aleph_\omega$, the next one by $\aleph_{\omega+1}$, etc., and the theory of ordinal numbers furnishes the means to extend this series farther and farther.

2. The continuum problem, the continuum hypothesis and the partial results concerning its truth obtained so far

So the analysis of the phrase "how many" leads unambiguously to quite a definite meaning for the question stated in the second line of this paper, namely, to find out which one of the $\aleph$'s is the number of points on a straight line or (which is the same) on any other continuum in Euclidean space. Cantor, after having proved that this number is certainly greater than $\aleph_0$, conjectured that it is $\aleph_1$, or (which is an equivalent proposition) that every infinite subset of the continuum has either the power of the set of integers or of the whole continuum. This is Cantor's continuum hypothesis.

[1] As to the question why there does not exist a set of all cardinal numbers, see footnote 14.

[2] In order to prove this theorem the axiom of choice (see *Fraenkel 1928*, p. 288 ff.) is necessary, but it may be said that this axiom is, in the present state of knowledge, exactly as well-founded as the system of the other axioms. It has been proved consistent, provided the other axioms are so (see *Gödel 1940*). It is exactly as evident as the other axioms for sets in the sense of arbitrary multitudes and, as for sets in the sense of extensions of definable properties; it also is demonstrable for those concepts of definability for which, in the present state of knowledge, it is possible to prove the other axioms, namely, those explained in footnotes 20 and 26.

But, although Cantor's set theory has now had a development of more than sixty years and the problem is evidently of great importance for it, nothing has been proved so far relative to the question what the power of the continuum is or whether its subsets satisfy the condition just stated, except (1) that the power of the continuum is not a cardinal number of a certain very special kind, namely, not a limit of denumerably many smaller cardinal numbers,[3] and (2) that the proposition just mentioned about the 517 subsets of the continuum is | true for a certain infinitesimal fraction of these subsets, the analytical[4] sets.[5] Not even an upper bound, however high, can be assigned for the power of the continuum. Nor is there any more known about the quality than about the quantity of the cardinal number of the continuum. It is undecided whether this number is regular or singular, accessible or inaccessible, and (except for König's negative result) what its character of cofinality[4] is. The only thing one knows, in addition to the results just mentioned, is a great number of consequences of, and some propositions equivalent to, Cantor's conjecture.[6]

This pronounced failure becomes still more striking if the problem is considered in its connection with general questions of cardinal arithmetic. It is easily proved that the power of the continuum is equal to $2^{\aleph_0}$. So the continuum problem turns out to be a question from the "multiplication table" of cardinal numbers, namely, the problem to evaluate a certain infinite product (in fact the simplest non-trivial one that can be formed). There is, however, not one infinite product (of factors > 1) for which only as much as an upper bound for its value can be assigned. All one knows about the evaluation of infinite products are two lower bounds due to Cantor and König (the latter of which implies a generalization of the aforementioned negative theorem on the power of the continuum), and some theorems concerning the reduction of products with different factors to exponentiations and of exponentiations to exponentiations with smaller bases or exponents. These theorems reduce[7] the whole problem of computing infinite products to the evaluation of $\aleph_\alpha^{\mathrm{cf}(\aleph_\alpha)}$ and the performance of certain fundamental operations on ordinal numbers, such as determining the limit of a series of them. $\aleph_\alpha^{\mathrm{cf}(\aleph_\alpha)}$, and therewith all products and powers, can easily be

[3]See *Hausdorff 1914*, p. 68. The discoverer of this theorem, J. König, asserted more than he had actually proved (see his *1905*).

[4]See the list of definitions at the end of this paper.

[5]See *Hausdorff 1935*, p. 32. Even for complements of analytical sets the question is undecided at present, and it can be proved only that they have (if they are infinite) either the power $\aleph_0$ or $\aleph_1$ or that of the continuum (see *Kuratowski 1933*, p. 246).

[6]See *Sierpiński 1934*.

[7]This reduction can be effected owing to the results and methods of *Tarski 1925*.

computed[8] if the "generalized continuum hypothesis" is assumed, i.e., if it is assumed that $2^{\aleph_\alpha} = \aleph_{\alpha+1}$ for every α, or, in other terms, that the number of subsets of a set of power $\aleph_\alpha$ is $\aleph_{\alpha+1}$. But, without making any undemonstrated assumption, it is not even known whether or not $m < n$ implies $2^m < 2^n$ (although it is trivial that it implies $2^m \leq 2^n$), nor even whether $2^{\aleph_0} < 2^{\aleph_1}$.

3. Restatement of the problem on the basis of an analysis of the foundations of set theory and results obtained along these lines

This scarcity of results, even as to the most fundamental questions in this field, may be due to some extent to purely mathematical difficulties; it seems, however (see Section 4 below), that there are also deeper reasons behind it and that a complete solution of | these problems can be obtained only by a more profound analysis (than mathematics is accustomed to give) of the meanings of the terms occurring in them (such as "set", "one-to-one correspondence", etc.) and of the axioms underlying their use. Several such analyses have been proposed already. Let us see then what they give for our problem. 518

First of all there is Brouwer's intuitionism, which is utterly destructive in its results. The whole theory of the $\aleph$'s greater than $\aleph_1$ is rejected as meaningless.[9] Cantor's conjecture itself receives several different meanings, all of which, though very interesting in themselves, are quite different from the original problem, and which lead partly to affirmative, partly to negative answers;[10] not everything in this field, however, has been clarified sufficiently. The "half-intuitionistic" standpoint along the lines of H. Poincaré and H. Weyl[11] would hardly preserve substantially more of set theory.

This negative attitude towards Cantor's set theory, however, is by no means a necessary outcome of a closer examination of its foundations, but only the result of certain philosophical conceptions of the nature of mathematics, which admit mathematical objects only to the extent in which they

[8]For regular numbers $\aleph_\alpha$ one obtains immediately:

$$\aleph_\alpha^{\mathrm{cf}(\aleph_\alpha)} = \aleph_\alpha^{\aleph_\alpha} = 2^{\aleph_\alpha} = \aleph_{\alpha+1}.$$

[9]See *Brouwer 1909*.

[10]See *Brouwer 1907*, I, 9; III, 2.

[11]See *Weyl 1932*. If the procedure of construction of sets described there (p. 20) is iterated a sufficiently large (transfinite) number of times, one gets exactly the real numbers of the model for set theory spoken of below in Section 4, in which the continuum hypothesis is true. But this iteration would hardly be possible within the limits of the half-intuitionistic standpoint.

are (or are believed to be) interpretable as acts and constructions of our own mind, or at least completely penetrable by our intuition. For someone who does not share these views there exists a satisfactory foundation of Cantor's set theory in its whole original extent, namely, axiomatics of set theory, under which the logical system of *Principia mathematica* (in a suitable interpretation) may be subsumed.

It might at first seem that the set-theoretical paradoxes would stand in the way of such an undertaking, but closer examination shows that they cause no trouble at all. They are a very serious problem, but not for Cantor's set theory. As far as sets occur and are necessary in mathematics (at least in the mathematics of today, including all of Cantor's set theory), they are sets of integers, or of rational numbers (i.e., of pairs of integers), or of real numbers (i.e., of sets of rational numbers), or of functions of real numbers (i.e., of sets of pairs of real numbers), etc.; when theorems about all sets (or the existence of sets) in general are asserted, they can always be interpreted without any difficulty to mean that they hold for sets of integers as well as for sets of real numbers, etc. (respectively, that there exist either sets of integers, or sets of real numbers, or ... etc., which have the asserted property). This concept of set, however, according to which a set is anything obtainable from the integers (or some other well-defined 519 | objects) by iterated application[12] of the operation "set of",[13] and not something obtained by dividing the totality of all existing things into two categories, has never led to any antinomy whatsoever; that is, the perfectly "naïve" and uncritical working with this concept of set has so far proved completely self-consistent.[14]

But, furthermore, the axioms underlying the unrestricted use of this concept of set, or, at least, a portion of them which suffices for all mathematical proofs ever produced up to now, have been so precisely formulated in axiomatic set theory[15] that the question whether some given proposition follows from them can be transformed, by means of logistic symbolism, into

[12]This phrase is to be understood so as to include also transfinite iteration, the totality of sets obtained by finite iteration forming again a set and a basis for a further application of the operation "set of".

[13]The operation "set of x's" cannot be defined satisfactorily (at least in the present state of knowledge), but only be paraphrased by other expressions involving again the concept of set, such as: "multitude of x's", "combination of any number of x's", "part of the totality of x's"; but as opposed to the concept of set in general (if considered as primitive) we have a clear notion of this operation.

[14]It follows at once from this explanation of the term "set" that a set of all sets or other sets of a similar extension cannot exist, since every set obtained in this way immediately gives rise to further application of the operation "set of" and, therefore, to the existence of larger sets.

[15]See, e.g., *Bernays 1937, 1941, 1942, 1942a, 1943, von Neumann 1925*; cf. also *von Neumann 1928a* and *1929, Gödel 1940.*

a purely combinatorial problem concerning the manipulation of symbols which even the most radical intuitionist must acknowledge as meaningful. So Cantor's continuum problem, no matter what philosophical standpoint one takes, undeniably retains at least this meaning: to ascertain whether an answer, and if so what answer, can be derived from the axioms of set theory as formulated in the systems quoted.

Of course, if it is interpreted in this way, there are (assuming the consistency of the axioms) a priori three possibilities for Cantor's conjecture: It may be either demonstrable or disprovable or undecidable.[16] The third alternative (which is only a precise formulation of the conjecture stated above that the difficulties of the problem are perhaps not purely mathematical) is the most likely, and to seek a proof for it is at present one of the most promising ways of attacking the problem. One result along these lines has been obtained already, namely, that Cantor's conjecture is not disprovable from the axioms of set theory, provided that these axioms are consistent (see Section 4).

It is to be noted, however, that, even if one should succeed in proving its undemonstrability as well, this would (in contradistinction, for example, to the proof for the transcendency of π) by no means settle the question definitively. | Only someone who (like the intuitionist) denies that the concepts and axioms of classical set theory have any meaning (or any well-defined meaning) could be satisfied with such a solution, not someone who believes them to describe some well-determined reality. For in this reality Cantor's conjecture must be either true or false, and its undecidability from the axioms as known today can only mean that these axioms do not contain a complete description of this reality; and such a belief is by no means chimerical, since it is possible to point out ways in which a decision of the question, even if it is undecidable from the axioms in their present form, might nevertheless be obtained. 520

For first of all the axioms of set theory by no means form a system closed in itself, but, quite on the contrary, the very concept of set[17] on which they are based suggests their extension by new axioms which assert the existence of still further iterations of the operation "set of". These axioms can also be formulated as propositions asserting the existence of very great cardinal numbers or (which is the same) of sets having these cardinal numbers.

[16]In case of the inconsistency of the axioms the last one of the four a priori possible alternatives for Cantor's conjecture would occur, namely, it would then be both demonstrable and disprovable by the axioms of set theory.

[17]Similarly also the concept "property of set" (the second of the primitive terms of set theory) can constantly be enlarged and, furthermore, concepts of "property of property of set" etc. be introduced whereby new axioms are obtained, which, however, as to their consequences for propositions referring to limited domains of sets (such as the continuum hypothesis) are contained in the axioms depending on the concept of set.

The simplest of these strong "axioms of infinity" assert the existence of inaccessible numbers (and of numbers inaccessible in the stronger sense) $> \aleph_0$. The latter axiom, roughly speaking, means nothing else but that the totality of sets obtainable by exclusive use of the processes of formation of sets expressed in the other axioms forms again a set (and, therefore, a new basis for a further application of these processes).[18] Other axioms of infinity have been formulated by P. Mahlo.[19] Very little is known about this section of set theory; but at any rate these axioms show clearly, not only that the axiomatic system of set theory as known today is incomplete, but also that it can be supplemented without arbitrariness by new axioms which are only the natural continuation of the series of those set up so far.

That these axioms have consequences also far outside the domain of very great transfinite numbers, which are their immediate object, can be proved; each of them (as far as they are known) can, under the assumption of consistency, be shown to increase the number of decidable propositions even in the field of Diophantine equations. As for the continuum problem, there is little hope of solving it by means of those axioms of infinity which can be set up on the basis of principles known today (the above-mentioned proof for the undisprovability of the continuum hypothesis, e.g., goes through for all of them without any change). But probably there exist others based on hitherto unknown principles; also there may exist, besides the ordinary 521 axioms, the axioms of infinity and | the axioms mentioned in footnote 17, other (hitherto unknown) axioms of set theory which a more profound understanding of the concepts underlying logic and mathematics would enable us to recognize as implied by these concepts.

Furthermore, however, even disregarding the intrinsic necessity of some new axiom, and even in case it had no intrinsic necessity at all, a decision about its truth is possible also in another way, namely, inductively by studying its "success", that is, its fruitfulness in consequences and in particular in "verifiable" consequences, i.e., consequences demonstrable without the new axiom, whose proofs by means of the new axiom, however, are considerably simpler and easier to discover, and make it possible to condense into one proof many different proofs. The axioms for the system of real numbers, rejected by the intuitionists, have in this sense been verified to some extent owing to the fact that analytical number theory frequently allows us to prove number-theoretical theorems which can subsequently be verified by elementary methods. A much higher degree of verification than that, however, is conceivable. There might exist axioms so abundant in

[18]See *Zermelo 1930*.

[19]See his *1911*, pp. 190–200, *1913*, pp. 269–276. From Mahlo's presentation of the subject, however, it does not appear that the numbers he defines actually exist.

their verifiable consequences, shedding so much light upon a whole discipline, and furnishing such powerful methods for solving given problems (and even solving them, as far as that is possible, in a constructivistic way) that quite irrespective of their intrinsic necessity they would have to be assumed at least in the same sense as any well-established physical theory.

4. Some observations about the question: In what sense and in which direction may a solution of the continuum problem be expected?

But are such considerations appropriate for the continuum problem? Are there really any strong indications for its unsolubility by the known axioms? I think there are at least two.

The first one is furnished by the fact that there are two quite differently defined classes of objects which both satisfy all axioms of set theory written down so far. One class consists of the sets definable in a certain manner by properties of their elements,[20] the other of the sets in the sense of arbitrary multitudes irrespective of if, or how, they can be defined. Now, before it is settled what objects are to be numbered, and on the basis of what one-to-one correspondences, one could hardly expect to be able to determine their number (except perhaps in case of some fortunate coincidence). If, however, someone believes that it is meaningless to speak of sets except in the sense of extensions of definable properties, or, at least, that no other sets exist, then, too, he can hardly expect more than a small fraction of the problems of set theory to be solvable without making use of this, in his opinion essential, characteristic of sets, namely, that they are | all derived from (or in a sense even identical with) definable properties. This characteristic of sets, however, is neither formulated explicitly nor contained implicitly in the accepted axioms of set theory. So from either point of view, if in addition one has regard to what was said above in Section 2, it is plausible that the continuum problem will not be solvable by the axioms set up so far, but, on the other hand, may be solvable by means of a new axiom

522

[20]Namely, definable "in terms of ordinal numbers" (i.e., roughly speaking, under the assumption that for each ordinal number a symbol denoting it is given) by means of transfinite recursions, the primitive terms of logic, and the ϵ-relation, admitting, however, as elements of sets and of ranges of quantifiers only previously defined sets. See my papers *1939a* and *1940*, where an exactly equivalent, although in its definition slightly different, concept of definability (under the name of "constructibility") is used. The paradox of Richard, of course, does not apply to this kind of definability, since the totality of ordinals is certainly not denumerable.

which would state or at least imply something about the definability of sets.[21]

The latter half of this conjecture has already been verified; namely, the concept of definability just mentioned (which is itself definable in terms of the primitive notions of set theory) makes it possible to derive the generalized continuum hypothesis from the axiom that every set is definable in this sense.[22] Since this axiom (let us call it "A") turns out to be demonstrably consistent with the other axioms, under the assumption of the consistency of these axioms, this result (irrespective of any philosophical opinion) shows the consistency of the continuum hypothesis with the axioms of set theory, provided that these axioms themselves are consistent.[23] This proof in its structure is analogous to the consistency proof for non-Euclidean geometry by means of a model within Euclidean geometry, insofar as it follows from the axioms of set theory that the sets definable in the above sense form a model for set theory in which furthermore the proposition A and, therefore, the generalized continuum hypothesis is true. But the definition of "definability" can also be so formulated that it becomes a definition of a concept of "set" and a relation of "element of" (satisfying the axioms of set theory) in terms of entirely different concepts, namely, the concept of "ordinal numbers", in the sense of elements ordered by some relation of "greater" and "less", this ordering relation itself, and the notion of "recursively defined function of ordinals", which can be taken as primitive and be described axiomatically by way of an extension of Peano's axioms.[24] (Note that this does not apply to my original formulation presented in the papers quoted above, because there the general concept of "set" with its element relation occurs in the definition of "definable set", although the definable sets remain the same if, afterwards, in the definition of "definability" the term "set" is replaced by "definable set".)

523 | A second argument in favor of the unsolubility of the continuum problem on the basis of the ordinary axioms can be based on certain facts (not known or not existing at Cantor's time) which seem to indicate that Cantor's

[21]D. Hilbert's attempt at a solution of the continuum problem (see his *1926*), which, however, has never been carried through, also was based on a consideration of all possible definitions of real numbers.

[22]On the other hand, from an axiom in some sense directly opposite to this one the negation of Cantor's conjecture could perhaps be derived.

[23]See my paper *1940* and note *1939a*. For a carrying through of the proof in all details, my paper *1940* is to be consulted.

[24]For such an extension see *Tarski 1924*, where, however, the general concept of "set of ordinal numbers" is used in the axioms; this could be avoided, without any loss in demonstrable theorems, by confining oneself from the beginning to recursively definable sets of ordinals.

conjecture will turn out to be wrong;[25] for a negative decision the question is (as just explained) demonstrably impossible on the basis of the axioms as known today.

There exists a considerable number of facts of this kind which, of course, at the same time make it likely that not all sets are definable in the above sense.[26] One such fact, for example, is the existence of certain properties of point sets (asserting an extreme rareness of the sets concerned) for which one has succeeded in proving the existence of undenumerable sets having these properties, but no way is apparent by means of which one could expect to prove the existence of examples of the power of the continuum. Properties of this type (of subsets of a straight line) are: (1) being of the first category on every perfect set,[27] (2) being carried into a zero set by every continuous one-to-one mapping of the line on itself.[28] Another property of a similar nature is that of being coverable by infinitely many intervals of any given lengths. But in this latter case one has so far not even succeeded in proving the existence of undenumerable examples. From the continuum hypothesis, however, it follows that there exist in all three cases not only examples of the power of the continuum,[29] but even such as are carried into themselves (up to denumerably many points) by *every* translation of the straight line.[30]

And this is not the only paradoxical consequence of the continuum hypothesis. Others, for example, are that there exist: (1) subsets of a straight line of the power of the continuum which are covered (up to denumerably many points) by *every* dense set of intervals, or (in other terms) which contain no undenumerable subset nowhere dense on the straight line,[31] (2) subsets of a straight line of the power of the continuum which contain no undenumerable zero set,[32] (3) subsets of Hilbert

[25]Views tending in this direction have been expressed also by N. Luzin in his *1935*, p. 129 ff. See also *Sierpiński 1935*.

[26]That all sets are "definable in terms of ordinals" if *all* procedures of definition, i.e., also quantification and the operation $\hat{x}$ with respect to *all* sets, irrespective of whether they have or can be defined, are admitted could be expected with more reason, but still it would not at all be justified to assume this as an axiom. It is worth noting that the proof that the continuum hypothesis holds for the definable sets, or follows from the assumption that all sets are definable, does not go through for this kind of definability, although the assumption that these two concepts of definability are equivalent is, of course, demonstrably consistent with the axioms.

[27]See *Sierpiński 1934a* and *Kuratowski 1933*, p. 269 ff.

[28]See *Luzin and Sierpiński 1918* and *Sierpiński 1934a*.

[29]For the 3rd case see *Sierpiński 1934*, p. 39, Theorem 1.

[30]See *Sierpiński 1935a*.

[31]See *Luzin 1914*, p. 1259.

[32]See *Sierpiński 1924*, p. 184.

space of the power of the continuum which contain no undenumerable subset of finite dimension,[33] (4) an infinite sequence A^i of decompositions of
524 any set M of the power of the continuum into continuum | many mutually exclusive sets A^i_x such that, in whichever way a set $A^i_{x_i}$ is chosen for each i, $\prod_i(M - A^i_{x_i})$ is always denumerable.[34] Even if in (1)–(4) "power of the continuum" is replaced by "$\aleph_1$", these propositions are very implausible; the proposition obtained from (3) in this way is even equivalent with (3).

One may say that many of the results of point-set theory obtained without using the continuum hypothesis also are highly unexpected and implausible.[35] But, true as that may be, still the situation is different there, insofar as in those instances (such as, e.g., Peano's curves) the appearance to the contrary can in general be explained by a lack of agreement between our intuitive geometrical concepts and the set-theoretical ones occurring in the theorems. Also, it is very suspicious that, as against the numerous plausible propositions which imply the negation of the continuum hypothesis, not one plausible proposition is known which would imply the continuum hypothesis. Therefore one may on good reason suspect that the role of the continuum problem in set theory will be this, that it will finally lead to the discovery of new axioms which will make it possible to disprove Cantor's conjecture.

Definitions of some of the technical terms

Definitions 4–12 refer to subsets of a straight line, but can be literally transferred to subsets of Euclidean spaces of any number of dimensions; definitions 13–14 refer to subsets of Euclidean spaces.

1. I call "character of cofinality" of a cardinal number m (abbreviated by "cf(m)") the smallest number n such that m is the sum of n numbers $< m$.
2. A cardinal number m is regular if cf(m) $= m$, otherwise singular.
3. An infinite cardinal number m is inaccessible if it is regular and has no immediate predecessor (i.e., if, although it is a limit of numbers $< m$, it is not a limit of fewer than m such numbers); it is inaccessible in the stronger sense if each product (and, therefore, also each sum) of fewer than m numbers $< m$ is $< m$. (See *Sierpiński and Tarski 1930*, *Tarski 1938*. From the generalized continuum hypothesis follows the equivalence of these two notions. This equivalence, however, is a

[33]See *Hurewicz 1932*.

[34]See *Braun and Sierpiński 1932*, p. 1, proposition (Q). This proposition and the one stated under (3) in the text are equivalent with the continuum hypothesis.

[35]See, e.g., *Blumenthal 1940*.

much weaker and much more plausible proposition. $\aleph_0$ evidently is inaccessible in both senses. As for finite numbers, 0 and 2 and no others are inaccessible in the stronger sense (by the above definition), which suggests that the same will hold also for the correct extension of the concept of inaccessibility to finite numbers.)

4. A set of intervals is dense if every interval has points in common with some interval of the set. (The endpoints of an interval are not considered as points of the interval.)
5. A zero set is a set which can be covered by infinite sets of intervals with arbitrarily small lengths-sum.
6. A neighborhood of a point P is an interval containing P.
7. A subset A of B is dense in B if every neighborhood of any point of B contains points of A.
8. A point P is in the exterior of A if it has a neighborhood containing no point of A.
9. A subset A of B is nowhere dense on B if those points of B which are in the exterior of A are dense in B. (Such sets A are exactly the subsets of the borders of the open sets in B, but the term "border-set" is unfortunately used in a different sense.)
10. A subset A of B is of the first category in B if it is the sum of denumerably many sets nowhere dense in B.
11. Set A is of the first category on B if the intersection $A \cdot B$ is of the first category in B. 525
12. A set is perfect if it is closed and has no isolated point (i.e., no point with a neighborhood containing no other point of the set).
13. Borel sets are defined as the smallest system of sets satisfying the postulates:
    (1) The closed sets are Borel sets.
    (2) The complement of a Borel set is a Borel set.
    (3) The sum of denumerably many Borel sets is a Borel set.
14. A set is analytic if it is the orthogonal projection of some Borel set of a space of next higher dimension. (Every Borel set therefore is, of course, analytic.)
15. Quantifiers are the logistic symbols standing for the phrases: "for all objects x" and "there exist objects x". The totality of objects x to which they refer is called their range.
16. The symbol "$\hat{x}$" means "the set of those objects x for which ...".

Veli Valpola

Kurt Gödel in his office at the Institute for Advanced Study, May 1958

Introductory note to *1949* and *1952*

These two papers represent Gödel's main contribution to relativistic cosmology. In the 1920s and the 1930s, the Friedmann–Robertson–Walker cosmological models had been introduced as the simplest solutions of the equations of Einstein's general theory of relativity that were consistent with the observed red-shift of distant galaxies. These models were spatially homogeneous and isotropic, and were expanding but were non-rotating. Gödel was the first to consider models that were rotating. The possible rotation of the universe has a special significance in general relativity because one of the influences that led Einstein to the theory in 1915 was Mach's principle. The exact formulation of the principle is rather obscure, but it is generally interpreted as denying the existence of absolute space. In other words, matter has inertia only relative to other matter in the universe. The principle is generally taken to imply that the local inertial frame defined by gyroscopes should be non-rotating with respect to the frame defined by distant galaxies.

Gödel showed that it was possible to have solutions of the Einstein field equations in which the galaxies were rotating with respect to the local inertial frame. He therefore demonstrated that general relativity does not incorporate Mach's principle. Whether or not this is an argument against general relativity depends on your philosophical viewpoint, but most physicists nowadays would not accept Mach's principle, because they feel that it makes an untenable distinction between the geometry of space-time, which represents the gravitational and inertial field, and other forms of fields and matter.

In the first of these papers (*1949*) Gödel presented a rotating solution that was not expanding but was the same at all points of space and time. This solution was the first to be discovered that had the curious property that in it it was possible to travel into the past. This leads to paradoxes such as "What happens if you go back and kill your father when he was a baby?" It is generally agreed that this cannot happen in a solution that represents our universe, but Gödel was the first to show that it was not forbidden by the Einstein equations. His solution generated a lot of discussion of the relation between general relativity and the concept of causality.

The second paper (*1952*) describes more reasonable rotating cosmological models that are expanding and that do not have the possibility of travel into the past. These models could well be a reasonable description of the universe that we observe, although observations of the isotropy of the microwave background indicate that the rate of rotation must be very low.

S. W. Hawking

⟦For a more detailed discussion of Gödel's cosmological models, the reader is referred to *Hawking and Ellis 1973*, pages 168–170, to *Malament 1985*, and to *Raychaudhuri 1979*, pages 92–95.⟧

An example of a new type of cosmological solutions of Einstein's field equations of gravitation (*1949*)

1. The main properties of the new solution

All cosmological solutions with non-vanishing density of matter known at present[1] have the common property that, in a certain sense, they contain an "absolute" time coordinate,[2] owing to the fact that there exists a one-parametric system of three-spaces everywhere orthogonal on the world lines of matter. It is easily seen that the non-existence of such a system of three-spaces is equivalent with a rotation of matter relative to the compass of inertia. In this paper I am proposing a solution (with a cosmological term $\neq 0$) which exhibits such a rotation. This solution, or rather the four-dimensional space S which it defines, has the further properties:

(1) S is homogeneous, i.e., for any two points P, Q of S there exists a transformation of S into itself which carries P into Q. In terms of physics this means that the solution is stationary and spatially homogeneous.

(2) There exists a one-parametric group of transformations of S into itself which carries each world line of matter into itself, so that any two world lines of matter are equidistant.

[1] See, for example, *Robertson 1933*.

[2] As to the philosophical consequences which have been drawn from this circumstance see *Jeans 1936* and my article *1949a*, forthcoming in the Einstein volume of the Library of Living Philosophers.

(3) S has rotational symmetry, i.e., for each point P of S there exists a one-parametric group of transformations of S into itself which carries P into itself.

(4) The totality of time-like and null vectors can be divided into $+$- and $-$-vectors in such a way that: (a) if ξ is a $+$-vector, $-\xi$ is a $-$-vector, (b) a limit of $+$- (or $-$-)vectors, if $\neq 0$, is again a $+$- (or $-$-)vector. That is, a positive direction of time can consistently be introduced in the whole solution.

After a direction of time has been introduced in this way, a temporal orientation is defined for the world line of every (real or possible) particle of matter or light, i.e., it is determined for any two neighboring points on it which one is earlier. On the other hand, however, no uniform temporal ordering of *all* point events, agreeing in direction with all these individual orderings, exists. This is expressed in the next property:

(5) It is not possible to assign a time coordinate t to each space-time point in such a way that t always increases, if one moves in a positive time-like direction; and this holds both for an open and a closed time coordinate.

(6) Every world line of matter occurring in the solution is an open line of infinite length, which never approaches any of its preceding points again; but there also exist closed time-like lines.[3] In particular, if P, Q are any two points on a world line of matter,[4] and P precedes Q on this line, there exists a time-like line connecting P and Q on which Q precedes P; i.e., it is theoretically possible in these worlds to travel into the past, or otherwise influence the past.

(7) There exist no three-spaces which are everywhere space-like and intersect each world line of matter in one point.

(8) If Σ is any system of mutually exclusive three-spaces, each of which intersects every world line of matter in one point,[5] then there exists a transformation which carries S and the positive direction of time into itself, but does *not* carry Σ into itself; i.e., an *absolute* time does not exist, even if it is not required to agree in direction with the times of all possible observers (where "absolute" means: definable without reference to individual objects, such as, e.g., a particular galactic system).

(9) Matter everywhere rotates relative to the compass of inertia with the angular velocity $2(\pi\kappa\rho)^{\frac{1}{2}}$, where ρ is the mean density of matter and κ Newton's gravitational constant.

[3]If the tangent of a line is discontinuous, the line is to be considered as time-like only if the corners can be so rounded off that the resulting line is everywhere time-like.

[4]"World line of matter" without further specification always refers to the world lines of matter occurring as such in the solution under consideration.

[5]Another hypothesis about Σ under which the conclusion holds is that Σ is one-parametric and oriented (where the orientation refers to the space whose points are the elements of Σ).

2. Definition of the linear element and proof that it satisfies the field equations

The linear element of S is defined by the following expression:[6]

$$a^2(dx_0{}^2 - dx_1{}^2 + (e^{2x_1}/2)dx_2{}^2 - dx_3{}^2 + 2e^{x_1}dx_0dx_2),$$

448 | where a is a positive number. The matrices of the g_{ik} and g^{ik}, therefore, are the two:

$$a^2 \cdot \begin{Vmatrix} 1 & 0 & e^{x_1} & 0 \\ 0 & -1 & 0 & 0 \\ e^{x_1} & 0 & e^{2x_1}/2 & 0 \\ 0 & 0 & 0 & -1 \end{Vmatrix}, \quad \frac{1}{a^2} \cdot \begin{Vmatrix} -1 & 0 & 2e^{-x_1} & 0 \\ 0 & -1 & 0 & 0 \\ 2e^{-x_1} & 0 & -2e^{-2x_1} & 0 \\ 0 & 0 & 0 & -1 \end{Vmatrix}.$$

Owing to the fact that only two of the forty $\partial g_{ik}/\partial x_l$ are $\neq 0$, namely $\partial g_{22}/\partial x_1$ and $\partial g_{02}/\partial x_1$, the $\Gamma_{i,kl}$ and Γ^i_{kl} can very easily be computed. One obtains the values:

$$\begin{aligned} &\Gamma_{0,12} = -\Gamma_{1,02} = \Gamma_{2,01} = (a^2/2)e^{x_1}, \\ &\Gamma_{1,22} = -\Gamma_{2,12} = -(a^2/2)e^{2x_1}, \\ &\Gamma^0_{01} = 1, \quad \Gamma^0_{12} = \Gamma^1_{02} = e^{x_1}/2, \\ &\Gamma^1_{22} = e^{2x_1}/2, \quad \Gamma^2_{01} = -e^{-x_1}. \end{aligned}$$

These $\Gamma_{i,kl}$ and Γ^i_{kl}, and those obtained from them by interchanging the last two (or the lower two) indices are the only ones that do not vanish.

Using for R_{ik} the formula[7]

$$R_{ik} = \frac{\partial}{\partial x_\sigma}\Gamma^\sigma_{ik} - \frac{1}{2}\frac{\partial^2 \log g}{\partial x_i \partial x_k} + \frac{1}{2}\Gamma^\sigma_{ik}\frac{\partial \log g}{\partial x_\sigma} - \Gamma^\rho_{\sigma i}\Gamma^\sigma_{\rho k},$$

[6]This quadratic form can also be written thus

$$a^2\left[(dx_0 + e^{x_1}dx_2)^2 - dx_1{}^2 - \frac{e^{2x_1}}{2}dx_2{}^2 - dx_3{}^2\right],$$

which makes it evident that, as required, its signature is everywhere -2. The three-space obtained by leaving out the term $-dx_3{}^2$ has a simple geometric meaning (see below). Essentially the | same three-space, but with the signature $+3$ and with more general values of the constants, has been investigated in connection with the theory of continuous groups, without any reference to relativity theory. See, for example, *Bianchi 1918*, p. 565. 448

[7]Note that physicists frequently denote with $-R_{ik}$ what is here denoted with R_{ik}, with a corresponding change of sign in the field equations.

and taking account of the fact that $\partial/\partial x_i$, except for $i = 1$, vanishes for every magnitude of the solution, and that $g = (a^8/2)e^{2x_1}$, we obtain

$$R_{ik} = \frac{\partial}{\partial x_1}\Gamma^1_{ik} + \Gamma^1_{ik} - \Gamma^\sigma_{\rho i}\Gamma^\rho_{\sigma k}.$$

This yields the values for the R_{ik}

$$R_{00} = 1, \quad R_{22} = e^{2x_1}, \quad R_{02} = R_{20} = e^{x_1};$$

all other R_{ik} vanish. Hence,

$$R = 1/a^2.$$

The unit vector u in the direction of the x_0-lines has the contravariant components $1/a$, 0, 0, 0 and, therefore, the covariant components a, 0, ae^{x_1}, 0.

Hence we obtain:

$$R_{ik} = 1/a^2 \cdot u_i u_k.$$

Since, furthermore, R is a constant, the relativistic field equations (with the x_0-lines as world lines of matter), i.e., the equations[8]

$$R_{ik} - \frac{1}{2}g_{ik}R = 8\pi\kappa\rho u_i u_k + \lambda g_{ik}$$

are satisfied (for a given value of ρ) if we put

$$1/a^2 = 8\pi\kappa\rho, \quad \lambda = -R/2 = -1/2a^2 = -4\pi\kappa\rho.$$

The sign of the cosmological constant here is the opposite of that occurring in Einstein's static solution. It corresponds to a positive pressure.

3. Proofs for the properties enumerated

That there exists no one-parametric system of three-spaces orthogonal on the x_0-lines follows immediately from the necessary and sufficient condition which a vector field v in a four-space must satisfy, if there is to exist

[8]The linear element is supposed to give time-like distances in seconds and space-like distances in light seconds. Therefore, the coefficient of $u_i u_k$ differs from the usual one by a factor c^2.

a system of three-spaces everywhere orthogonal on the vectors of the field. This condition requires that the skew symmetric tensor

$$a_{ikl} = v_i \left(\frac{\partial v_k}{\partial x_l} - \frac{\partial v_l}{\partial x_k} \right) + v_k \left(\frac{\partial v_l}{\partial x_i} - \frac{\partial v_i}{\partial x_l} \right) + v_l \left(\frac{\partial v_i}{\partial x_k} - \frac{\partial v_k}{\partial x_i} \right)$$

should vanish identically. The components of the corresponding vector

$$w^j = \frac{\epsilon^{jikl}}{6 \cdot \sqrt{g}} a_{ikl},$$

however, in our case (i.e., for $v_i = u_i$) have the values 0, 0, 0, $\sqrt{2}/a^2$. The non-vanishing of w^3 shows, moreover, that there exist no surfaces orthogonal on the x_0-lines in the subspaces $x_3 =$ constant.

If v is the unit vector representing the velocity of matter, the vector w (which evidently is always orthogonal to v) is twice the angular velocity of matter in a local inertial system in whose origin matter is at rest at the moment considered.[9] Hence, property (9) follows at once.

The properties (1) and (2) follow from the directly verifiable fact that the space S admits the following four systems of transformations into itself,

(I)	$x_0 = x_0{}' + b$	(II)	$x_2 = x_2{}' + b$
	$x_i = x_i{}'$ for $i \neq 0$		$x_i = x_i{}'$ for $i \neq 2$
(III)	$x_3 = x_3{}' + b$	(IV)	$x_1 = x_1{}' + b$
	$x_i = x_i{}'$ for $i \neq 3$		$x_2 = x_2{}' e^{-b}$
			$x_0 = x_0{}'$
			$x_3 = x_3{}'$,

where b is an arbitrary real number.

A division of the time-like and null vectors into +- and −-vectors as required by (4) can be effected by defining ξ to be a +- or a −-vector according as to whether the inner product $(\xi u) = g_{ik}\xi^i u^k$ is $>$ or < 0.

In order to prove (3) we introduce new coordinates r, ϕ, t, y (where r, ϕ, t are cylindrical coordinates in the subspaces $x_3 =$ constant, and y, up to a constant factor, is $= x_3$) by the following formulas of transformation, 449 | which are easily solvable with respect to the x_i,

$$e^{x_1} = \text{ch}2r + \cos\phi\,\text{sh}2r$$

$$x_2 e^{x_1} = \sqrt{2}\sin\phi\,\text{sh}2r$$

[9]This is an immediate consequence of the definition of a local inertial system, which requires that $g_{ik} = \pm\delta_k{}^i$ and $\partial g_{ik}/\partial x_l = 0$ for every i, k, l.

$$\operatorname{tg}\left(\frac{\phi}{2}+\frac{x_0-2t}{2\sqrt{2}}\right)=e^{-2r}\operatorname{tg}\frac{\phi}{2}, \text{ where } \left|\frac{x_0-2t}{2\sqrt{2}}\right|<\frac{\pi}{2}$$

$$x_3 = 2y.$$

This leads[10] to the expression for the linear element,

$$4a^2(dt^2 - dr^2 - dy^2 + (\operatorname{sh}^4 r - \operatorname{sh}^2 r)d\phi^2 + 2\sqrt{2}\operatorname{sh}^2 r d\phi dt),$$

which directly exhibits the rotational symmetry, since the g_{ik} do not depend on ϕ.

Property (6) now follows easily: if c is defined by $\operatorname{sh} c = 1$ (i.e., $c = \log(1+\sqrt{2}))$, then for any $R > c$ we have $\operatorname{sh}^4 R - \operatorname{sh}^2 R > 0$; hence, the circle defined by $r = R$, $t = y = 0$ is everywhere time-like (the positive direction of time, by the above definition, being that of increasing ϕ). Hence, the line defined by

$$r = R, \quad y = 0, \quad t = -\alpha\phi \qquad (0 \leq \phi \leq 2\pi)$$

for sufficiently small α also will be everywhere time-like. However, the initial point Q of this line (i.e., the point corresponding to $\phi = 0$) and the end point P (i.e., that corresponding to $\phi = 2\pi$) are situated on the t-line: $r = R$, $y = \phi = 0$, and P precedes Q on this line if $\alpha > 0$. Repeating this procedure, any point preceding Q on its t-line can be reached, and because of the homogeneity of the solution the same can be done for every point.

Property (7), in view of (2) and (4), is an immediate consequence of (6). For, a three-space satisfying the two conditions stated in (7) in conjunction with time measured along the world lines of matter in their positive direction would yield a coordinate system with the property that the 0th coordinate always increases if one moves in a positive time-like direction, in contradiction to (6), which implies that all coordinates of the initial and the end point of a time-like line are equal in certain cases.

Property (5), for an open time coordinate, is an immediate consequence of the existence of closed time-like lines; for a closed time coordinate it follows from the fact that the subspaces t = constant would contradict property (7) (as can easily be shown owing to the simple connectivity of S).

In order to prove property (8), let U be an element of Σ; then U intersects the subspace S_0 of S defined by $x_3 = 0$ in a surface V (for it has one point

[10]This computation is rather cumbersome. It is simpler to derive both forms of the linear element independently from each other from the geometrical meaning of S given below. The first form is obtained by taking for the x_1x_2-space of the coordinate system the point set corresponding to any two-parametric subgroup of the multiplicative group of the hyperbolic quaternions as defined in footnote 14.

in common with each x_0-line situated on S_0). Now, according to what was proved, V cannot be orthogonal on all x_0-lines in S_0. So let l be an x_0-line in S_0 on which it is not orthogonal, and P the point of intersection of V and l. Then by rotating S_0 around l (and every S_b defined by $x_0 = b$ by the same angle around the x_0-line obtained from l by the translation $x'_3 = x_3 + b$), U goes over into a three-space different from U, but passing through P, hence not contained in Σ, since the elements of Σ were assumed to be mutually exclusive. Hence Σ goes over into a system different from Σ.

4. Some additional theorems and considerations about the solution

I am mentioning without proof that, disregarding the connectivity in the large (which can be changed by identifying the points of certain point sets with each other), the solution given and Einstein's static universe are the only spatially homogeneous cosmological solutions with non-vanishing density of matter and equidistant world lines of matter.[11]

The space S has a simple geometric meaning. It is the direct product of a straight line and the three-space S_0, defined by $x_3 = 0$; and S_0 is obtained from a space R of constant positive curvature and signature $+\ -\ -$ by stretching the metric[12] in the ratio $\sqrt{2}:1$ in the direction of a system of time-like Clifford parallels.[13]

This definition of S_0 also leads to an elegant representation of its group of transformations. To this end we map the points of R on the hyperbolic quaternions $u_0 + u_1 j_1 + u_2 j_2 + u_3 j_3$ of positive absolute value[14] by means

[11]There exist stationary homogeneous solutions in which the world lines of matter are not equidistant. They lead, however, into difficulties in consequence of the inner friction which would arise in the "gas" whose molecules are the galaxies, unless the irregular motion of the galaxies is zero and stays so.

[12]By "stretching the metric in the ratio μ in the direction of the lines of a system π" I mean that a new distance PQ' of neighboring points is introduced by the equation $(PQ')^2 = PR^2 + (\mu \cdot RQ)^2$, where R is the foot of the perpendicular drawn from P on the line of π passing through Q; or in other terms: $(ds')^2 = ds^2 + (\mu^2 - 1)(v_i dx_i)^2$, where v is the field of the tangent vectors of unit length of the lines of π.

[13]That is, a system of pairwise equidistant straight lines which for each point of space contains exactly one line passing through it.

[14]Here the u_i are real numbers and the units j_n are defined by $j_1 = i_1$, $j_2 = i \cdot i_2$, $j_3 = i \cdot i_3$, where the i_n are the units of the ordinary quaternions and i the imaginary unit, which is assumed to commute with all i_n. The term "hyperbolic quaternions" occurs in the literature in a different sense, but the number system just defined evidently is what should be so called. For: $\mathrm{norm}(u) = u \cdot \overline{u} = {u_0}^2 + {u_1}^2 - {u_2}^2 - {u_3}^2$, and moreover, the multiplicative group of these quaternions, if quaternions differing by a real factor are identified, is isomorphic with the group of transformations of the Lobatchefskian plane into itself. That the metric of R remains invariant under the transformations given in the text follows immediately from the equation $\mathrm{norm}(uv) = \mathrm{norm}(u) \cdot \mathrm{norm}(v)$.

of projective coordinates[15] $u_0 u_1 u_2 u_3$ so chosen that Klein's fundamental quadric takes on the form $| {u_0}^2 + {u_1}^2 - {u_2}^2 - {u_3}^2$. Then, any motion $u \to u'$ of R into itself can be represented in the form $u' = p \cdot u \cdot q$, where p and q are hyperbolic quaternions of positive norm. A system π of Clifford parallels can be represented by $\sigma^\alpha \cdot u$, where σ is a hyperbolic quaternion depending on π alone, and the individual lines of π are obtained by assigning a fixed value to u and varying α from $-\infty$ to $+\infty$. It follows that those motions of R into itself which leave π (and the orientation of its lines) invariant are represented by $u' = \sigma^\beta \cdot u \cdot q$, where β varies over all real numbers and q over all hyperbolic quaternions of positive norm. These motions, however, evidently form the four-parametric continuous group of transformations which carry S_0 into itself. The lines of π, of course, are the world lines of matter.

450

Evidently, in whatever ratio μ the metric of R is stretched in the direction of the lines of π, the resulting space R' has rotational symmetry. Therefore, the contracted Riemann tensor of $R' \times l$ (l being a straight line), if the coordinate system in the point considered is orthonormal, and its first basis vector $e^{(0)}$ has the direction of the π-lines, its last one $e^{(3)}$ the direction of l, has the form

$$\begin{pmatrix} a & & & \\ & b & 0 & \\ & 0 & b & \\ & & & 0 \end{pmatrix}$$

where a and b are functions of μ. Computation shows that $\mu = \sqrt{2}$ yields $b = 0$, i.e., $R_{ik} = a \cdot {e_i}^{(0)} {e_k}^{(0)}$, which makes it possible to satisfy the field equations in the manner described above.

As to the physical meaning of the solution proposed in this paper, it is clear that it yields no red shift for distant objects. For, by using the transformation (I) defined in the proof of the properties (1) and (2), one proves immediately that light signals sent from one particle of matter (occurring in the solution) to another one arrive with the same time intervals in which they are sent. For the period of rotation one obtains $2 \cdot 10^{11}$ years, if for ρ the value of 10^{-30}g/cm^3 is substituted. Assuming galactical systems were formed by condensation of matter originally distributed uniformly, and taking for the ratio of contraction $1:200$ (which is suggested by the

[15]It is to be noted, however, that there exist different topological forms of spaces of constant positive curvature and signature -1, and that that form which can be represented in projective coordinates in a one-to-one manner does not lead exactly to the space S defined before, but rather to a space obtained from S by identifying any two points which are situated on the same line of the system π and whose distance on that line is equal to a certain constant. A corresponding difference subsists for the groups of transformations.

observed average ratio of 1:200 between diameter and distance of galaxies), one obtains (using the law of conservation of angular momentum) for the average period of rotation of galactic systems $5 \cdot 10^6$ years. This number is of the correct order of magnitude, but, in view of the fact that this would have to be approximately the period of rotation in the outer parts of the nebulae, the observed value is found to be considerably larger.[16] Of course such comparison with observation has very little significance before an expansion has been combined with the rotation. Moreover, an explanation would have to be found for the apparent irregularity of the distribution of the axes of rotation of the galaxies. This, however, is perhaps not impossible, since there exist various circumstances which would tend to blur the original order, or make it appear blurred, especially if the axes of rotation of matter in different places (unlike in the solution described above) were not parallel with each other. The radius of the smallest time-like circles, in the solution given in this paper, is of the same order of magnitude as the world radius in Einstein's static universe.

[16]From the numerical data which E. Hubble (*1934*, p. 74) gives about two galaxies of medium size follow periods of rotation of $2 \cdot 10^7$ and $7 \cdot 10^7$ years at a distance of about half the radius from the center. The period of rotation of the Andromeda Nebula in the central region is estimated at $1.5 \cdot 10^7$ years.

Introductory note to *1949a*

This paper, written for a collection intended to honor and to discuss the work of Einstein, appears to be the only published piece by Gödel that deals with philosophical issues not directly concerned with mathematics. In it Gödel argues, on the basis of the very interesting cosmological solutions of Einstein's general-relativistic field equations obtained by him (*1949*), that those philosophers are right who have denied the "objectivity of change".[a]

A caution seems in order concerning the use in the title of this paper of the phrase "idealistic philosophy". The word "idealism" has been used historically in connection with a very diverse class of metaphysical views, whose common characteristic is the claim that what is ultimately "real" is something fundamentally "mental". By no means all such philosophies have denied the objectivity of change—for change may be attributed to minds or their contents. On the other hand, the contention that change is not objective, but is in some sense a "mere appearance", need not be associated with the view that all that is real is mental; and, indeed, it is far from plain in Gödel's paper that the latter is his own view, since he bases his argument on the physical possibility ("compatibility with the laws of nature") of worlds in which temporal relations have the bizarre characteristics he describes: thus his conclusion seems to be, not that the world of physics is grounded in something "mental", but that our conception of the world as *changing* is subjective or illusory—a contribution of our minds.

[a]One point in Gödel's discussion of his cosmological solutions perhaps deserves comment, although it does not substantially affect the argument of the paper. Gödel defends these solutions against a charge of absurdity by the consideration that the "time travel" that is physically possible in the "worlds" they describe would not be *practically* possible. Others have in fact rejected these solutions as "unphysical" because of the possibility of time travel. But it is hard to see the cogency of this rejection—or the need for the defense Gödel offers. Objections of the type "What if I were to go back and, for example, murder my own younger self?" admit a perfectly straightforward answer: in a cosmos of the sort in question, that act would simply not be possible. It would seem, in fact, that such a cosmos would have to be regarded as fully deterministic—or fully determinate; but Gödel's own argument against "the objectivity of change" leads in any case to determinateness as characteristic of things. And, after all, *classical* physics was generally conceived as deterministic. However obnoxious this notion has been to some philosophies, the objection "But I can always *choose* whether or not, for instance, to lift my arm" has never really carried any weight; and the objection raised in the context of time travel, although in some sense more poignant, is fundamentally of the same type.

This conclusion has a distinct relation to the position of Kant, to whom Gödel refers and who himself *repudiated* metaphysical idealism, but asserted what he called the "transcendental ideality" of time (as well as of space). The force of this assertion was that spatial and temporal attributes fundamentally characterize, not things "as they are in themselves", but a certain relation of those things to us—to our faculties of perception and representation. In particular, it is the special constitution of these latter faculties, according to Kant, that is responsible for the general structures of space and time that form the subject of geometry and of what may be called "pure chronometry"; in this sense, Kant characterizes these disciplines as concerned with the "pure form" of our "intuitive" (that is, our receptive or sensitive) faculty. As to the issue of metaphysical idealism, Kant rejects altogether any claims to knowledge of what things are apart from our experience (what they are "in themselves"); but *within* experience—that is, within the entire field of what can in any way be known—the structures of space and time by his doctrine are fully objective; as "forms" that condition the very possibility of perception, they constitute a universal framework for all objective scientific knowledge. Thus, affirming the "transcendental ideality" of space and time, Kant as emphatically asserts their "empirical reality". Furthermore, on his doctrine, these universal and empirically real structures can be *known* independently of experience, just because they are effects of our own constitution and are *conditions* of all possible experience.

On this latter point, it is evident that Gödel cannot adhere to Kant's view, since his own examples depart radically from the structure Kant thought necessary a priori for science, and since, far from claiming a grounding in something like Kant's "pure intuition", he emphasizes the "astonishing" and "strange" character of the results that form the basis of his argument, and their departure from "the intuitive idea" of an absolute and objective lapse of time. Further light is thrown on the question of Gödel's own conception of the relation of his view to Kant's by an as yet unpublished manuscript (found in his *Nachlass*, and bearing the title "Some observations about the relationship between theory of relativity and Kantian philosophy"), which discusses that relation in some detail and which makes explicit that a central difference from Kant concerns just this point: Kant, in Gödel's opinion, overemphasized in his epistemological discussion the dependence of spatiotemporal structure upon our faculty of representation, and was led by this into two errors—he concluded, erroneously, that the temporal properties of things (perhaps one should rather say, "of events") must be the same for all human beings (since human beings all have the same species of representational faculty); and he failed to see that geometry is at least in one sense an empirical science.

Thus, so far as Kantian philosophy is concerned, the principal analogy that Gödel has in mind between it and relativity theory concerns the strong sense in which temporal properties become (in general) well defined only relative to certain structures within the world: the world-lines of bodies. In the case of the bodies of sentient beings, these world-lines are also the loci of their immediate sensual contact with the reality outside themselves—so that in this special case the relation can be said to be "to the sensibility of the observer". This relativity of course affects the notion of "change" or "passage", centrally emphasized in the present paper. But to be relative is not to be illusory: in this paper, Gödel speaks of "an unequivocal proof for the view of those philosophers who ... consider change as an illusion due to our special mode of perception", whereas in the manuscript referred to (which does not explicitly mention "idealistic philosophy") he puts great stress upon the *objective* character of the relations in question.

In view of Gödel's well-known, long-standing, and deep interest in philosophical matters, it is cause for great regret that what we thus far possess of his reflections on such matters is so meager. The apparent discrepancy just noted between the present paper and the unpublished manuscript makes one wish both to know their comparative dates (if that can be determined) and to obtain further light—perhaps reconciling what appears discrepant—upon the metaphysical view here too briefly adumbrated. There is even greater need for clarification of the epistemological side of these views: we have so far in print only a brief comment, quoted (from a letter) in *Greenberg 1980* (page 250), suggesting that Gödel did believe that we have in some sense an a priori "physical intuition" of spatial structure "in the small"; and some enigmatic but intriguing remarks (*Gödel 1964*, pages 271–272, and *Wang 1974*, pages 84–85) about the relation to reality of human knowledge. It is very much to be hoped that the materials found in the Gödel *Nachlass* will help to illuminate our understanding of his philosophical position.

Howard Stein

A remark about the relationship between relativity theory and idealistic philosophy (*1949a*)

One of the most interesting aspects of relativity theory for the philosophical-minded consists in the fact that it gave new and surprising insights into the nature of time, of that mysterious and seemingly self-contradictory[1] being which, on the other hand, seems to form the basis of the world's and our own existence. The very starting point of special relativity theory consists in the discovery of a new and very astonishing property of time, namely the relativity of simultaneity, which to a large extent implies[2] that of succession. The assertion that the events A and B are simultaneous (and, for a large class of pairs of events, also the assertion that A happened before B) loses its objective meaning, in so far as another observer, with the same claim to correctness, can assert that A and B are not simultaneous (or that B happened before A).

Following up the consequences of this strange state of affairs, one is led to conclusions about the nature of time which are very far reaching indeed. In short, it seems that one obtains an unequivocal proof for the view of those philosophers who, like Parmenides, Kant, and the modern idealists, deny the objectivity of change and consider change as an illusion or an

558 appearance due to our special mode of perception.[3] The argu|ment runs as follows: Change becomes possible only through the lapse of time. The existence of an objective lapse of time,[4] however, means (or, at least, is equivalent to the fact) that reality consists of an infinity of layers of "now"

[1]Cf., e.g., *McTaggart 1908*.

[2]At least if it is required that any two point events are either simultaneous or one succeeds the other, i.e., that temporal succession defines a complete linear ordering of all point events. There exists an absolute partial ordering.

[3]Kant, in the *Critique of pure reason* (*1787*, p. 54), expresses this view in the following words: "Those affections which we represent to ourselves as changes, in beings with other forms of cognition, would give rise to a perception in which the idea of time, and therefore also of change, would not occur at all." This formulation agrees so well with the situation subsisting in relativity theory that one is almost tempted to add: such as, e.g., a perception of the inclination relative to each other of the world lines of matter in Minkowski space.

[4]One may take the standpoint that the idea of an objective lapse of time (whose essence is that only the present really exists) is meaningless. But this is no way out of the dilemma: for by this very opinion one would take the idealistic viewpoint as to

which come into existence successively. But, if simultaneity is something relative in the sense just explained, reality cannot be split up into such layers in an objectively determined way. Each observer has his own set of "nows", and none of these various systems of layers can claim the prerogative of representing the objective lapse of time.[5]

This inference has been pointed out by some, although by surprisingly few, philosophical writers, but it has not remained | unchallenged. And actually to the argument in the form just presented it can be objected that the complete equivalence of all observers moving with different (but uniform) velocities, which is the essential point in it, subsists only in the abstract space-time scheme of special relativity theory and in certain empty worlds of general relativity theory. The existence of matter, however, as well as the particular kind of curvature of space-time produced by it, largely destroys the equivalence of different observers[6] and distinguishes some of

559

the idea of change, exactly as those philosophers who consider it as self-contradictory. For in both views one denies that an objective lapse of time is a possible state of affairs, *a fortiori* that it exists in reality, and it makes very little difference in this context, whether our idea of it is regarded as meaningless or as self-contradictory. Of course, for those who take either one of these two viewpoints the argument from relativity theory given below is unnecessary, but even for them it should be of interest that perhaps there exists a second proof for the unreality of change based on entirely different grounds, especially in view of the fact that the assertion to be proved runs so completely counter to common sense. A particularly clear discussion of the subject independent of relativity theory is to be found in *Mongré 1898*.

[5]It may be objected that this argument only shows that the lapse of time is something relative, which does not exclude that it is something objective, whereas idealists maintain that it is something merely imagined. A relative lapse of time, however, if any meaning at all can be given to this phrase, would certainly be something entirely different from the lapse of time in the ordinary sense, which means a change in the existing. The concept of existence, however, cannot be relativized without destroying its meaning completely. It may furthermore be objected that the argument under consideration only shows that time lapses in different ways for different observers, whereas the lapse of time itself may nevertheless be an intrinsic (absolute) property of time or of reality. A lapse of time, however, which is not a lapse in some definite way seems to me as absurd as a colored object which has no definite colors. But, even if such a thing were conceivable, it would again be something totally different from the intuitive idea of the lapse of time to which the idealistic assertion refers.

[6]Of course, according to relativity theory all observers are equivalent in so far as the laws of motion and interaction for matter and field are the same for all of them. But this does not exclude that the structure of the world (i.e., the actual arrangement of matter, motion, and field) may offer quite different aspects to different observers, and that it may offer a more "natural" aspect to some of them and a distorted one to others. The observer, incidentally, plays no essential rôle in these considerations. The main point, of course, is that the [four-dimensional] world itself has certain distinguished directions, which directly define certain distinguished local times.

them conspicuously from the rest. namely. those which follow in their motion the mean motion of matter.[7] Now in all cosmological solutions of the gravitational equations (i.e., in all possible universes) known at present the local times of all *these* observers fit together into one world time. so that apparently it becomes possible to consider this time as the "true" one. which lapses objectively. whereas the discrepancies of the measuring results of other observers from this time may be conceived as due to the influence which a motion relative to the mean state of motion of matter has on the measuring processes and physical processes in general.

From this state of affairs. in view of the fact that some of the known cosmological solutions seem to represent our world correctly. James Jeans has concluded[8] that there is no reason to abandon the intuitive idea of an absolute time lapsing objectively. I do not think that the situation justifies 560 this conclu|sion and am basing my opinion chiefly[9] on the following facts and considerations:

There exist cosmological solutions of another kind[10] than those known at present, to which the aforementioned procedure of defining an absolute time is not applicable. because the local times of the special observers used above cannot be fitted together into one world time. Nor can any other procedure which would accomplish this purpose exist for them: i.e.. these worlds possess such properties of symmetry that for each possible concept of simultaneity and succession there exist others which cannot be distinguished from it by any intrinsic properties. but only by reference to individual objects. such as. e.g.. a particular galactic system.

[7]The value of the mean motion of matter may depend essentially on the size of the regions over which the mean is taken. What may be called the "true mean motion" is obtained by taking regions so large that a further increase in their size does not any longer change essentially the value obtained. In our world this is the case for regions including many galactic systems. Of course a true mean motion in this sense need not necessarily exist.

[8]Cf. *Jeans 1936*. pp. 22–23.

[9]Another circumstance invalidating Jeans' argument is that the procedure described above gives only an approximate definition of an absolute time. No doubt it is possible to refine the procedure so as to obtain a precise definition. but perhaps only by introducing more or less arbitrary elements (such as. e.g.. the size of the regions or the weight function to be used in the computation of the mean motion of matter). It is doubtful whether there exists a precise definition which has so great merits that there would be sufficient reason to consider exactly the time thus obtained as the true one.

[10]The most conspicuous physical property distinguishing these solutions from those known at present is that the compass of inertia in them everywhere rotates [in the same direction] relative to matter. which in our world would mean that it rotates relative to the totality of galactic systems. These worlds. therefore. can fittingly be called "rotating universes". In the subsequent considerations I have in mind a particular kind of rotating universes which have the additional properties of being static and spatially homogeneous. and a cosmological constant < 0. For the mathematical representation of these solutions. cf. my forthcoming *1949* [and. for a general discussion of rotating universes. my *1952*].

Consequently, the inference drawn above as to the non-objectivity of change doubtless applies at least in these worlds. Moreover it turns out that temporal conditions in these universes (at least in those referred to in the end of footnote 10) show other surprising features, strengthening further the idealistic viewpoint. Namely, by making a round trip on a rocket ship in a sufficiently wide curve, it is possible in these worlds to travel into any region of the past, present, and future, and back again, exactly as it is possible in other worlds to travel to distant parts of space.

This state of affairs seems to imply an absurdity. For it enables one, e.g., to travel into the near past of those places where | he has himself lived. 561
There he would find a person who would be himself at some earlier period of his life. Now he could do something to this person which, by his memory, he knows has not happened to him. This and similar contradictions, however, in order to prove the impossibility of the worlds under consideration, presuppose the actual feasibility of the journey into one's own past. But the velocities which would be necessary in order to complete the voyage in a reasonable length of time[11] are far beyond everything that can be expected ever to become a practical possibility. Therefore it cannot be excluded a priori, on the ground of the argument given, that the space-time structure of the real world is of the type described.

As to the conclusions which could be drawn from the state of affairs explained for the question being considered in this paper, the decisive point is this: that for *every* possible definition of a world time one could travel into regions of the universe which are past according to that definition.[12] This again shows that to assume an objective lapse of time would lose every justification in these worlds. For, in whatever way one may assume time to be lapsing, there will always exist possible observers to whose experienced lapse of time no objective lapse corresponds (in particular also possible

[11]Basing the calculation on a mean density of matter equal to that observed in our world, and assuming one were able to transform matter completely into energy, the weight of the "fuel" of the rocket ship, in order to complete the voyage in t years (as measured by the traveller), would have to be of the order of magnitude of $10^{22}/t^2$ times the weight of the ship (if stopping, too, is effected by recoil). This estimate applies to $t \ll 10^{11}$. Irrespective of the value of t, the velocity of the ship must be at least $1/\sqrt{2}$ of the velocity of light.

[*Translation of the author's addition to the German edition* (*1955*): A second reason for excluding *a priori* the universes mentioned above could be found in the possibility of "telegraphing a message into one's own past". But the practical difficulties in doing so would hardly seem to be trifling. Moreover, the boundary between difficulties in practice and difficulties in principle is not at all fixed. What was earlier a practical difficulty in atomic physics has today become an impossibility in principle, in consequence of the uncertainty principle; and the same could one day happen also for those difficulties that reside not in the domain of the "too small", but of the "too large".]

[12]For this purpose incomparably smaller velocities would be sufficient. Under the assumptions made in footnote 11 the weight of the fuel would have to be at most of the same order of magnitude as the weight of the ship.

observers whose whole existence objectively would be simultaneous). But, if the experience of the lapse of time can exist without an objective lapse of time, no reason can be given why an objective lapse of time should be assumed at all.

It might, however, be asked: Of what use is it if such conditions prevail in certain *possible* worlds? Does that mean anything for the question interesting us whether in *our* world there | exists an objective lapse of time? I think it does. For: (1) Our world, it is true, can hardly be represented by the particular kind of rotating solutions referred to above (because these solutions are static and, therefore, yield no red-shift for distant objects); there exist however also *expanding* rotating solutions. In such universes an absolute time also might fail to exist,[13] and it is not impossible that our world is a universe of this kind. (2) The mere compatibility with the laws of nature[14] of worlds in which there is no distinguished absolute time, and ⟦in which⟧, therefore, no objective lapse of time can exist, throws some light on the meaning of time also in those worlds in which an absolute time *can* be defined. For, if someone asserts that this absolute time is lapsing,

562

[13]At least if it required that successive experiences of one observer should never be simultaneous in the absolute time or (which is equivalent) that the absolute time should agree in direction with the times of all possible observers. Without this requirement an absolute time always exists in an expanding (and homogeneous) world. Whenever I speak of an "absolute" time, this of course is to be understood with the restriction explained in footnote 9, which also applies to other possible definitions of an absolute time.

[*Translation of the author's addition to the German edition* (*1955*): By an "absolute time" I understand a world time that can be defined without reference to particular objects and that satisfies the requirement formulated at the beginning of this footnote. More precisely, this should be called a "possible absolute time", since several can exist within *one* world, even though that is only exceptionally the case in spatially homogeneous universes.]

[14]The solution considered above only proves the compatibility with the general form of the field equations in which the value of the cosmological constant is left open; this value, however, which at present is not known with certainty, evidently forms part of the laws of nature. But other rotating solutions might make the result independent of the value of the cosmological constant (or rather of its vanishing or non-vanishing and of its sign, since its numerical value is of no consequence for this problem). At any rate these questions would first have to be answered in an unfavorable sense before one could think of drawing a conclusion like that of Jeans mentioned above. *Note added 2 September 1949*: I have found in the meantime that for *every* value of the cosmological constant there do exist solutions in which there is no world time satisfying the requirement of footnote 13.

[*Translation of the author's addition to the German edition* (*1955*): The second law of thermodynamics would also seem to be compatible with the solutions above. For within them a positive direction can be defined for all time-like lines in a unique and continuous way. Furthermore, the probability of any material system returning *exactly* to a former state is vanishingly small; and, if that happens only approximately, it merely means that somewhere two examples of the same system (in general having different entropies) exist simultaneously side by side. Of course, the initial conditions in such worlds cannot be chosen entirely freely.]

he accepts as a consequence that whether or not an objective lapse of time exists (i.e., whether or not a time in the ordinary sense of the word exists) depends on the particular way in which matter and its motion are arranged in the world. This is not a straightforward contradiction; nevertheless, a philosophical view leading to such consequences can hardly be considered as satisfactory.

Rotating universes in general relativity theory (*1952*)

⟦The introductory note to this paper and to *1949* is found on page 189, immediately preceding *1949*.⟧

In this lecture I am setting forth the main results (for the most part without proofs) to which my investigations on rotating universes have led me so far.

1. Definition of the type of rotatory solutions to be considered

I am starting from the relativistic field equations:[1]

$$R_{ik} - \frac{1}{2} g_{ik} R = T_{ik} - \lambda g_{ik} \tag{1}$$

and am assuming that:

1) the relative velocity of masses (i.e., galactic systems) close to each other is small compared with c;
2) no other forces except gravitation come into play.

Under these assumptions T_{ik} takes on the form:

$$T_{ik} = \rho v_i v_k \tag{2}$$

where:

$$\rho > 0, \tag{3}$$

$$g^{ik} v_i v_k = -1, \tag{4}$$

and, of course:

$$\text{The signature of } g_{ik} \text{ is } +2. \tag{5}$$

The local angular velocity of matter relative to the compass of inertia

[1] I am supposing that such measuring units are introduced as make $c = 1$, $8\pi k/c^2 = 1$.

can be represented by the following vector ω (which is always orthogonal on v):

$$\omega^i = \frac{\epsilon^{iklm}}{12(-g)^{1/2}} a_{klm}, \tag{6}$$

where the skew-symmetric tensor a_{klm} is defined by:

$$a_{klm} = v_k \left(\frac{\partial v_l}{\partial x_m} - \frac{\partial v_m}{\partial x_l} \right) + v_l \left(\frac{\partial v_m}{\partial x_k} - \frac{\partial v_k}{\partial x_m} \right) + v_m \left(\frac{\partial v_k}{\partial x_l} - \frac{\partial v_l}{\partial x_k} \right). \tag{7}$$

That ω represents the angular velocity relative to the compass of inertia is seen as follows: In a coordinate system which, in its origin, is geodesic and normal, and in whose origin matter is at rest (i.e., for which in O: $\partial g_{ik}/\partial x_l = 0$, $g_{ik} = \eta_{ik}$, $v^4 = 1$, $v^i = 0$ for $i \neq 4$),[2] one obtains for ω^i in O:

$$\omega^1 = \frac{1}{2} \left(\frac{\partial v^3}{\partial x_2} - \frac{\partial v^2}{\partial x_3} \right) = \frac{1}{2} \left(\frac{\partial}{\partial x_2} \left(\frac{v^3}{v^4} \right) - \frac{\partial}{\partial x_3} \left(\frac{v^2}{v^4} \right) \right), \quad etc. \tag{8}$$

$$\omega^4 = 0. \tag{9}$$

| In such a coordinate system, however, since parallel displacement (in its origin) means constancy of the components, the angular velocity relative to the compass of inertia, in O, is given by the same expressions as in Newtonian physics, i.e. the right-hand sides of (8) are its components. *Evidently ω is the only vector the first 3 components of which, in the particular coordinate systems defined, coincide with the angular velocity computed as in Newtonian physics and the 4th component is* 0. 176

Any Riemann 4-space with some ρ, v_i defined in it, which everywhere satisfies the conditions (1)–(5) and permits of no extension free from singularities, and for which, moreover, ω is continuous and $\neq 0$ in every point, represents a rotating universe. However, in the sequel I am chiefly concerned with solutions satisfying the following three further postulates (suggested both by observation and theory):

I. The solution is to be homogeneous in space (i.e., for any two world lines of matter l, m there is to exist a transformation of the solution into itself which carries l into m).

II. Space is to be finite (i.e., the topological space whose points are the world lines of matter is to be closed, i.e., compact).

III. ρ is not to be a constant.

[2] A coordinate system satisfying the first two conditions may fittingly be called a "local inertial system".

Postulate III is indispensable also for rotating universes, since it can be proved that a *red-shift which, for small distances, increases linearly with the distance implies an expansion, no matter whether the universe rotates or not.*[3]

As to the question of the existence of rotating solutions satisfying the postulates I, II, III, cf. §5.

2. Some general properties of these solutions

In view of III the equation $\rho =$ constant defines a one-parameter system of 3-spaces. *In rotating universes these 3-spaces of constant density cannot be orthogonal on the world lines of matter.* This follows immediately from the fact that $a_{klm} = 0$ is the necessary and sufficient condition for the existence of *any* system of 3-spaces orthogonal on a vector field v.

The inclination of the world lines of matter toward the spaces of constant density yields *a directly observable necessary and sufficient criterion for the rotation of an expanding spatially homogeneous and finite universe:* namely, *for sufficiently great distances, there must be more galaxies in one half of the sky than in the other half.*

In the first approximation, i.e., for solutions differing little from one spatially isotropic, the magnitude of this effect is given by the following theorem: *If N_1, N_2 are the numbers of galaxies in the two hemispheres into which a spatial sphere*[4] *of radius r (small compared with the world*

177 *radius R) is decomposed by a | plane orthogonal on ω, then:*

$$\frac{|N_1 - N_2|}{N_1 + N_2} = \frac{9}{8} \cdot \frac{|\omega| rRh}{c^2}, \tag{10}$$

where h is Hubble's constant $(=\dot{R}/R)$.

For plausible values of the constants (where ω is estimated from the velocity of rotation of the galaxies[5]) this effect is extremely small. But the uncertainty in the knowledge of the constants is too great for drawing any definitive conclusions.

The group of transformations existing owing to I evidently carries each of the spaces $\rho =$ constant into itself, and therefore (the case of isotropy being

[3]Provided, of course, that the atomic constants do not vary in time and space, or, to be more exact, provided that the dimensionless numbers definable in terms of the constants of nature (such as e^2/hc) are the same everywhere.

[4]I.e., one situated in a 3-space orthogonal on v at the point under consideration.

[5]Cf. my *1949*, p. 450.

excluded) can only have 3 or 4 parameters.[6] The number 4 (i.e., the case of rotational symmetry) cannot occur either. *There exist no rotationally symmetric rotating universes satisfying the conditions stated in* §1.[7] The only symmetry around one point which can occur is that of one rotation by π. This case will be referred to as the symmetric one.

In any case the group of transformations must be 3-parameter. Since moreover, owing to II, it must be compact, and since (as can easily be shown) it cannot be commutative in rotating universes,[8] it follows that *the group of transformations of any rotating solution of the type characterized in* §1 *must be isomorphic* (*as a group of transformations*) *with the right* (*or the left*) *translations of a* 3-*space of constant positive curvature, or with these translations plus certain rotations by an angle* π. Hence also the topological connectivity of space must be that of a spherical or elliptical 3-space.

The metric g_{ik} can be decomposed (relative to the world lines of matter) into a space-metric $\overline{g_{ik}}$ and a time-metric $\overline{\overline{g_{ik}}}$, by defining the spatial distance of two neighboring points P_1, P_2 to be the orthogonal distance of the two world lines of matter passing through P_1, P_2, and the temporal distance to be the orthogonal projection of P_1P_2 on one of these two lines. This decomposition evidently is exactly that which (in the small) holds for the observers moving along the world lines of matter. It has the following properties:

$$\overline{\overline{g_{ik}}} = -v_i v_k, \quad \overline{g_{ik}} = g_{ik} + v_i v_k, \qquad (11)$$

$$\text{Det}\,(\overline{g_{ik}}) = \text{Det}\,(\overline{\overline{g_{ik}}}) = 0.$$

If the coordinate system is so chosen that the x_4-lines are the world lines of matter and the x_4-coordinate measures the length of these lines, $\overline{g_{ik}}$ takes on the form:

$$\overline{g_{ik}} = \left\| \begin{matrix} h_{ik} & 0 \\ 0 & 0 \end{matrix} \right\| \qquad (12)$$

| (where h_{ik} is positive definite) and the Hubble-constant in the space-direction dx^i (orthogonal on v), as measured by an observer moving along with matter, becomes equal to: 178

$$\frac{1}{2}\frac{\dot{h}_{ik}dx^i dx^k}{h_{ik}dx^i dx^k}, \quad \text{where } \dot{h}_{ik} = \frac{\partial h_{ik}}{\partial x_4}.$$

[6] There exists, in every space $\rho =$ constant, a positive definite metric which is carried into itself, namely the metric h_{ik} defined below.

[7] This even is true irrespective of postulate II (the finiteness of space).

[8] The reason is that the curl of a vector field invariant under a transitive commutative group vanishes identically.

The surface $\dot{h}_{ik}x_i x_k = 1$ in the 3-dimensional subspace, orthogonal on v, of the tangent space, may be called the ellipsoid of expansion or, more generally, the quadric of expansion.

The theorem about the nonexistence of rotationally symmetric solutions,[9] *under the additional hypothesis that the universe contains no closed time-like lines* (cf. §3), can be strengthened to the statement that *the quadric of expansion, at no moment of time, can be rotationally symmetric around* ω. In particular it can never be a sphere, i.e., the expansion is necessarily coupled with a deformation. This even is true for *all* solutions satisfying I–III and gives another directly observable property of the rotating universes of this type.

Moreover the asymmetry of the expansion around ω opens up a possibility for the explanation of the spiral structure of the galaxies. For, if under these circumstances a condensation is formed, the chances are that it will become an oblong body rotating around one of its smaller axes; and such a body, because its outer parts will rotate more slowly, will, in the course of time, be bent into a spiral. It remains to be seen whether a quantitative elaboration of this theory of the formation of spirals will lead to agreement with observation.

3. Rotation and time-metric

The formulae (6), (7), (11) show that it is, in the first place, the time-metric (relative to the observers moving along with matter) which determines the behavior of the compass of inertia. In fact *a necessary and sufficient condition for a spatially homogeneous universe to rotate is that the local simultaneity of the observers moving along with matter be not integrable* (i.e., do not define a simultaneity in the large). This property of the time-metric in rotating universes is closely connected with the possibility of closed time-like lines.

The latter anomaly, however, occurs only if the angular velocity surpasses a certain limit. This limit, roughly speaking, is that value of $|\omega|$ for which the maximum linear velocity caused by the rotation becomes equal to c; i.e., it is approximately c/R if, at the moment considered, the space-metric in the 3-space $\rho =$ constant does not differ too much from a space of the constant curvature $1/R^2$. *The precise necessary and sufficient condition for the nonexistence of closed time-like lines* (provided that the 179 one-parameter manifold of the spaces $\rho = |$ constant is not closed) *is that*

[9]This theorem makes it very likely that there exist no rotating spatially homogeneous and expanding solutions whatsoever in which the ellipsoid of expansion is *permanently* rotationally symmetric around ω.

the metric in the spaces of constant density be space-like.[10] This holds for solutions satisfying all conditions stated in §1.

For these solutions, also, *the nonexistence of closed time-like lines is equivalent with the existence of a "world-time"*, where by a world-time we mean an assignment of a real number t to every space-time point, which has the property that t always increases if one moves along a time-like line in its positive direction.[11] If in addition any two 3-spaces of simultaneity are equidistant and the difference of t is their distance, one may call it a *metric world-time.* If the spaces of constant density are space-like, a metric world-time can be defined by taking these 3-spaces as spaces of simultaneity. Evidently (up to transformations $\bar{t} = f(t)$) this is the only world-time invariant under the group of transformations of the solution.

4. Behavior of the angular velocity in the course of the expansion

No matter whether postulates I–III are satisfied or not, the temporal change of ω is described by the following theorem: *In a coordinate system in which the x_4-lines are the world lines of matter, $g_{44} = -1$ everywhere, and moreover $g_{i4} = 0$ (for $i \neq 4$) on the X_4-axis, one has along the whole X_4-axis:*

$$\omega^i(-g)^{1/2} = \omega^i h^{1/2} = \text{constant} \qquad (i = 1, 2, 3). \tag{13}$$

The proof can be given in a few lines: Evidently $v^4 = 1$, $v^i = 0$ (for $i \neq 4$) everywhere; hence: $v_i = g_{i4}$. Substituting these values of v_i in (7), one obtains on X_4:

$$a_{4ik} = \frac{\partial g_{4k}}{\partial x_i} - \frac{\partial g_{4i}}{\partial x_k}, \qquad a_{123} = 0. \tag{14}$$

But $\partial g_{4i}/\partial x_4 = 0$ (because the x_4-lines are geodesics and $g_{44} = -1$). Hence by (14), $\partial a_{klm}/\partial x_4 = 0$ on X_4. Hence by (6) also, $\partial(\omega^i(-g)^{1/2})/\partial x_4 = 0$ on X_4.

The equation (13) means two things:

[10] This condition, too, means that at the border separating the two cases the linear velocity caused by the rotation becomes equal to c, if by this linear velocity is understood the velocity of matter relative to the orthogonals on the spaces of constant density.

[11] A time-like vector is positive if it is contained in the same half of the light-cone as the vector v.

A. that the vector ω (or, to be more exact, the lines l_ω whose tangent everywhere has the direction ω) permanently connects the same particles with each other;

B. that the absolute value $|\omega|$ increases or decreases in proportion to the contraction or expansion of matter orthogonal on ω, where this contraction or expansion is measured by the area of the intersection of an infinitesimal spatial cylinder[4] around l_ω (permanently including the same particles) with a surface orthogonal on l_ω.

Since in the proof of (13) nothing was used except the fact that the 180 world | lines of matter are geodesics (and in particular the homogeneity of space was not used), (13), and therefore A, B, also describe the behavior of the angular velocity, if condensations are formed under the influence of gravitation;[12] i.e., $|\omega|$, under these circumstances, increases by the same law as in Newtonian mechanics.

The direction of ω, even in a homogeneous universe, need not be displaced parallel to itself along the world lines of matter. *The necessary and sufficient condition for it to be displaced parallel at a certain moment is that it coincide with one of the principal axes of the quadric of expansion.* For, if P, Q are two neighboring particles connected by ω, then, only under the condition just formulated, the direction PQ at the given moment, will be at rest relative to the compass of inertia (in order to see this one only has to introduce the local inertial system defined in §1 (cf. footnote 2) and then argue exactly as in Newtonian physics). Since however (because of A) the direction of ω coincides *permanently* with the direction of PQ, the same condition applies for the direction of ω. This condition however, in general, is not satisfied (only in the symmetric case it is always satisfied).

The fact that the direction of ω need not be displaced parallel to itself might be the reason for the irregular distribution of the directions of the axes of rotation of the galaxies (which at first sight seems to contradict an explanation of the rotation of the galaxies from a rotation of the universe). For, if the axis of rotation of the universe is not displaced parallel, the direction of the angular momentum of a galaxy will depend on the moment of time at which it was formed.

5. Existence theorems

It can be shown that, *for any value of* λ (*including* 0), *there exist* ∞^8 *rotating solutions satisfying all conditions stated in* §1. *The same is true if in addition it is required that a world-time should exist* (*or should not*

[12]Of course, only as long as the gas and radiation pressure remain small enough to be neglected.

exist). The value of the angular velocity is quite arbitrary, even if ρ and the mean world radius (at the moment under consideration) are given. In particular, there exist rotating solutions with $\lambda = 0$ which differ arbitrarily little from the spatially isotropic solution with $\lambda = 0$.

Thus the problem arises of distinguishing, by properties of symmetry or simplicity, certain solutions in this vast manifold of solutions. E.g., one might try to require that the universe should expand from one point and contract to one point.

6. Method of proof

The method of proof by which the results given above were obtained is based on postulate I of §1. This postulate implies that all world lines of matter (and all orthogonals on the spaces of constant density) are equivalent with each other. It is, therefore, sufficient to confine the consideration to one | such world line (or one such orthogonal). This reduces the problem to a system of *ordinary* differential equations. 181

Moreover, this system of differential equations can be derived from a Hamiltonian principle, i.e., it is a problem of analytical mechanics with a finite number of degrees of freedom. The equations of relativity theory, however, assign definite values to the integrals of energy and momentum, so that the relativistic problem is a little more special than the corresponding one of analytical mechanics.

The symmetric case, by means of the integrals of momentum, can be reduced to a problem with three degrees of freedom (g_1, g_2, g_3), whose Lagrangian function reads as follows:

$$\left\{\sum_{i<k}\frac{\dot{g}_i\,\dot{g}_k}{g_i g_k} + \frac{1}{g}\left[2\sum_i g_i^2 - \left(\sum_i g_i\right)^2\right] + \frac{V^2}{g_1(g_2-g_3)^2}\right\} g^{1/2} + 2\left(1+\frac{V^2}{g_1}\right)^{1/2}, \tag{15}$$

where $g = g_1 g_2 g_3$ and V is a constant which determines the velocity of rotation. The general case can be reduced to a system of differential equations of the 8th order.

7. Stationary rotating solutions

It might be suspected that the desired particular solutions (cf. §5 above) will have a close relationship to the stationary homogeneous solutions, and it is therefore of interest to investigate these, too. By a stationary homogeneous solution we mean one whose group, for any two points P, Q of the whole 4-space, contains transformations carrying P into Q.

These solutions can all be determined and expressed by elementary functions. One thus obtains the following results:

1. *There exist no stationary homogeneous solutions with* $\lambda = 0$.

2. *There exist rotating stationary homogeneous solutions with finite space, no closed time-like lines, and* $\lambda > 0$; in particular also such as differ arbitrarily little from Einstein's static universe.

The world lines of matter in these solutions, however, are not equidistant: neighboring particles of matter, relative to the compass of inertia, rotate around each other, not in circles, but in ellipses (or, to be more exact, in rotating ellipses).

Introductory note to *1958* and *1972*

1. Preliminary remarks; history of the paper

Gödel's "*Dialectica* paper" appeared in 1958, in German, in honor of P. Bernays' 70th birthday; it is reproduced, following this introductory note, together with a translation into English by Stefan Bauer-Mengelberg and Jean van Heijenoort. The ideas in this paper date back at least as far as 1941, since Gödel lectured at that time on his interpretation at Princeton and Yale. (In Gödel's *Nachlass* there is the text of a lecture, "In what sense is intuitionistic logic constructive?", given at Yale University on 15 April 1941.)

In this volume there is also a second version of *Gödel 1958*, which has not previously appeared in print. This, too, has a long history; in the form reproduced here it dates from circa 1972. Between the first and the second version, Gödel's interest shifted from the mathematical result to its philosophical aspects; as he wrote in a letter to Bernays of 16 May 1968, "In those days, after all, I set no particular store by the philosophical aspect; rather, it was chiefly the mathematical result that was important to me, while now it is the other way around."[a] Nevertheless, *1958* is already presented by Gödel as a foundational contribution, not as a technical one.

Gödel never managed to express his ideas on the philosophical aspects of the interpretation to his own satisfaction, as is evidenced by the vicissitudes of the second version of the *Dialectica* paper (henceforth cited as *Gödel 1972*). In 1965 Bernays informed Gödel of the plan to publish an English translation (by Leo F. Boron) of *1958*, again in *Dialectica*. Gödel then undertook to revise his paper for that occasion; and in January 1967 he expressed satisfaction with the result.

In 1968, however, on rereading the philosophical introduction to the original paper, Gödel became dissatisfied with it and rewrote it completely. Equally dissatisfied with this revision, he gave up the idea of rewriting the introduction and decided to add a series of notes (**a**–**n**) instead (Gödel to Bernays, 16 May and 17 December 1968). In 1970 the revised version, after much active prompting and help from Bernays and D. S. Scott, was sent to the printer. When the proof sheets arrived, Gödel was not pleased with notes **i** and **h**, and rewrote them almost completely,

[a] "Ich legte ja damals keinen besonderen Wert auf das Philosophische, sondern es kam mir hauptsächlich auf das mathematische Resultat an, während es jetzt umgekehrt ist."

Richard Arens

Albert Einstein and Kurt Gödel at the Institute for Advanced Study, about 1954

but never returned the proof sheets. Apparently he continued to make minor revisions and corrections as late as 1972—at least there are some handwritten corrections on the proof sheets which concern points raised by Bernays in a letter to Gödel of 16 March 1972.

Gödel had been in poor health since 1968, and this may have had much to do with the hesitations and doubts that are manifested in the style of the additional notes in *1972*, which lack the sureness of touch of Gödel's earlier work (see *Kreisel 1980*). A less subjective reason for the difficulties lies in the material itself: it is quite difficult to formulate the philosophical gain achieved in *1958* and *1972*.

The presentation of the mathematical results also leaves much to be desired; the system T is given in outline only, in the main body of text; the description is then expanded and/or modified in the additional notes for *1972*. Similarly, the description of interpretations of T is not at all clear-cut. Accordingly, after discussing the aims of Gödel's paper in Section 2 below, we devote a (long) Section 3 to the technical background of Gödel's system T and his main result, while Section 4 is devoted to interpretations of T; Section 5 describes further work that has been prompted by *Gödel 1958*.

Our commentary on these papers concentrates on *1972*; references to *1958* are explicitly noted in the text. We use *Troelstra 1973* as a source of technical background information.

References to Gödel's notes: We list Gödel's notes 1–11 and **a**–**n** with an indication of where they are referred to in our introduction: 3 (2.2), 5 (4.1), 6 (2.1, 2.2, 4.1), 7 (3.1, 3.2), 8 (3.2), 10 (3.1); **b** (2.1, 2.2), **c** (2.2, 3.3), **d** (5.1), **e** (2.1), **g** (4.2), **h** (2.1, 3.3, 4.1–3), **i** (3.1–3), **j** (3.1), **l** (3.2), **m** (3.1), **n** (3.2, 3.3, 4.3).

2. Aims of Gödel's paper

2.1 Gödel originally devised his interpretation for technical applications; specifically, the underivability of $\neg\neg\forall x(A(x) \vee \neg A(x))$ in intuitionistic predicate logic (*Kleene 1973*, note 7). In correspondence, G. Kreisel wrote that Gödel wanted to establish that intuitionistic proofs of existential theorems provide explicit realizations.

However, in *1958* Gödel presents his results as a contribution to a liberalized version of Hilbert's program: to justify classical systems, in particular arithmetic, in terms of notions as intuitively clear as possible. Hilbert wanted to find these intuitively clear notions in the domain of 'finitary mathematics'.

In Hilbert's sense, this may be described as mathematics of a purely combinatorial nature, dealing with configurations of finite, discrete, concretely representable objects that can be surveyed (grasped) in all their parts. Elementary-school arithmetic may be regarded as typically finitary in Hilbert's sense: it deals with natural numbers and certain *specific* operations on them, such as addition and multiplication, which have purely combinatorial character. On the other hand, the *general* concept of a function from $\mathbb{N}$ to $\mathbb{N}$ is not finitary (see Gödel's note **b**). Even more abstract in character, and therefore further removed from finitary mathematics, is the use of abstract (intuitionistic) proofs in the explanation[b] of the intuitionistic logical operators (see the opening paragraph of *1958* or of *1972*, and notes **e** and **h**). In this explanation the meaning of the logical operators is given by describing proofs of logically compound statements in terms of the constituent statements. Two typical clauses are the following:

(1) p proves $A \vee B$ if p is either of the form $(0, p')$, where p' is a proof of A, or of the form $(1, p')$, where p' is a proof of B.

(2) p proves $A \rightarrow B$ if p is a construction (method) which, when applied to any proof q of A, yields a proof $p(q)$ of B. ('p proves $\neg A$' is a special case: p transforms any hypothetical proof of A into a proof of an absurdity.)

Thus $A \vee \neg A$ is not valid for this interpretation. Accepting these explanations for intuitionistic logic provides us with a justification and consistency proof for intuitionistic first-order arithmetic **HA** ('Heyting's arithmetic') comparable to the technically trivial consistency proof for classical first-order arithmetic **PA** ('Peano arithmetic') that simply consists in interpreting **PA** in the intended (standard) model and observing that the logical rules preserve truth. Combining the indicated justification for **HA** with Gödel's own 'negative translation' (*1933e*) that reduces[c] **PA** to **HA**, one obtains a justification of **PA** in terms of intuitionistic principles. As Gödel observes in note 6, even if we do not have a sufficiently clear idea of the notion of a constructive proof, we do not doubt that the laws of intuitionistic logic hold for it. (Conversely, *if* one is worried about the coherence of the explanations of intuitionistic logic in terms of abstract proofs, but accepts the classical notion of truth, then **HA** $\subset$ **PA** is an immediate justification of **HA**.)

2.2 Gödel argues that, since the finitistic methods considered are not sufficient to carry out Hilbert's program, one has to admit at least some abstract notions in a consistency proof; the necessity of this is

[b]Gödel calls this Heyting's explanation, though "Heyting–Kolmogorov" might be more appropriate. See our introductory notes to *1933e* and *1933f* in Volume I.

[c]The systems of classical and intuitionistic arithmetic considered in *1933e* are not identical with **PA** and **HA** of the recent literature, but that is not essential.

shown both by the second incompleteness theorem and by our experience with known consistency proofs, all of which appeal at some point to an abstract notion (see the second paragraph of *1958* or *1972*, and notes 3 and **c**; as to the use of the term 'finitary', see also **b**). In a letter to Bernays, dated 25 July 1969, Gödel said that the restriction to 'visualizable' objects in Hilbert's finitism was quite unnatural.

However, Gödel did not want to go as far as admitting Heyting's abstract notion of constructive proof; hence he tried to replace the notion of constructive proof by something more definite, less abstract (that is, more nearly finitistic), his principal candidate being a notion of 'computable functional of finite type' which is to be accepted as sufficiently well understood to justify the axioms and rules of his system T, an essentially logic-free theory of functionals of finite type (see note 6).

2.3 The method used by Gödel consists in associating with each A in the language of arithmetic a formula of the form $\exists \boldsymbol{x} \forall \boldsymbol{y} A_D(\boldsymbol{x}, \boldsymbol{y})$, where A_D is quantifier-free and $\boldsymbol{x}$, $\boldsymbol{y}$ are strings of variables for functionals of finite type. This association is such that, if **HA** proves A, then T proves $A_D(\boldsymbol{t}, \boldsymbol{y})$ for a suitable sequence of terms $\boldsymbol{t}$ (not containing $\boldsymbol{y}$). For a quantifier-free A, $A_D \equiv A$; thus consistency follows if we accept T, since, if T is correct, one cannot derive $1 = 0$.

We are not asked to think of the reinterpretation of the logical operators of A, involved in the transition to A_D, as particularly fundamental; the main point is that it permits a translation of A in logic-free terms (see our comments in 3.2 below).

2.4 In order to gain a better understanding of what exactly Gödel achieved, it helps to go into the technical background and content of Gödel's paper. We do so at length in the next two sections; in 4.4 we shall return to an assessment of Gödel's reduction.

3. Gödel's main result

This section is quite long and technically the most complicated one in this introduction. The best procedure for the reader is perhaps to alternate between Gödel's text and our explanations, starting with the present section.

3.1 The basic languages and systems

As noted earlier, Gödel's description of his system T is rather sketchy: it is supposed to be a quantifier-free theory of computable finite-type functionals, with a primitive notion of decidable (intensional) equality for all finite types (note 7) and in which functionals can be introduced by explicit definition and recursion. Additional specifications are contained in note **i**. To explain them in definite and precise terms, we begin by describing a version **T** of T that seems to correspond well to Gödel's intentions in *1958* and the main text of *1972*; from this we extract a subsystem $\mathbf{T}_0$ suggested by note **i** and sufficient for Gödel's principal result. The motivation for our choice is found in subsection 3.3.

We shall use $\wedge$, $\vee$, $\rightarrow$, $\neg$, $\forall$, $\exists$ for the logical operators (Gödel has $\supset$, ( ) for $\rightarrow$, $\forall$). Intuitionistic arithmetic **HA** (Gödel's 'H'; see notes 10, **j** and **i**5) is taken to be formulated with number variables (x, y, z, u, v, w), the Peano axioms for zero (0) and successor (S^+), the induction axiom schema (or the corresponding rule), and constants for all primitive recursive functions, with their defining equations as axioms.[d]

The type structure $\mathcal{T}$ (on which **T** is based) is generated from type 0 (natural numbers) by the rule: if σ, $\tau_1, \ldots, \tau_n$ are types, then so is $(\sigma, \tau_1, \ldots, \tau_n)$. Intuitively, $(\sigma, \tau_1, \ldots, \tau_n)$ consists of n-place functions which, applied to arguments of types $\tau_1, \ldots, \tau_n$, give a value of type σ. These are called *functionals* in general, since, for $\tau_i \neq 0$, the arguments are themselves functions.

For Gödel's type (σ, τ) we shall write $(\tau)\sigma$ or $\tau \rightarrow \sigma$. Then Schönfinkel's device of reducing n-place functions to unary functions by means of the isomorphism $X^{Y \times Z} \cong (X^Y)^Z$ permits us to think of the type $(\sigma, \tau_1, \ldots, \tau_n)$ as an abbreviation for $(\tau_1) \ldots (\tau_n)\sigma$. The restriction to unary functionals is customary in the literature and technically convenient. We also use 1 for the type (0)0 and 2 for (1)0.

The language $\mathcal{L}$ of **T** contains variables for each type σ of $\mathcal{T}$(x^σ, y^σ, z^σ, u^σ, v^σ, w^σ, possibly subscripted) and individual constants for certain types, to be specified below. The informal interpretation is that the variables of type σ, for $\sigma \neq 0$, range over the computable functionals of that type; this will be analyzed in more detail in Section 4 below.

For each type $\sigma \in \mathcal{T}$ there is a primitive binary predicate constant $=_\sigma$ in $\mathcal{L}$ for equality at type σ.

The letters t and s (possibly with sub- or superscripts) will be used for terms. 't is of type σ' can be written as $t \in \sigma$ or as t^σ; where there is no ambiguity, type superscripts are suppressed. If $t \in (\sigma)\tau$, $t' \in \sigma$, then $\mathrm{App}_{\sigma,\tau}(t, t')$ ("t applied to t'") is a term of type τ, for which we shall simply write $t(t')$ or even tt'. The notation $t_1 t_2 \ldots t_n$ is short for $t_1(t_2)(t_3) \ldots (t_n)$, which corresponds to Gödel's $t_1(t_2, \ldots, t_n)$.

[d] As observed by Gödel in note **m**, one can dispense with negation as a primitive, defining $\neg A$ as $A \rightarrow 1 = 0$. In fact, in **HA** one can also dispense with disjunction, defining it as $A \vee B := \exists x[(x = 0 \rightarrow A) \wedge (x \neq 0 \rightarrow B)]$. In note **i**5 Gödel observes that, because of the presence of the predecessor function, one of the Peano axioms is in fact redundant.

When suppressing type specifications for compound expressions, we shall always tacitly assume types to be 'fitting', that is, in a term tt', t must be of type $(\sigma)\tau$, t' of type σ, for certain σ, $\tau \in \mathcal{T}$.

Prime formulas of $\mathcal{L}$ are term equations: $t^\sigma =_\sigma s^\sigma$, or simply $t = s$ without type indications; compound formulas are constructed by means of the propositional operators $\wedge$, $\vee$, $\rightarrow$; $\neg A$ is an abbreviation for $A \rightarrow S^+0 = 0$.

The logical basis of **T** is the intuitionistic (many-sorted) propositional calculus with equality, where equality is assumed to be decidable for each type σ, that is, $t =_\sigma s \vee t \neq_\sigma s$. Equivalently, one might have taken classical propositional logic as the basis, since the decidability of the prime formulas entails the decidability of all compound formulas by intuitionistic logic. Gödel adds a rule of substitution: if we have derived $A(x^\sigma)$, we may infer $A(t^\sigma)$; but this rule is not needed if, instead, all schemas formulated with free variables are replaced by schemas formulated for arbitrary terms.

Before continuing our description of **T**, we introduce some further abbreviations, which will enable us to deal with finite strings of variables and terms. We shall use $\boldsymbol{u}$, $\boldsymbol{v}$, $\boldsymbol{w}$, $\boldsymbol{x}$, $\boldsymbol{y}$, $\boldsymbol{z}$, $\boldsymbol{U}, \ldots, \boldsymbol{Z}$ for finite sequences of variables, and $\boldsymbol{t}$, $\boldsymbol{s}$, $\boldsymbol{S}$ for finite sequences of terms.

Let $\boldsymbol{x} \equiv (x_1, \ldots, x_n)$, $\boldsymbol{y} \equiv (y_1, \ldots, y_m)$. Then $\forall \boldsymbol{x}$, $\exists \boldsymbol{x}$ abbreviate $\forall x_1 \ldots \forall x_n$, $\exists x_1 \ldots \exists x_n$ respectively; $\forall \boldsymbol{xy}$ stands for $\forall \boldsymbol{x} \forall \boldsymbol{y}$; etc. If $\boldsymbol{t} \equiv (t_1, \ldots, t_n)$ and $\boldsymbol{s} \equiv (s_1, \ldots, s_m)$, then $\boldsymbol{ts}$ in a formula $A(\boldsymbol{ts})$ stands for a finite sequence $(t_1 s_1 \ldots s_m, \ldots, t_n s_1 \ldots s_m)$; $\boldsymbol{t} = \boldsymbol{s}$ is a sequence of equations $t_1 = s_1, \ldots, t_n = s_n$ (n has to be equal to m).

We often write $t[x]$ or $t[x_1, \ldots, x_n]$ in order to refer to the (possibly empty) sets of free occurrences of the variables x, or $x_1, \ldots, x_n$, in t. Then $t[t_1, \ldots, t_n]$ denotes the result of simultaneously substituting $t_1, \ldots, t_n$ for $x_1, \ldots, x_n$, respectively.

The non-logical axioms and rules for **T** contain the (Peano) axioms for zero (0) and successor (S^+) and the induction rule

IND–R $\qquad A(0), A(x^0) \rightarrow A(S^+x^0) \Rightarrow A(x^0).$

Besides 0 and S^+, **T** should also contain (i) constants for functionals introduced by explicit definition (combinatorial completeness), that is, whenever t is a term built by application from variables $x_1, \ldots, x_n$ and constants already introduced, there is a constant ϕ such that

$$\phi x_1 \ldots x_n = t,$$

and (ii) constants for functionals defined by primitive recursion.

(i) can be guaranteed by having λ-abstraction in **T**, or, as is often technically more convenient, by having combinator constants $K_{\sigma,\tau}$ of type $(\sigma)(\tau)\sigma$ and $S_{\rho,\sigma,\tau}$ of type $((\rho)(\sigma)\tau)((\rho)\sigma)(\rho)\sigma$ such that

$$(1) \qquad K_{\sigma,\tau}xy = x, \quad S_{\rho,\sigma,\tau}xyz = xz(yz);$$

it is well known that these permit us to define a term $\lambda x.t[x]$, for each term t, such that $(\lambda x.t[x])(t') = t[t']$ (see *Troelstra 1973*, 1.6.8).

As to (ii), it suffices to add for each σ a 'recursor' R_σ of type $(\sigma)((\sigma)(0)\sigma)(0)\sigma$ such that

$$(2) \qquad R_\sigma xy0 = x, \quad R_\sigma xy(S^+z) = y(R_\sigma xyz)z.$$

From the R_σ one can define (sequences of) constants for simultaneous recursion (see *Troelstra 1973*, 1.7).

Following Gödel's note **i**4, let $\mathbf{T}_0$ (with language $\mathcal{L}_0$) be the subsystem of $\mathbf{T}$ obtained by restricting prime formulas to numerical equalities (that is, now $=_0$ is the only primitive notion of equality). Of course, the equality axioms (1) and (2) for higher types cannot be expressed in $\mathbf{T}_0$ as they stand; instead, one assumes the corresponding replacement schemas; for an arbitrary term $s[x]$ of type 0 we have

$$(3) \qquad \begin{array}{ll} s[Ktt'] = s[t], & s[Stt't''] = s[tt''(t't'')], \\ s[Rtt'0] = s[t], & s[Rtt'(S^+t'')] = s[t'(Rtt't'')t'']. \end{array}$$

The decidability of equality for type 0 can be proved; for higher types the decidability assumption is dropped.

This completes our description of one precise version $\mathbf{T}$ of Gödel's T and a significant subsystem $\mathbf{T}_0$ of $\mathbf{T}$.

3.2 Gödel's translation and interpretation

Gödel's translation ^{D} (denoted by $'$ in Gödel's paper) assigns to each A in the language of **HA** a formula A^D of the form $\exists \boldsymbol{x}\forall \boldsymbol{y} A_D(\boldsymbol{x}, \boldsymbol{y})$, where $A_D \in \mathcal{L}_0$ (that is, A_D is quantifier-free). Gödel's principal result may then be stated as follows:

(I) If $\mathbf{HA} \vdash A$, then for some sequence of terms $\boldsymbol{t}$, $\mathbf{T} \vdash A_D(\boldsymbol{t}, \boldsymbol{y})$ ($\boldsymbol{y}$ not in $\boldsymbol{t}$).

In fact, already $\mathbf{T}_0 \vdash A_D(\boldsymbol{t}, \boldsymbol{y})$. If $\mathbf{T}$ ($\mathbf{T}_0$) is regarded as embodying evident principles, this can be regarded as a consistency proof for **HA**. To obtain a similar result for **PA**, one first applies the translation from *Gödel 1933e*.

It is not too easy to explain the intuitive content of A^D in a few words; so we shall first formally define the translation and afterwards return to this question. A^D is defined for arbitrary formulas by induction on their logical complexity.

$(=)^D$ For A prime, $\boldsymbol{x}$, $\boldsymbol{y}$ are empty sequences and $A^D \equiv A_D \equiv A$.

For all other clauses, let $A^D \equiv \exists \boldsymbol{x} \forall \boldsymbol{y} A_D(\boldsymbol{x}, \boldsymbol{y})$, $B^D \equiv \exists \boldsymbol{u} \forall \boldsymbol{v} B_D(\boldsymbol{u}, \boldsymbol{v})$. Then

$(\wedge^D)$ $(A \wedge B)^D := \exists \boldsymbol{xu} \forall \boldsymbol{vy}(A_D \wedge B_D)$,
$(\vee^D)$ $(A \vee B)^D := \exists z^0 \boldsymbol{xu} \forall \boldsymbol{vy}[(z = 0 \wedge A_D) \vee (z = 1 \wedge B_D)]$,
$(\forall^D)$ $(\forall z A(z))^D := \exists \boldsymbol{x} \forall z \boldsymbol{y} A_D(\boldsymbol{x}z, \boldsymbol{y})$,
$(\exists^D)$ $(\exists z A(z))^D := \exists z \boldsymbol{x} \forall \boldsymbol{y} A_D(z, \boldsymbol{x}, \boldsymbol{y})$,
$(\rightarrow^D)$ $(A \rightarrow B)^D := \exists \boldsymbol{UY} \forall \boldsymbol{xv}(A_D(\boldsymbol{x}, \boldsymbol{Yxv}) \rightarrow B_D(\boldsymbol{Ux}, \boldsymbol{v}))$.

In the above definition certain obvious requirements on the variables are to be met; for example, in $(\wedge^D)$ the sequences $\boldsymbol{x}$, $\boldsymbol{u}$, $\boldsymbol{v}$, $\boldsymbol{y}$ must all be kept disjoint, if necessary by renaming variables. As to the clause $(\vee^D)$, see Gödel's note 1.[e]

Suppose $\mathbf{PA} \vdash A$ and let A' be a prenex form of $\neg A$; with the help of the 'negative translation' (*Gödel 1933e*) one easily sees that $\mathbf{HA} \vdash \neg A'$; then by (I) there are terms $\boldsymbol{t}$ such that $\mathbf{HA} \vdash \neg A'_D(\boldsymbol{x}, \boldsymbol{tx})$. This is Kreisel's no-counterexample interpretation[f] for $\mathbf{PA}$ (*Kreisel 1951*, *1952*; see also *Troelstra 1973*, 3.5.3).

In *1958* and *1972* the definition of A^D is given only for formulas A of the language of $\mathbf{HA}$, that is, z is of type 0 in the clauses $(\forall^D)$, $(\exists^D)$. But, formally, these clauses make just as good sense when extended to variables z of any finite type. We write $\mathcal{L}^*$, $\mathcal{L}_0^*$ for the languages of $\mathbf{T}$, $\mathbf{T}_0$ extended with quantifiers for all finite types. A^D is thus defined for all A in $\mathcal{L}^*$.

The following may serve to motivate the definition of A^D. It is obvious how we can constructively verify statements $\exists \boldsymbol{x} \forall \boldsymbol{y} A(\boldsymbol{x}, \boldsymbol{y})$, for A quantifier-free, namely by providing an explicit $\boldsymbol{t}$ such that $A(\boldsymbol{t}, \boldsymbol{y})$ holds (the constructive meaning of a quantifier-free statement is taken to be immediate). If we wish to assign a similar interpretation to *arbitrary* A, we should look for a statement $A^D \equiv \exists \boldsymbol{x} \forall \boldsymbol{y} A_D(\boldsymbol{x}, \boldsymbol{y})$ (classically) equiv-

[e]The clause $(\vee^D)$ can also be formulated as

$$(A \vee B)^D := \exists z^0 \boldsymbol{xu} \forall \boldsymbol{vy}[(z = 0 \rightarrow A_D) \wedge (z \neq 0 \rightarrow B_D)],$$

which corresponds precisely with the definition of $\vee$ given in footnote d.

[f]This treatment of the no-counterexample interpretation (n.c.i.) introduces type-2 functionals *via* the higher types. Such a detour can be avoided by a direct treatment, as shown by Tait. An adequate system of type-2 functionals is given in *Tait 1965a*. A nice treatment is in *Schwichtenberg 1977*. The no-counterexample interpretation, and hence also Gödel's interpretation, may be viewed as an extension and generalization of Herbrand's theorem to arithmetic (see e.g. *Girard 1982*); in this connection it is perhaps interesting that Gödel already realized in the early forties (in 1943 or before) that Herbrand's proof of his theorem was in need of correction.

alent to A. Gödel's definition of A^D accomplishes this in such a way that explicit realization of existential quantifiers and choices of disjuncts for disjunctions are encoded in the verifying $\boldsymbol{x}$, as may be seen by inspecting the clauses $(\vee^D)$ and $(\exists^D)$. As to the required equivalence $A \leftrightarrow A^D$, note that $A \leftrightarrow A^D$ for A prime, and intuitionistically $(A \wedge B)^D \leftrightarrow A^D \wedge B^D$, $(A \vee B)^D \leftrightarrow A^D \vee B^D$ and $(\exists z A(z))^D \leftrightarrow \exists z A(z)^D$; $(\forall z A)^D \leftrightarrow \forall z A^D$ holds if we accept

$$\text{AC} \qquad \forall \boldsymbol{x} \exists \boldsymbol{y} B(\boldsymbol{x}, \boldsymbol{y}) \rightarrow \exists Y \forall \boldsymbol{x} B(\boldsymbol{x}, Y\boldsymbol{x}).$$

Finally, clause $(\rightarrow)^D$ can be motivated[g] by the following sequence of steps (*Spector 1962*):

$$[\exists \boldsymbol{x} \forall \boldsymbol{y} A_D(\boldsymbol{x}, \boldsymbol{y}) \rightarrow \exists \boldsymbol{u} \forall \boldsymbol{v} B_D(\boldsymbol{u}, \boldsymbol{v})] \overset{\text{(i)}}{\longleftrightarrow}$$
$$[\forall \boldsymbol{x}(\forall \boldsymbol{y} A_D(\boldsymbol{x}, \boldsymbol{y}) \rightarrow \exists \boldsymbol{u} \forall \boldsymbol{v} B_D(\boldsymbol{u}, \boldsymbol{v}))] \overset{\text{(ii)}}{\longleftrightarrow}$$
$$[\forall \boldsymbol{x} \exists \boldsymbol{u}(\forall \boldsymbol{y} A_D(\boldsymbol{x}, \boldsymbol{y}) \rightarrow \forall \boldsymbol{v} B_D(\boldsymbol{u}, \boldsymbol{v}))] \overset{\text{(iii)}}{\longleftrightarrow}$$
$$[\forall \boldsymbol{x} \exists \boldsymbol{u} \forall \boldsymbol{v}(\forall \boldsymbol{y} A_D(\boldsymbol{x}, \boldsymbol{y}) \rightarrow B_D(\boldsymbol{u}, \boldsymbol{v}))] \overset{\text{(iv)}}{\longleftrightarrow}$$
$$[\forall \boldsymbol{x} \exists \boldsymbol{u} \forall \boldsymbol{v} \exists \boldsymbol{y}(A_D(\boldsymbol{x}, \boldsymbol{y}) \rightarrow B_D(\boldsymbol{u}, \boldsymbol{v}))] \overset{\text{(v)}}{\longleftrightarrow}$$
$$\exists UY \forall xv(A_D(\boldsymbol{x}, Y\boldsymbol{x}\boldsymbol{v}) \rightarrow B_D(U\boldsymbol{x}, \boldsymbol{v})).$$

All steps (i)–(iv) are classically justified, in fact (i) and (iii) are intuitionistic; (v) consists in a double application of AC. A further discussion of the assumptions involved in (ii) and (iv) will be taken up in 3.3 below.

It should be pointed out that Gödel does not claim any fundamental significance for ^{D} as such; his principal interest lies in achieving a conceptual reduction of intuitionistic arithmetic to 'nearly finitistic' notions; that the translation used is perhaps very tortuous is irrelevant to this aim.

The verification of the main result (I) proceeds by induction on the length of deductions in **HA**, that is, by showing that, for suitable $\boldsymbol{t}$,

$$(*) \qquad \mathbf{T} \vdash A_D(\boldsymbol{t}, \boldsymbol{y})$$

holds for each axiom A of **HA** and is preserved by the rules of inference. Details are given in *Troelstra 1973.* There are only two delicate points involved. One comes when verifying $(*)$ for instances of induction in **HA**. It may be seen that what is needed in **T** for this purpose is the generalization of the rule of induction mentioned by Gödel in note **i1**.

[g]Gödel's own motivation, in the text, is presented differently.

This rule is in fact derivable in **T**, as asserted by Gödel (**i**3); for a proof, see *Troelstra 1973*, 1.7.

A second point occurs in verifying (*) for the seemingly innocuous axioms of the form $A \to A \wedge A$. Here we need not only the decidability of prime formulas but the stronger fact that there exist characteristic terms for them, that is to say that for each prime formula $A(\boldsymbol{x})$ with free variables among the $\boldsymbol{x}$ there is a term t_A such that $t_A\boldsymbol{x} = 0 \leftrightarrow A(\boldsymbol{x})$. To see how the need for t_A arises, let us consider the interpretation of $A \to A \wedge A$. Let $A^D \equiv \exists \boldsymbol{x} \forall \boldsymbol{y} A_D$; then $(A \to A \wedge A)^D$ becomes

$$[\exists \boldsymbol{x} \forall \boldsymbol{y} A_D \to \exists \boldsymbol{x}' \boldsymbol{x}'' \forall \boldsymbol{y}' \boldsymbol{y}'' (A_D(\boldsymbol{x}', \boldsymbol{y}') \wedge A_D(\boldsymbol{x}'', \boldsymbol{y}''))]^D \equiv$$
$$\exists \boldsymbol{y} \boldsymbol{x}' \boldsymbol{x}'' \forall \boldsymbol{x} \boldsymbol{y}' \boldsymbol{y}'' (A_D(\boldsymbol{x}, \boldsymbol{y}\boldsymbol{x}\boldsymbol{y}'\boldsymbol{y}'') \to A_D(\boldsymbol{x}'\boldsymbol{x}, \boldsymbol{y}') \wedge A_D(\boldsymbol{x}''\boldsymbol{x}, \boldsymbol{y}'')).$$

Now, if $t_A\boldsymbol{x}\boldsymbol{y} = 0 \leftrightarrow A_D(\boldsymbol{x}, \boldsymbol{y})$, we can take

$$\boldsymbol{x}' \equiv \lambda \boldsymbol{x} . \boldsymbol{x}, \quad \boldsymbol{x}'' \equiv \lambda \boldsymbol{x} . \boldsymbol{x}, \quad \boldsymbol{y}\boldsymbol{x}\boldsymbol{y}'\boldsymbol{y}'' = \begin{cases} \boldsymbol{y}' & \text{if } t_A\boldsymbol{x}\boldsymbol{y}' \neq 0 \\ \boldsymbol{y}'' & \text{if } t_A\boldsymbol{x}\boldsymbol{y}' = 0. \end{cases}$$

The definition of $\boldsymbol{y}$ is an instance of definition-by-cases, which is readily justified with the help of the recursor constants. Note that the choice of interpretation here is in some respects arbitrary; we might equally well have taken

$$\boldsymbol{y}\boldsymbol{x}\boldsymbol{y}'\boldsymbol{y}'' = \boldsymbol{y}'' \quad \text{if } t_A\boldsymbol{x}\boldsymbol{y}'' \neq 0, \quad \boldsymbol{y}\boldsymbol{x}\boldsymbol{y}'\boldsymbol{y}'' = \boldsymbol{y}' \quad \text{otherwise.}$$

This is to say, the straightforward construction of terms $\boldsymbol{t}$ such that (*) holds, by induction on the length of derivations in **HA**, is not canonical at this step.

If A is arithmetical, A_D will be quantifier-free with equality of type 0 only (that is, A_D belongs to $\mathcal{L}_0$), and the existence of t_{A_D} satisfying $t_{A_D} = 0 \leftrightarrow A_D$ is in fact provable in $\mathbf{T}_0$. The rule for disjunctive definition as stated in Gödel's note **i**2 is in fact equivalent to the existence of characteristic functions plus the following more restricted rule for definition-by-cases of a functional f,

$$f\boldsymbol{x} = t_1 \quad \text{if } t = 0, \quad f\boldsymbol{x} = t_2 \quad \text{otherwise}$$

(all free variables of t_1, t_2, t contained in $\boldsymbol{x}$).

Next we wish to discuss T as it actually figures in Gödel's text. We believe that **T** as described in 3.1 corresponds pretty well to Gödel's intentions. Certainly in *1958* Gödel wanted T to have a decidable primitive notion of equality for each finite type (note 7), since the decidability of prime formulas together with intuitionistic logic justifies the decidability of all formulas of T, and the formulas may therefore be regarded as essentially 'logic-free'. (No abstract interpretation of the logical operations is involved, as there is, for example, in the proof-interpretation of $\rightarrow$.)

It is doubtful whether in *1958* Gödel had already realized the need for characteristic functions in connection with the axiom $A \rightarrow A \wedge A$. In a letter of 1970 J. Diller explicitly drew Gödel's attention to the role of characteristic functions. Gödel reacted in a letter to Bernays (14 July 1970) as follows: "I do not understand what it means to say that in my proof of the formula $p \supset p \wedge p$ a passage (which is not possible) to the characteristic term of a formula is required. What is required is the decidability of intensional equations between functions."[h]

Gödel probably regarded the existence of (computable) equality functionals as a concomitant of the decidability of equality; from the passage just quoted one cannot tell whether Gödel realized that in general (that is, at higher types) the decidability of equality does not entail, axiomatically, the presence of equality functionals at higher types. In any case the proof sheets of *1972* contain an earlier, crossed-out version of footnote **i2** which reads:

> 2. The principle of *disjunctive definition*, added to Axiom 5 in the present version of the paper, is the following: A function f may be defined by stipulating
>
> $$A \supset f(x) = t_1, \quad \neg A \supset f(x) = t_2,$$
>
> where t_1, t_2 are terms and A is a formula, both containing only previously defined functions and no variables except those of the sequence x. This principle is needed for the proofs that the axioms 1 and 4 of H and the deduction rule 6 of H hold in the interpretation defined below. It can be derived if *equality functions* with the axioms $G(f, g) = 0 \equiv f = g$ are introduced as primitive terms at all types. At any rate either disjunctive definition, or the axiom for equality functions, must be added to the axioms mentioned in the first edition.

Afterwards Gödel must have realized that A_D for *arithmetical* A required only equations between numerical terms, and he accordingly rewrote note **i2** to the version we find in the text.[i] Diller and Nahm, on

[h] "Ich verstehe nicht was es heissen soll, dass in meinem Beweis der Formel $p \supset p \wedge p$ ein (nicht möglicher) Übergang zum characteristischen Term einer Formel nötig sei. Was nötig ist, ist die Entscheidbarkeit von intensionalen Gleichungen zwischen Funktionen."

[i] Thus, in the words of G. Kreisel, the whole issue of decidability of equality and characteristic functions for equality at higher types turned out to be a "red herring", at least for Gödel's purposes.

the other hand, had attempted to extend Gödel's result to higher-type arithmetic formulated in the language $\mathcal{L}^*$, that is, $\mathcal{L}$ with quantifiers $\forall x^\sigma$, $\exists x^\sigma$ added, where equality functionals at all types (or equivalently, characteristic functions at all types) are necessary. In *1974* they gave a variant $^\wedge$ of the translation ^{D} which also achieved reduction to $\exists\forall$-form, for which an interpretation theorem similar to (I) could be proved and for which no appeal to decidability of prime formulas and characteristic functions was necessary (more about this in Section 5.4).

Gödel's note **i4** shows that he realized that one can dispense with all references to equality at higher types for Theorem I, provided one reformulates the defining axioms for the functional constants as replacement schemas for terms, as we did (3.1(3)) in formulating $\mathbf{T}_0$. Clearly, in note **i4** Gödel had something like our $\mathbf{T}_0$ in mind. He also observes that one can in fact dispense with the propositional operators, reducing everything to a term-equation calculus; in the context of primitive recursive arithmetic **PRA** this fact was first proved by Goodstein (*1945*; see also his *1957*). *Hilbert and Bernays 1934* (Chapter 7) showed that propositional combinations of term equations can be replaced by a single term equation, but not that the addition of the propositional operators with their usual rules is in fact a conservative addition to the term-equation calculus.

In note **n**, Gödel refers to the system Σ_2 of *Spector 1962* minus axiom F, corresponding to the system $\mathbf{WE\text{-}HA}^\omega$ as described below in 3.3. However, the extensionality rule of Spector's system does not seem to fit in with Gödel's decidedly *intensional* view of equality; and in fact Gödel's own paraphrase of $\Sigma_2-\{F\}$, when read in combination with note **i4**, rather points to the system $\mathbf{HA}_0^\omega$ ('quantified $\mathbf{T}_0$'; see 3.4 below).

We do not feel quite certain of the correct reading of note 8. One possible interpretation is the following: since, for Gödel, T has functionals with n arguments as a primitive notion, one can introduce a constant functional P of higher type by $(P(x_1, \ldots, x_n))(y_1, \ldots, y_m) = t$, where all free variables of t are contained in $\{x_1, \ldots, x_n, y_1, \ldots, y_m\}$; with the notation $\lambda\boldsymbol{x}$ for simultaneous abstraction for a sequence of variables one might write $P = \lambda\boldsymbol{x}(\lambda\boldsymbol{y}.t)$. So P is of type $((\tau, \tau_1, \ldots, \tau_m), \sigma_1, \ldots, \sigma_n)$ (Gödel's notation), where σ_i is the type of x_i, τ_j of y_j, and τ of t. This has no analogue in the usual formulation of **PRA**, since there only functions with values of type 0 are introduced. This formal difference disappears if we apply the Schönfinkel method for reducing everything to unary functions, both to **PRA** and to T. On this reading, Gödel's note refers only to a fairly superficial formal difference between **PRA** and T.

3.3 Extensions to finite type arithmetic; constructive content of ^{D}

It is quite illuminating to see what happens to Gödel's result (I) if one tries to extend it from **HA** to various systems of intuitionistic higher-type arithmetic. Let $\mathcal{L}_0^*$ and $\mathcal{L}^*$ be the languages of $\mathbf{T}_0$ and $\mathbf{T}$, respectively, extended with higher-type quantifiers. Let $\mathbf{HA}_0^\omega$ be the extension of $\mathbf{T}_0$ to $\mathcal{L}_0^*$ with intuitionistic predicate logic; $\mathbf{HA}^\omega$ is the corresponding extension of $\mathbf{T}$ to $\mathcal{L}^*$, but without assuming decidability of equality at higher types; $\textbf{I-HA}^\omega$ is $\mathbf{HA}^\omega$ with characteristic functions E_σ for equality at type σ, $\sigma \in \mathcal{T}$, added. E_1 is certainly not definable in $\mathbf{HA}^\omega$, since all type-2 functionals of $\mathbf{HA}^\omega$ are continuous, while $\lambda y^1.E(\lambda x^0.0)y$ is not continuous in y^1.

An extensional variant $\textbf{E-HA}^\omega$ in the language $\mathcal{L}_0^*$ is obtained by *defining* equality at higher types as extensional equality, assuming the usual equality axioms for this defined equality at all types. A weakly extensional variant $\textbf{WE-HA}^\omega$ has, instead of the substitution schema $t_1^\sigma = t_2^\sigma \wedge A[t_1] \rightarrow A[t_2]$, the weaker *rule*

$$t_1 x_1 \ldots x_n = t_2 x_1 \ldots x_n, A(t_1) \Rightarrow A(t_2),$$

where $x_1, \ldots, x_n$ is a string of variables, not occurring free in t_1, t_2, such that $t_1 x_1 \ldots x_n$, $t_2 x_1 \ldots x_n$ are of type 0. Schematically

$$\mathbf{HA} \subset \mathbf{HA}_0^\omega \begin{cases} \subset \mathbf{HA}^\omega \begin{cases} \subset \textbf{I-HA}^\omega \\ \subset \textbf{E-HA}^\omega \end{cases} \\ \qquad\qquad\qquad\qquad \| \\ \subset \textbf{WE-HA}^\omega \subset \textbf{E-HA}^\omega . \end{cases}$$

As shown in *Rath 1978*, $\mathbf{HA}^\omega$ is in fact conservative over $\mathbf{HA}_0^\omega$.

If $\mathbf{H}$ is one of the systems $\mathbf{HA}_0^\omega$, $\mathbf{HA}^\omega$, $\textbf{I-HA}^\omega$, $\textbf{WE-HA}^\omega$, $\textbf{E-HA}^\omega$, qf-$\mathbf{H}$ denotes the fragment without quantifiers ("quantifier-free"), so qf-$\mathbf{HA}_0^\omega = \mathbf{T}_0$, and $\mathbf{T} \subset$ qf-$\textbf{I-HA}^\omega$.

For $\mathbf{H}$ any one of $\textbf{I-HA}^\omega$, $\textbf{WE-HA}^\omega$ or $\mathbf{HA}_0^\omega$, we can show straightforwardly the following extensions of Gödel's (I):

(II) $\mathbf{H} \vdash A(\boldsymbol{x}) \Rightarrow$ qf-$\mathbf{H} \vdash A_D(\boldsymbol{t}\boldsymbol{x}, \boldsymbol{y}, \boldsymbol{x})$ for a suitable sequence $\boldsymbol{t}$ of closed terms;

(III) $\mathbf{H} \vdash A \Rightarrow \mathbf{H} \vdash A^D$ (soundness, as a corollary to II).

Soundness must necessarily fail for $\mathbf{HA}^\omega$, as shown by Howard (*Troelstra 1973*, 3.5.6), since soundness for systems in the language $\mathcal{L}^*$ would entail $\forall y^1 z^1(\neg\neg y^1 = z^1 \vee \neg y^1 = z^1)$, i.e., a weak form of decidability, which cannot be proved in $\mathbf{HA}^\omega$; and soundness also fails for $\mathbf{E\text{-}HA}^\omega$, since no functional of $\mathbf{E\text{-}HA}^\omega$ can satisfy the ^{D} translation of $\forall z^{(1)0}x^1y^1(x = y \to zx = zy)$ (Howard, in *Troelstra 1973*, Appendix). Of course, there is an indirect method for interpreting $\mathbf{E\text{-}HA}^\omega$: first interpret this theory in $\mathbf{HA}_0^\omega$ by hereditarily restricting all quantifiers in $\mathbf{HA}_0^\omega$ to elements of type σ which respect extensional equality, and then apply the *Dialectica* interpretation; this is the road taken in *Luckhardt 1973*.

$\mathbf{WE\text{-}HA}^\omega$ as an intermediate possibility is not very attractive: the deduction theorem does not hold for this theory. (However, it does hold for qf-$\mathbf{WE\text{-}HA}^\omega$; in this respect the comments in *Troelstra 1973*, 1.6.12, are misleading and partly wrong. See also Gödel's note **n**.)

Let us now return to a discussion of the equivalence $A \leftrightarrow A^D$. If the sole aim of the translation were to associate with each A a classically equivalent (modulo AC) A' of the form $\exists \boldsymbol{x}\forall \boldsymbol{y}A''(\boldsymbol{x}, \boldsymbol{y})$, any standard recipe for rewriting A in prenex normal form, followed by a number of applications of AC so as to bring all $\exists$ in front ("Skolemization"), would do the job. However, A^D is designed so as to keep as close as possible to an intuitionistic reading of A, by minimizing the non-intuitionistic steps in the translation from A to A^D. The only clause which needs inspection in this connection is $(\to^D)$, since the other clauses involve only AC and transitions valid by intuitionistic logic. As noted already in 3.2, the crucial steps in the definition of $(A \to B)^D$ are (ii) and (iv). Step (ii) is an instance of the schema

$$\mathrm{IP}' \qquad (\forall \boldsymbol{x} C_1 \to \exists y C_2) \to \exists y(\forall \boldsymbol{x} C_1 \to C_2)$$

(C_1, C_2 quantifier-free, y not free in C_1, $\boldsymbol{x}$ not free in C_2). The notation IP′ derives from "Independence of Premiss", since, assuming $\forall \boldsymbol{x} C_1 \to \exists y C_2$, IP′ requires that we can *a priori* indicate y, independently of the truth of $\forall \boldsymbol{x} C_1$; an intuitionistic reading of $\forall \boldsymbol{x} C_1 \to \exists y C_2$ requires only that, once a proof of $\forall \boldsymbol{x} C_1$ is given, we can find a y (possibly depending on the given proof).

Step (iv) can be justified on the basis of the following generalization of "Markov's principle":

$$\mathrm{M}' \qquad \neg\forall \boldsymbol{v} C \to \exists \boldsymbol{v} \neg C \quad (C \text{ quantifier-free})$$

((iv) is justified if $B_D(\boldsymbol{u}, \boldsymbol{v})$ holds; if not, apply M′ with $A_D(\boldsymbol{x}, \boldsymbol{y})$ for C). Markov's principle as accepted by the Russian constructivist school is the special case where $\boldsymbol{v}$ is of type 0 and C primitive recursive.

For the interpretation in terms of abstract proofs, IP′ and M′ are usually not accepted as valid, and AC only in special cases (in particular where $\boldsymbol{x}$ is of type 0). However, nothing in the intuitionist point of view prevents us from giving a more special interpretation to the logical operators in the context of a given language, *provided* that interpretation is itself meaningful and intelligible from an intuitionistic point of view (which is certainly the case for ^{D}). For the same reasons, however, one should question Gödel's remark (at the end of note **h**) that his interpretation is *more* constructive than the proof-interpretation because it validates Markov's schema. In fact, Markov's schema is false for some perfectly coherent intuitionistic theories such as the theory of lawless sequences (see *Troelstra 1977*, Chapter II), while Gödel himself, at the end of note **c**, regards choice sequences as coming close to being finitistic.

Not only does the *Dialectica* interpretation validate Markov's principle, but in fact any instance of M′, IP′ and AC. As a result, we obtain the following strengthening of (II) and (III), for the same systems **H** as before:

(IV) $\mathbf{H} + \mathrm{IP}' + \mathrm{M}' + \mathrm{AC} \vdash A \leftrightarrow A^D$, and
$\mathbf{H} + \mathrm{IP}' + \mathrm{M}' + \mathrm{AC} \vdash A \Rightarrow \text{qf-}\mathbf{H} \vdash A_D(\boldsymbol{t}, y)$ for suitable $\boldsymbol{t}$.

The first half of (IV) also holds for $\mathbf{H} \equiv \mathbf{E\text{-}HA}^\omega$ or $\mathbf{HA}^\omega$, and is implicit in *Kreisel 1959* (2.11,3.51). The second half of (IV) was explicitly stated in *Yasugi 1963* for $\mathbf{WE\text{-}HA}^\omega$.

4. Models for T

For the sake of definiteness, we shall here identify Gödel's T with our **T** of 3.1. Three models for **T** are mentioned in Gödel's paper: the computable functionals, the hereditarily recursive operations, and the term model. We shall now briefly discuss each of them in turn.

4.1 The computable functionals of finite type

These are informally described by Gödel as follows: the computable functions of type 0 are the natural numbers; a computable function of type $(\sigma)\tau$ is a well-defined mathematical procedure which, applied to a computable function of type σ, yields a computable function of type τ. Here "well-defined mathematical procedure" must be taken as an understood primitive notion, and Gödel stresses (note 6) the parallel with the role of "constructive proof" in Heyting's explanations of the logical operators.

The precise wording of the description of this notion together with the explanatory note 6 caused Gödel a good deal of trouble. In a letter to Bernays, dated 17 December 1968, Gödel, referring to note 1 of *1958* (corresponding to note 6 of *1972*), observes that "it seems to be very difficult (or impossible) to make this more precise and yet to maintain it to its full extent".

In the description of computable function, the proof sheets have, instead of "this general fact is constructively evident ... without any further explanation [6]", the following crossed-out passage: "this general fact is intuitionistically demonstrable. This definition of 'computable function of type $(t_0, t_1, \ldots, t_k)$' must be accepted as having a clear meaning without any further explanation [6] provided one already has clear ideas of the meanings of the phrases 'computable function of type t_i' for $i = 0, 1, \ldots, k$." (Gödel's [6] is note 6 in *1972* as reproduced here.)

As Bernays observed in a letter to Gödel of 12 December 1970, this makes it appear as if the notion of computable function *depends* on the general notion of intuitionistic proof, contrary to Gödel's intentions. Consequently, Gödel replaced "intuitionistically demonstrable" by "constructively evident", but he remained dissatisfied as is shown by his letters to Bernays (that of 26 December 1972, in particular). His note **h**, discussed below, may be seen as an attempt to interpret T with the help of a narrower concept of proof that is more obviously independent of the general notion of intuitionistic proof.

As Gödel realized, it was not possible to avoid a certain "impredicativity" in the notion of a finite-type function. "Impredicative" here refers to the fact that, e.g., functionals of type (0)0 could be defined *via* functionals of much more complex types—just as an intuitionistic proof of a statement may perhaps refer to proofs of more complex statements. It is this fact which makes it difficult to formulate the epistemological gain obtained in replacing the general concept of intuitionistic proof by "computable function of finite type" or by one of the interpretations discussed in 4.2 and 4.3 below.

4.2 The hereditarily recursive operations *HRO*

For each $\sigma \in \mathcal{T}$, define V_σ, the set of (Gödel numbers of) hereditarily recursive operations as follows

$$V_0 := N, \quad V_{(\sigma)\tau} := \{x : \forall y \in V_\sigma \exists z \in V_\tau(\{x\}(y) \simeq z)\}.$$

Here $\{\cdot\}$ are the "Kleene-brackets" for partial recursive function application; i.e., with Kleene's T-predicate and result-extracting function U,

$$\{x\}(y) \simeq z \leftrightarrow \exists u(T(x, y, u) \wedge U(u) = z).$$

Equality at each type σ is interpreted as equality of Gödel numbers. The reader should note that with increasing complexity of the type σ the arithmetical complexity (level in the arithmetical hierarchy) of the predicates V_σ increases without bound, and that $x \in V_\sigma$ as a predicate of x is certainly not recursive for $\sigma \neq 0$. Thus the only "reductive gain" of the interpretation of **T** in terms of *HRO* is that arbitrary arithmetical predicates are explained in terms of the more special predicates $x \in V_\sigma$. For details, see *Troelstra 1973*, 2.4.8. Clearly it is this model to which Gödel refers in notes **g** and **h** as based on Turing's notion of a computable function.

4.3 The term model

Many variants of this can be given; we choose a simple version that is easy to describe and is suitable for illustrating Gödel's intentions.

Redexes are terms of one of the following forms: Ktt', $Stt't''$, $Rtt'0$, $Rtt'(S^+t'')$, where t, t', t'' do not contain subterms that are redexes; these *convert* respectively to t, $tt''(t't'')$, t, $t'(Rtt't'')t''$. It can be shown that each closed term t can be *reduced* to a unique redex-free term (the *normal form* of t), by successively converting redexes occurring as subterms. We use NF for the set of closed terms in normal form.

The most straightforward method for proving this is by means of so-called[j] 'computability' predicates (*Tait 1967*). Let t, t' range over closed terms, and put

$$\mathrm{Comp}_0(t) := t \text{ is of type 0 and reduces to normal form,}$$
$$\mathrm{Comp}_{(\sigma)\tau}(t) := t \text{ reduces to normal form}$$
$$\text{and } \forall t' \in \mathrm{Comp}_\sigma(tt' \in \mathrm{Comp}_\tau).$$

The proof then proceeds by noting that all applicative combinations of computable terms are computable, and that all constants are computable; afterwards one can prove the normal form to be unique.

For the model of **T** one takes the $t \in NF$ of type σ to be the objects of type σ. Equality is interpreted as literal identity (hence it is recursive!) and application $\mathrm{App}(s,t)$ as the (unique) normal form of $s(t)$.

In view of Gödel's choice of terminology ("reductive proof") in note **n1**, it is tempting to think he had something like a term model, defined via reductions, in mind. But there is no conclusive evidence for this. Though we may assume that *Tait 1967* was known to Gödel, he does not refer to it in the paper, nor does he ever refer to Tait's work in his letters to Bernays. Nevertheless, the model NF may be used in an effort to understand Gödel's intentions in note **h**—at the same time revealing problematic aspects.

[j] The terminology 'hereditarily normalizable', instead of 'computable', would have been more appropriate.

So let us attempt to interpret note **h** in terms of NF, taking $\mathbf{T}_0$ for T. We may then think of $\mathbf{T}_0$ as an equation calculus of finite-type functionals, and take Gödel's T′ to be $\mathbf{T}_0$ interpreted in NF. 'Reductively provable' for an equation between closed terms $t = s$ would then mean that the equation can be verified by reducing t, s to their normal forms t^N, s^N and finding that $t^N \equiv s^N$. This is indeed decidable, and the proof-procedure is defined in advance (see note **h**1,2).

Of course, the proofs of $\mathbf{T}_0$ are not reductive proofs as such; but according to Gödel, it should be possible to justify all of them on the basis of the notion of reductive proof alone (note **h**3: "no other *concept* of proof ... occurs in ..."). The simplest way to interpret this claim is to verify that NF is indeed a model for $\mathbf{T}_0$ (see *Troelstra 1973*, 2.5). This can be done in a fairly straightforward manner, though it is not entirely trivial; the obvious metamathematical argument relies on induction plus the existence of a unique normal form for each closed term. In fact, we have for equations between closed terms t and s

$$\mathbf{HA}_0^\omega \vdash t = s \overset{(1)}{\Longleftrightarrow} \mathbf{T}_0 \vdash t = s \overset{(2)}{\Longleftrightarrow} t^N \equiv s^N$$

($\Leftrightarrow t = s$ true in $NF \Leftrightarrow t = s$ reductively provable, by definition). Here (1) holds by the soundness of Gödel's interpretation ((II) in Section 3.3), (2) by the fact that NF is a model for $\mathbf{T}_0$.

One might also think of a stricter interpretation of Gödel's claim, namely, that it ought to be possible to justify all proofs in $\mathbf{T}_0$ by a method that would explicitly transform any proof of a closed term equation in $\mathbf{T}_0$ into a reductive proof. (For proofs with free variables in the conclusion, we ought then to require that this be possible for each substitution of closed terms for the free variables of the conclusion.)

A step in such a transformation could be the replacement of an application of the induction rule $A(0)$, $A(n) \rightarrow A(S^+n) \Rightarrow A(\overline{m})$ for a numeral $\overline{m}$, by m applications of modus ponens to proofs of $A(0)$, $A(0) \rightarrow A(1), \ldots, A(\overline{m}-1) \rightarrow A(\overline{m})$ (see Gödel's observation on induction in note **n**).

Though it is possible to carry through this stricter interpretation, it is certainly not trivial.

4.4 Assessment of the reduction achieved by Gödel

We now return to the issues left hanging in 2.4. The "impredicativity" of **T** has already been mentioned. As will have become clear from 4.2 and 4.3, narrowing down the intended interpretation does not remove this feature: we may make the description of the intended model of **T** very definite and concrete, but the "impredicativity" then pops up in the arguments needed to show that the interpretations for **T** are correct.

Not only the V_σ, but also the arithmetized version of the predicates Comp_σ, run through all stages of the arithmetical hierarchy as the complexity of σ increases. Gödel's aim was to replace the abstract intuitionistic logical notions by a notion of functional, as concrete as possible; he succeeded in fact in eliminating the logic except for the logic hidden in the precise definition of the intended class of functionals. (In 4.1, we got rid of the logic by accepting "computable functional" as a primitive; in 4.2 and 4.3 the remaining logic resides in the predicates $x \in V_\sigma$, respectively $x \in \mathrm{Comp}_\sigma$.)

If we look at the generalization of Gödel's result, for example to $\mathbf{HA}_0^\omega$, we see that the logic is "absorbed", under the interpretation, by the notion of higher-type functional. In short, there is some reductive gain though it is not clear-cut; we think it falls short of Gödel's aims.[k]

5. Later research flowing from *Gödel 1958*

In this section we intend to present a brief survey of research more or less directly inspired by Gödel's paper.[l] As principal themes we distinguish

(1) extensions of Gödel's main result to other systems,
(2) investigations of the functionals needed for the interpretation,
(3) metamathematical applications,
(4) the study of related interpretations.

To each of these topics we devote a subsection.

5.1 Extensions of Gödel's main result

Already in *Kreisel 1959* it was observed that Gödel's result is easily extended to intuitionistic arithmetic with transfinite induction.

Spector (*1962*) extended Gödel's result to analysis formalized with function variables by adding to T a new definition principle, the schema

[k]We should mention, however, that Gödel, as late as 1974, expressed himself in the following terms, writing about *1958*: "...the most direct way of arriving at an intuitionistic interpretation of T ... does *not* pass through Heyting's logic, or the general intuitionistic concept of proof or implication, but rather through much narrower (and in principle decidable) concepts of 'provable' and 'implies'. Thus the implicit use of 'implication' and 'demonstrability' in the definition of 'computable function of finite type' does *not* give rise to any circularity" (from a letter, or draft of a letter, to a Mr. Sawyer, then a graduate student at the University of Pittsburgh; we do not know whether the letter was actually sent or not).

[l]Some papers connected with the *Dialectica* interpretation and Gödel's T not explicitly reviewed below are: *Kreisel 1959a, 1959b, Grzegorczyk 1964, Diller and Schütte 1971, Vesley 1972, Schwichtenberg 1973, 1975, 1979, Goodman 1976* and *Scott 1978*.

of bar recursion at all finite types. Bar recursion (BR) is closely related to the axiom schema BI of bar induction at all finite types, and BR can in fact be justified with the help of BI. Brouwer's "bar theorem" is equivalent to BI_0, bar induction of type 0. Gödel, in note **d**, refers to Spector's work, but does not distinguish clearly between bar recursion and bar induction. The justification of BI_0 mentioned by Gödel refers to the method of elimination of choice sequences,[m] which reduces BI_0 to the theory of a single generalized inductive definition.

Spector's work was later refined by Howard (*1968*) and Luckhardt (*1973*). Howard (*1972*) also extended Gödel's result to a theory of abstract constructive ordinals (again a theory of a single generalized inductive definition, namely an abstract version of Kleene's recursive ordinals), thereby realizing a possibility suggested by Gödel at the end of note **d**.

Quite recently Friedrich (*1984*, *1985*) has carried the interpretation through for analysis extended with a game quantifier.

Girard (*1971, 1972*) was the first to treat classical analysis and finite type theory formulated with *set* (*predicate*) *variables* of finite type and full comprehension. Girard had to invent several new technical devices, such as the introduction of a type structure with variable types and the use of 'reducibility candidates', in establishing normalization for his system of functionals. The latter idea led to direct proofs of normalization and cut-elimination for analysis and the theory of types (*Girard 1972*, *Prawitz 1971*, *Martin-Löf 1971*).

Maaß (*1976*) gave a treatment of predicative analysis. Koletsos (*1985*) extended Gödel's interpretation to Girard's β-logic.

5.2 Investigations of extensions of Gödel's T

Gödel's interpretation provides a consistency proof for arithmetic modulo the assumption that closed terms of type zero have a unique numerical value, which precludes a proof of $0 = 1$. If one does not want to rely on the insight that the computable functionals are a model of T, one can try to prove the assumption by analyzing the computation of terms of T.

Thus many investigations have been devoted to showing that terms of T and some of its extensions can always be reduced to normal form. In particular, for closed terms of type 0 this entails that they can be shown to be equal to a numeral, that is, all closed terms of type 0 can be evaluated.

[m]The result as stated in Gödel's source, *Kreisel 1965*, is not quite correct. See *Kreisel and Troelstra 1970* for a corrected version.

As to the methods for proving normalization, they are principally of two kinds: (a) by defining suitable computability predicates (*Dragalin 1968*, *Tait 1967*, *1971*, *Girard 1971*, *1972*, *Luckhardt 1973*, *Vogel 1977*) and (b) by ordinal assignments to terms (*Hinata 1967*, *Diller 1968*,[n] *Howard 1970*, *1980*, *1981*, *1981a*). The method of assigning ordinals less than ϵ_0 to terms of T can be used as another route to Gentzen's theorem that the consistency of **PA** can be established in primitive recursive arithmetic plus quantifier-free ϵ_0-induction. In *Tait 1965* Gentzen's result is obtained via an assignment of infinite terms to functionals of T; the infinite terms of type zero are shown, by means of quantifier-free ϵ_0-induction, to have a unique numerical value. Infinite terms are also used in *Howard 1972*.

Sanchis (*1967*) and Diller (*1968*) establish normalization of the type-zero terms of T by means of bar induction (Diller also gave an ordinal assignment, as noted above); Hanatani (*1975*) uses cut-elimination for a system like $\mathbf{HA}^\omega$.

5.3 Metamathematical applications

Here we give some examples only. One of the first applications is in *Kreisel 1959*. There a constructive interpretation of formulas of analysis is given by combining the translation ^{D} with specific models for T. A typical result is the following: if a formula A of analysis does not contain $\exists$ or $\vee$, then $A \leftrightarrow (A^D)^*$ holds classically. Here $(A^D)^*$ is obtained from $A^D \equiv \exists \boldsymbol{x} \forall \boldsymbol{y} A_D$ by letting $\boldsymbol{y}$ range over the continuous functionals and $\boldsymbol{x}$ over the recursive continuous functionals (see e.g. *Troelstra 1973*, 2.6.5).

Conservative extension results can be obtained from the axiomatization of $A^D \leftrightarrow A$ (IV in 3.3 above); thus e.g. $\mathbf{H} + \mathrm{IP}' + \mathrm{M}' + \mathrm{AC}$ is conservative over $\mathbf{H}$ for $\exists\forall$-formulas if $\mathbf{H} \equiv \mathbf{I\text{-}HA}^\omega$, $\mathbf{WE\text{-}HA}^\omega$ or $\mathbf{HA}_0^\omega$. Other examples are given in *Troelstra 1973* (3.5.14, 3.6.6, 3.6.18 (iii)).

One of the best-known applications of ^{D} is to show closure under Markov's rule (*Troelstra 1973*, 3.8.3); recently, the more elegant and more widely applicable[o] method of Friedman (*1978*) and Dragalin (*1980*) has become available.

[n] Diller's assignment is not optimal, that is to say, he uses ordinals beyond ϵ_0.

[o] It remains to be seen whether the new method of proof yields better results in extracting *effective* bounds from *classical* proofs of Π_2^0-statements. In this connection see the discussion in *Kreisel and Macintyre 1982*. On the use of the no-counterexample interpretation and the *Dialectica* interpretation for the extraction of explicit bounds from classical proofs, see also *Girard 1982* and *Kreisel 1982*.

Examples of applications in the proof theory of classical systems are Parsons' use (*1970*) of the *Dialectica* interpretation for an analysis of subsystems of arithmetic and Feferman's use (*1971* and *1977*, 8.6.2) of the interpretation in the study of subsystems of classical analysis by means of **T** relativized to non-constructive functionals.

5.4 Related interpretations

Towards the end of Section 3.2 we mentioned the variant interpretation due to Diller and Nahm (*1974*). For this interpretation, one widens the notion of "quantifier-free" by permitting bounded numerical quantifiers $\forall x^0 < t$ in addition to $\wedge$, $\vee$, $\rightarrow$, $\neg$, and then one associates with each A of $\mathbf{HA}^\omega$ a translation $A^\wedge$ of the form $\exists \boldsymbol{x} \forall \boldsymbol{y} A_\wedge(\boldsymbol{x}, \boldsymbol{y})$, where $A_\wedge$ is quantifier-free in the wider sense. An interpretation result completely similar to (I) in 3.2 can then be proved (see *Troelstra 1973*, 3.5.17).

The Diller–Nahm interpretation was extended to analysis in *Diller and Vogel 1975*, and to systems with self-applicable operators in *Beeson 1978*.

Stein (*1976, 1978, 1980*) interpolated an infinite sequence of interpretations between $^\wedge$ and modified realizability, a functional interpretation originally devised by Kreisel to show underivability of Markov's principle (*Kreisel 1959*; see *Troelstra 1973*, 3.4). Moreover, all these interpretations can be seen as special cases of a single interpretation ^{M} in a language with "set-types" (see also *Rath 1978*, *Diller 1979*). For theories with decidable prime formulas, $^\wedge$ is equivalent to ^{D}.

Normalization and cut-elimination, ^{D}, $^\wedge$, modified realizability and realizability interpretations all give explicit realizations for numerical existential statements proved in intuitionistic arithmetic. It can be shown that all these methods can be made to yield the same realizations (*Mints 1974, 1975, 1979*; *Stein 1976, 1980, 1981*)—a fact which is by no means obvious, since the choice of terms for the *Dialectica* interpretation is not always canonical (see the discussion in Section 3.2 above).

In conclusion, we mention a variant of the *Dialectica* interpretation of a different nature, due to Shoenfield (*1967*). This variant is directly applicable to classical first-order arithmetic **PA** formalized using $\forall$, $\vee$, $\neg$. To each A of the language of **PA** a formula A^s of the form $\forall \boldsymbol{x} \exists \boldsymbol{y} A_s(\boldsymbol{x}, \boldsymbol{y})$ is assigned, where $\exists \boldsymbol{y}$ now abbreviates $\neg \forall \boldsymbol{y} \neg$, with A_s quantifier-free. A^s is said to be *valid* if, for some term sequence $\boldsymbol{t}$, $A_s(\boldsymbol{x}, \boldsymbol{t})$ is derivable in (a version of) Gödel's T. Then A^s is valid for each A provable in **PA**. This reduction bypasses the "negative translation" of **PA** into **HA** (*Gödel 1933e*). See also *Troelstra 1973*, 3.5.18.

5.5 Concluding remarks

As the preceding survey will have made clear, Gödel's paper led to many interesting results of a technical nature. But also, notwithstanding the fact that Gödel did not quite achieve his own aims (see 4.4), the work connected with his paper has taught us several facts of philosophical interest. For example, the issue of the interpretation of $A \to A \wedge A$ made us aware of the role of decidable equality and of the contrast between intensionally and extensionally conceived functionals (see 3.2, 3.3 and 5.4).

The most important insight is perhaps that the use of logic can be

Über eine bisher noch nicht benützte Erweiterung des finiten Standpunktes (*1958*)

P. Bernays hat wiederholt darauf hingewiesen,[1] dass angesichts der Tatsache der Unbeweisbarkeit der Widerspruchsfreiheit eines Systems mit geringeren Beweismitteln als denen des Systems selbst eine Überschreitung des Rahmens der im Hilbertschen Sinn finiten Mathematik nötig ist, um die Widerspruchsfreiheit der klassischen Mathematik, ja sogar um die der klassischen Zahlentheorie zu beweisen. Da die finite Mathematik als die der *anschaulichen* Evidenz definiert ist,[2] so bedeutet das (wie auch von Bernays (*1935*, p. 62 und 69) explizit formuliert wurde), dass man für den Widerspruchsfreiheitsbeweis der Zahlentheorie gewisse *abstrakte* Begriffe braucht. Dabei sind unter abstrakten (oder nichtanschaulichen) Begriffen solche zu verstehen, die wesentlich von zweiter oder höherer Stufe sind, das heisst, die nicht Eigenschaften oder Relationen *konkreter Objekte* (z. B. von Zeichenkombinationen) beinhalten, sondern sich auf *Denkgebilde* (z. B. Beweise, sinnvolle Aussagen usw.) beziehen, wobei in den Beweisen Einsichten über die letzteren gebraucht werden, die sich nicht aus den kombinatorischen (raumzeitlichen) Eigenschaften der sie darstellenden Zeichenkombinationen, sondern nur aus deren *Sinn* ergeben.

[1] Vgl. z. B.: *Bernays 1941a*, p. 144, 147; ferner: *Hilbert und Bernays 1939*, §5; und: *Bernays 1954*, p. 10.

[2] Vgl. die Hilbertsche Formulierung in *Hilbert 1926*, p. 171–173.

replaced by the use of higher-type functionals, for a very limited set of types. The study of the functionals in T and its extensions has made it clear that formal proofs and functionals are in many ways similar. Thus, the work reported in 5.2 above has taught us, among other things, that normalization of functionals is, essentially, the same as normalization of proofs.

A. S. Troelstra[p]

[p]This commentary owes much to discussions with my friend and colleague Justus Diller. I gratefully acknowledge also the help and extensive comments from J. Dawson, G. Kreisel and especially S. Feferman.

On a hitherto unutilized extension of the finitary standpoint (*1958*)

P. Bernays has pointed out on several occasions[1] that, since the consistency of a system cannot be proved using means of proof weaker than those of the system itself, it is necessary to go beyond the framework of what is, in Hilbert's sense, finitary mathematics if one wants to prove the consistency of classical mathematics, or even that of classical number theory. Consequently, since finitary mathematics is defined as the mathematics in which evidence rests on what is *intuitive*,[2] certain *abstract* notions are required for the proof of the consistency of number theory (as was also explicitly formulated by Bernays in his *1935*, pages 62 and 69). Here, by abstract (or nonintuitive) notions we must understand those that are essentially of second or higher order, that is, notions that do not involve properties or relations of *concrete objects* (for example, of combinations of signs), but that relate to *mental constructs* (for example, proofs, meaningful statements, and so on); and in the proofs we make use of insights, into these mental constructs, that spring not from the combinatorial (spatiotemporal) properties of the sign combinations representing the proofs, but only from their *meaning*.

[1]See, for example, *Bernays 1941a*, pp. 144 and 147; see also *Hilbert and Bernays 1939*, §5, and *Bernays 1954*, p. 10.

[2]See Hilbert's formulation in his *1926*, pp. 171–173.

Obwohl in Ermanglung eines präzisen Begriffs der anschaulichen, beziehungsweise abstrakten, Evidenz ein strenger Beweis für die Bernayssche Feststellung nicht vorliegt, so kann doch über ihre Richtigkeit praktisch kein Zweifel bestehen, insbesondere seit dem Gentzenschen Beweis für die For-
281 malisierbarkeit aller Rekur|sionen nach Ordinalzahlen $< \epsilon_0$ in der Zahlentheorie. Denn die Gültigkeit des Rekursionsschlusses für ϵ_0 kann sicher nicht unmittelbar anschaulich gemacht werden, wie das zum Beispiel bei ω^2 möglich ist. Das heisst genauer, man kann die verschiedenen strukturellen Möglichkeiten, die für absteigende Folgen bestehen, nicht mehr übersehen und hat daher keine anschauliche Erkenntnis von der Notwendigkeit des Abbrechens jeder solchen Folge. Insbesondere kann durch schrittweises Übergehen von kleineren zu grösseren Ordinalzahlen eine solche *anschauliche* Erkenntnis nicht realisiert werden, sondern bloss eine abstrakte Erkenntnis mit Hilfe von Begriffen höherer Stufe. Das letztere wird durch den abstrakten Begriff der "Erreichbarkeit"[3] geleistet, welcher durch die inhaltliche Beweisbarkeit der Gültigkeit einer gewissen Schlussweise definiert ist. Auch ist es im Rahmen der für uns anschaulichen Mathematik nicht möglich, den Induktionsschluss nach einer hinreichend grossen Ordinalzahl auf eine Kette anderer Einsichten zurückzuführen. Vielmehr führt jeder Versuch, das zu tun, zu Induktionen von im wesentlichen derselben Ordnung. Ob die Notwendigkeit abstrakter Begriffe bloss durch die praktische Unmöglichkeit, kombinatorisch allzu komplizierte Verhältnisse an-
282 schaulich vorzustellen,[4] bedingt ist oder prinzipielle Gründe hat, | lässt sich nicht ohne weiteres entscheiden. Im zweiten Fall müsste nach Präzisierung der fraglichen Begriffe ein strenger Beweis für das Bestehen jener Notwendigkeit möglich sein.

[3]W. Ackermann erklärt zwar in *1951*, p. 407, dass "erreichbar" einen anschaulichen Sinn habe, wenn Beweisbarkeit als formale Beweisbarkeit nach gewissen Regeln verstanden wird. Aber darauf ist zu erwidern, dass aus dieser anschaulichen Tatsache die Gültigkeit des Schlusses durch transfinite Induktion für eine vorgelegte Eigenschaft nur mit Hilfe abstrakter Begriffe (oder mit Hilfe transfiniter Induktion in der Metamathematik) folgt. Allerdings ist der Begriff "erreichbar", zumindest für Induktionen bis ϵ_0, durch schwächere abstrakte Begriffe ersetzbar (vgl. *Hilbert und Bernays 1939*).

[4]Man beachte, dass eine adäquate beweistheoretische Charakterisierung einer durch Absehen von dieser Schranke *idealisierten* anschaulichen Evidenz Schlussweisen enthalten wird, die *für uns* nicht anschaulich sind und die sehr wohl eine Reduktion des induktiven Schlusses auf den einer wesentlich kleineren Ordnung gestatten könnten. Eine andere Möglichkeit, den ursprünglichen finiten Standpunkt zu erweitern, für die dasselbe gilt, besteht darin, dass man abstrakte Begriffe, die auf nichts anderes als auf finite Begriffe und Gegenstände, und zwar in kombinatorisch finiter Weise, Bezug nehmen, mit zur finiten Mathematik rechnet und diesen Prozess iteriert. Solche Begriffe sind zum Beispiel diejenigen, welche in der Reflexion auf den Inhalt schon konstruierter finiter For-
282 malismen involviert sind. Ein | dieser Idee entsprechender Formalismus wurde von G. Kreisel aufgestellt. Vgl. seinen Vortrag auf dem Internationalen Mathematikerkongress in Edinburgh, 1958 ⟦*Kreisel 1960*⟧. Man beachte, dass bei dieser Art der Erweiterung des Finitismus das abstrakte Element in einer wesentlichen schwächeren Form auftritt als bei der weiter unten besprochenen oder in der intuitionistischen Logik.

In the absence of a precise notion of what it means to be evident, either in the intuitive or in the abstract realm, we have no strict proof of Bernays' assertion; practically speaking, however, there can be no doubt that it is correct, in particular after Gentzen proved that all recursions on ordinals less than ϵ_0 can be formalized in number theory. For, the validity of inference by recursion up to ϵ_0 surely cannot be made immediately intuitive, as it can up to, say, ω^2. More precisely, we can no longer survey the various structural possibilities that obtain for descending sequences, and therefore we cannot intuitively recognize that every such sequence will necessarily terminate. In particular, we cannot acquire such knowledge *intuitively* by passing stepwise from smaller to larger ordinals; we can only gain knowledge abstractly by means of notions of higher type. This is achieved by means of the abstract notion of 'accessibility',[3] which is defined by our being able to give an informally understood proof that a certain kind of inference is valid. Moreover, within the framework of that part of mathematics which is intuitive to us, inference by induction up to a sufficiently large ordinal cannot be reduced to a chain of other insights. Rather, every attempt to do so leads to inductions of essentially the same order. It cannot be determined out of hand whether the need for abstract notions is due merely to the practical impossibility of our intuitively imagining states of affairs that are all too complex from the combinatorial point of view[4] or whether there are theoretical reasons for it. In the latter case it would have to be possible, once the notions in question have been made precise, to give a strict proof that this need exists.

[3]To be sure, W. Ackermann tells us in his *1951*, p. 407, that 'accessible' will be intuitively meaningful if provability is understood as formal provability according to certain rules. But to this one must reply that, from this intuitive fact, the validity of inference by transfinite induction for a given property can be demonstrated only by means of abstract notions (or by means of transfinite induction in metamathematics). The notion 'accessible' can, however, be replaced, at least for inductions up to ϵ_0, by weaker abstract notions (see *Hilbert and Bernays 1939*).

[4]Note that, if we were to give an adequate proof-theoretic characterization of *idealized* intuitive evidence while ignoring this limitation, we would use kinds of inference that, *for us*, are not intuitive and that might very well allow us to reduce the inductive inference to one of a substantially lower order. The same holds of another possible extension of the original finitary standpoint; it consists in adjoining to finitary mathematics abstract notions that relate, in a combinatorially finitary way, only to finitary notions and objects, and then iterating this procedure. Among such notions are, for example, those that are involved when we reflect on the content of finitary formalisms that have already been constructed. A formalism embodying this idea was set up by G. Kreisel. See his *1960*. Note that, when finitism is extended in this way, the abstract element appears in an essentially weaker form than in the extension discussed below or in intuitionistic logic.

Jedenfalls lehrt die Bernayssche Bemerkung, zwei Bestandteile in der finiten Einstellung unterscheiden, nämlich erstens das konstruktive Element, welches darin besteht, dass von mathematischen Objekten nur insoweit die Rede sein darf, als man sie aufweisen oder durch Konstruktion tatsächlich herstellen kann; zweitens das spezifisch finitistische Element, welches darüber hinaus fordert, dass die Objekte, über welche man Aussagen macht, mit welchen die Konstruktionen ausgeführt werden und welche man durch sie erhält, "anschaulich" sind, das heisst letzten Endes raum-zeitliche Anordnungen von Elementen, deren Beschaffenheit abgesehen von Gleichheit und Verschiedenheit irrelevant ist. (Im Gegensatz dazu sind jene Objekte in der intuitionistischen Logik sinnvolle Aussagen und Beweise.)

Es ist die zweite Forderung, welche fallen gelassen werden muss. Dieser Tatsache wurde bisher dadurch Rechnung getragen, dass man Teile der intuitionistischen Logik und Ordinalzahltheorie zur finiten Mathematik adjungierte. Im folgenden wird gezeigt, dass man statt dessen für den Widerspruchsfreiheitsbeweis der Zahlentheorie auch den Begriff der berechenbaren Funktion endlichen Typs über den natürlichen Zahlen und gewisse sehr elementare Konstruktionsprinzipien für solche Funktionen verwenden kann. Dabei wird der Begriff "berechenbare Funktion vom Typus t" folgendermassen erklärt: 1. Die berechenbaren Funktionen vom Typus 0 sind die natürlichen Zahlen. 2. Wenn die Begriffe "berechenbare Funktion vom Typus t_0", "berechenbare Funktion vom Typus t_1", ..., "berechenbare Funktion vom Typus t_k" (wobei $k \geq 1$) bereits definiert sind, so wird eine berechenbare Funktion vom Typus $(t_0, t_1, \dots t_k)$ definiert als eine immer ausführbare (und als solche konstruktiv erkennbare) Operation, 283 welche jedem k-tupel | berechenbarer Funktionen der Typen $t_1, t_2, \dots t_k$ eine berechenbare Funktion vom Typus t_0 zuordnet. Dieser Begriff[5] ist als unmittelbar verständlich[6] zu betrachten, vorausgesetzt dass man die

[5]Man kann darüber im Zweifel sein, ob wir eine genügend deutliche Vorstellung vom Inhalt dieses Begriffs haben, aber nicht darüber, ob die weiter unten angegebenen Axiome für ihn gelten. Derselbe scheinbar paradoxe Sachverhalt besteht auch für den der intuitionistischen Logik zugrunde liegenden Begriff des inhaltlich richtigen Beweises. Wie die nachfolgenden Überlegungen und die intuitionistisch interpretierte Theorie der rekursiven Funktionen und Funktionale zeigen, sind diese beiden Begriffe innerhalb gewisser Grenzen als Grundbegriffe durcheinander ersetzbar. Dabei ist zu beachten, dass, wenn der Begriff der berechenbaren Funktion nicht implizit den Begriff des Beweises enthalten soll, die Ausführbarkeit der Operationen unmittelbar aus der Kette der Definitionen ersichtlich sein muss, wie das für alle Funktionen des weiter unten angegebenen Systems T der Fall ist.

[6]A. M. Turing hat bekanntlich mit Hilfe des Begriffs einer Rechenmaschine eine Definition des Begriffs einer berechenbaren Funktion erster Stufe gegeben. Aber wenn dieser Begriff nicht schon vorher verständlich gewesen wäre, hätte die Frage, ob die Turingsche Definition adäquat ist, keinen Sinn.

In any case Bernays' remark teaches us to distinguish two components in the finitary attitude; namely, first, the constructive element, which consists in our being allowed to speak of mathematical objects only in so far as we can exhibit them or actually produce them by means of a construction; second, the specifically finitistic element, which makes the further demand that the objects about which we make statements, with which the constructions are carried out and which we obtain by means of these constructions, are 'intuitive', that is, are in the last analysis spatiotemporal arrangements of elements whose characteristics other than their identity or nonidentity are irrelevant. (By contrast, in intuitionistic logic these objects are meaningful statements and proofs.)

It is the second requirement that must be dropped. This fact has hitherto been taken into account by our adjoining to finitary mathematics parts of intuitionistic logic and the theory of ordinals. In what follows we shall show that, for the consistency proof of number theory, we can use, instead, the notion of computable function of finite type on the natural numbers and certain rather elementary principles of construction for such functions. Here the notion 'computable function of type t' is defined as follows:

(1) the computable functions of type 0 are the natural numbers;

(2) if the notions 'computable function of type t_0', 'computable function of type t_1', ..., 'computable function of type t_k' (with $k \geq 1$) have already been defined, then a computable function of type $(t_0, t_1, \ldots, t_k)$ is defined as an operation, always performable (and constructively recognizable as such), that to every k-tuple of computable functions of types $t_1, \ldots, t_k$ assigns a computable function of type t_0. This notion[5] is to be regarded as immediately intelligible,[6] provided the notions 'computable function of type t_i' $(i = 0, 1, \ldots, k)$ are already understood. If we then regard the type t as a variable, we arrive at the notion, required for the consistency proof, of a computable function of finite type t.

[5]One may doubt whether we have a sufficiently clear idea of the content of this notion, but not that the axioms given below hold for it. The same apparently paradoxical situation also obtains for the notion, basic to intuitionistic logic, of a proof that is informally understood to be correct. As the considerations presented below and the intuitionistically interpreted theory of recursive functions and functionals show, these two notions are, within certain limits, interchangeable as primitive notions. If the notion of computable function is not to implicitly contain the notion of proof, we must see to it that it is immediately apparent from the chain of definitions that the operations can be performed, as is the case for all functions in the system T specified below.

[6]As is well known, A. M. Turing, using the notion of a computing machine, gave a definition of the notion of computable function of the first order. But, had this notion not already been intelligible, the question whether Turing's definition is adequate would be meaningless.

Begriffe "berechenbare Funktion vom Typus t_i" $(i = 0, 1, \ldots k)$ bereits verstanden hat. Indem man dann den Typus t als Variable betrachtet, gelangt man zu dem für den Widerspruchsfreiheitsbeweis benötigten Begriff einer berechenbaren Funktion endlichen Typs t.

Als evidente Axiome sind, neben den Axiomen der Identität (auch für Funktionen[7]), dem 3. und 4. Peanoschen Axiom und der Substitutionsregel für freie Variable, keine anderen nötig als erstens solche, die es gestatten, Funktionen durch Gleichsetzung mit einem aus Variablen und vorher definierten Konstanten aufgebauten Term und durch einfache Induktion nach einer Zahlvariablen zu definieren, und zweitens den Schluss der vollständigen Induktion nach einer Zahlvariablen anzuwenden. Das heisst die Axiome dieses Systems (es werde T genannt) sind formal fast 284 dieselben[8] wie die der primitiv rekursiven Zahlentheorie, nur dass | die Variablen (ausser denen, auf die Induktion angewendet wird), sowie auch die definierten Konstanten, einen beliebigen endlichen Typus über den natürlichen Zahlen haben können. Der Einfachheit halber wird im folgenden der zweiwertige Aussagenkalkül, angewendet auf Gleichungen, hinzugenommen, obwohl die Wahrheitsfunktionen durch zahlentheoretische Funktionen ersetzbar sind. Gebundene Variable werden nicht zugelassen. Das System T ist von gleicher Beweisstärke wie ein System der rekursiven Zahlentheorie, in dem vollständige Induktion für alle Ordinalzahlen $< \epsilon_0$ (in der gewöhnlichen Darstellung) zugelassen wird.

Die Zurückführung der Widerspruchsfreiheit der klassischen Zahlentheorie auf die des Systems T gelingt mit Hilfe der folgenden Interpretation der Heytingschen Zahlentheorie, auf welche ja die klassische zurückführbar ist:[9]

Es wird jeder Formel F der intuitionistischen Zahlentheorie[10] (deren freie Variable in ihrer Gesamtheit mit x bezeichnet werden) eine Formel F' der Gestalt $(\exists y)(z)A(y, z, x)$ zugeordnet, wobei y und z endliche Reihen von Variablen irgendwelcher Typen sind, und $A(y, z, x)$ ein quantorenfreier Ausdruck mit keinen andern als den in x, y, z vorkommenden Variablen. Die Variablen der Reihen x, y, z, deren Gliederzahl auch 0 sein kann, sind sämtlich untereinander verschieden. Mit xy wird die aus x und y in dieser Reihenfolge zusammengesetzte Reihe bezeichnet.

[7] Identität zwischen Funktionen ist als intensionale oder Definitionsgleichheit zu verstehen.

[8] Bei der Definition durch Gleichsetzung mit einem Term tritt insofern ein Unterschied auf, als man eine Funktion P höheren Typs auch durch $[P(x_1, x_2, \ldots x_n)](y_1, y_2, \ldots y_m) = E$ definieren kann. Aber dieser Unterschied fällt weg, falls mehrstellige Funktionen in der von A. Church angegebenen Weise durch einstellige ersetzt werden.

[9] Vgl. *Gödel 1933e*.

[10] Die Zahlentheorie soll so formalisiert sein, dass keine Aussagen- oder Funktionsvariable vorkommen. Die Axiome des Aussagenkalküls sind als Schemata für alle möglichen Einsetzungen zu betrachten.

Besides the axioms of identity (including those for functions[7]), Peano's third and fourth axioms, and the rule of substitution for free variables, we need no other axioms ⟦for the notion of computable function⟧ than the following equally evident ones: (1) axioms that allow us to define functions by setting them equal to a term constructed from variables and previously defined constants, as well as by simple induction on a number variable; (2) axioms that allow us to use inference by mathematical induction on a number variable. That is, the axioms of this system (let it be T) are formally almost the same[8] as those of primitive recursive number theory, the only exception being that the variables (other than those on which induction is carried out), as well as the defined constants, can be of any finite type over the natural numbers. For the sake of simplicity, we shall, in what follows, avail ourselves of the two-valued propositional calculus, applied to equations, even though truth functions could be replaced by number-theoretic ones. Bound variables are not admitted. The system T is of the same proof-theoretic strength as a system of recursive number theory in which induction is permitted up to any ordinal less than ϵ_0 (in the usual representation).

The consistency of classical number theory can be reduced to that of the system T by means of the following interpretation of Heyting's number theory, to which, of course, classical number theory is reducible:[9]

To each formula F of intuitionistic number theory[10] (x standing for all of its free variables) we assign a formula F' of the form $(\exists y)(z)A(y,z,x)$, where y and z are finite sequences of variables of any type and $A(y,z,x)$ is a quantifier-free expression containing no other variables than those occurring in x, y and z. The variables of the (possibly empty) sequences x, y and z are understood to be pairwise distinct. Let xy be the sequence obtained when x is immediately followed by y.

[7]Identity between functions is to be understood as intensional or definitional equality.

[8]When we define a function by setting it equal to a term, a difference does occur, since we can also define a function P of higher type by the stipulation

$$[P(x_1, x_2, \ldots, x_n)](y_1, y_2, \ldots, y_m) = E.$$

But this difference vanishes if we replace many-place functions by one-place functions in the way specified by A. Church.

[9]See *Gödel 1933e*.

[10]Number theory is assumed to be formalized so that no propositional or functional variables occur. The axioms of the propositional calculus are to be regarded as schemas in which all possible substitutions are permitted.

Ferner werden folgende Bezeichnungen verwendet:

1. v, w sind endliche Reihen von Variablen irgendwelcher Typen; s, t sind Zahlvariable; u ist eine Reihe von Zahlvariablen.

2. V ist eine Reihe von Variablen, deren Anzahl und Typen dadurch bestimmt sind, dass jede von ihnen auf y als Argumentreihe angewendet werden kann und dass die Reihe der so erhaltenen Werte (welche mit $V(y)$ bezeichnet wird) hinsichtlich der Anzahl und der Typen ihrer Glieder mit der Reihe v übereinstimmt.

285 | 3. Analog wird die Variablenreihe Y (bzw. Z, bzw. $\overline{Z}$) hinsichtlich der Anzahl und der Typen ihrer Glieder durch die Argumentreihe s (bzw. yw, bzw. y) und durch die mit der Reihe der Werte gleichtypige Reihe y (bzw. z, bzw. z) bestimmt.

Funktionen mit 0 Leerstellen und Werten vom Typus τ werden mit Objekten vom Typus τ identifiziert, eingliedrige Variablenreihen mit Variablen.

Die Zuordnung von F' zu F geschieht durch Induktion nach der Anzahl k der in F enthaltenen logischen Operatoren. (Die bei der Wahl der Symbole für die gebundenen Variablen zu beachtenden Bedingungen und die heuristische Begründung der Definitionen werden nach den Formeln gegeben.)

I. Für $k = 0$ sei $F' = F$.

II. Es sei

$$F' = (\exists y)(z)A(y, z, x)$$

und

$$G' = (\exists v)(w)B(v, w, u)$$

bereits definiert; dann ist *per definitionem*:

1. $(F \wedge G)' = (\exists yv)(zw)[A(y, z, x) \wedge B(v, w, u)]$.
2. $(F \vee G)' = (\exists yvt)(zw)[t = 0 \wedge A(y, z, x) \,.\vee.\, t = 1 \wedge B(v, w, u)]$.
3. $[(s)F]' = (\exists Y)(sz)A\big(Y(s), z, x\big)$.
4. $[(\exists s)F]' = (\exists sy)(z)A(y, z, x)$.
5. $(F \supset G)' = (\exists VZ)(yw)[A\big(y, Z(yw), x\big) \supset B\big(V(y), w, u\big)]$.
6. $(\neg F)' = (\exists \overline{Z})(y)\neg A\big(y, \overline{Z}(y), x\big)$.

s ist eine beliebige Zahlvariable. Vor Anwendung der Regeln 1–5 sind nötigenfalls die gebundenen Variablen der Formeln F' und G' so umzubenennen, dass sie sämtlich untereinander und von den Variablen der Reihen x, u sowie auch von s verschieden sind. Ferner sind die durch Anwendung der Regeln 2, 3, 5, 6 neu eingeführten gebundenen Variablen der Reihen t, Y, V, Z, $\overline{Z}$ so zu wählen, dass sie untereinander und von den in den betreffenden Formeln schon vorkommenden Variablen verschieden sind.

Man beachte, dass 6. aus 5. folgt, falls $\neg p$ durch $p \supset .0 = 1$ definiert wird. Zu 5. gelangt man, indem man (für die auftretenden Spezialfälle) 286 die Aussage $(\exists x)H(x) \supset (\exists y)R(y)$ (bzw. $(y)R(y) \supset$ | $(x)H(x)$) mit der Existenz von für alle Argumentreihen vom Typus der Variablenreihe x definierten berechenbaren Funktionen identifiziert, welche jedem Beispiel

We also use the following notation:

(1) v and w are finite sequences of variables of any type; s and t are number variables; u is a sequence of number variables;

(2) V is a sequence of variables whose number and types are determined thus: each of these variables can take y as an argument sequence, and the sequence of values thus obtained (let it be $V(y)$) agrees with the sequence v in the number and types of its terms;

(3) similarly, the sequence Y (or Z, or $\overline{Z}$) is determined, as far as the number and types of its terms are concerned, by the argument sequence s (or yw, or y, respectively), as well as by the sequence y (or z, or z, respectively) whose types are those of the sequence of values.

Zero-place functions whose values are of type τ are identified with objects of type τ, and one-term sequences of variables are identified with variables.

The assignment of F' to F proceeds by induction on the number k of logical operators contained in F. (The conditions to be observed in choosing symbols for the bound variables, as well as the heuristic justification of the definitions, will be given after the formulas.)

I. For $k = 0$, let $F' = F$.

II. Let

$$F' = (\exists y)(z)A(y, z, x)$$

and

$$G' = (\exists v)(w)B(v, w, u)$$

be already defined; then we have by definition

1. $(F \wedge G)' = (\exists yv)(zw)[A(y, z, x) \wedge B(v, w, u)]$.
2. $(F \vee G)' = (\exists yvt)(zw)[t = 0 \wedge A(y, z, x) \,.\!\vee\!.\, t = 1 \wedge B(v, w, u)]$.
3. $[(s)F]' = (\exists Y)(sz)A(Y(s), z, x)$.
4. $[(\exists s)F]' = (\exists sy)(z)A(y, z, x)$.
5. $(F \supset G)' = (\exists VZ)(yw)[A(y, Z(yw), x) \supset B(V(y), w, u)]$.
6. $(\neg F)' = (\exists \overline{Z})(y)\neg A(y, \overline{Z}(y), x)$.

Here s is any number variable. Before applying Rules 1–5, we rename, if necessary, the bound variables of formulas F' and G' so that they will all be distinct from one another and from the variables of the sequences x and u, as well as from s. Further, the bound variables of the sequences t, Y, V, Z and $\overline{Z}$ that are newly introduced when Rules 2, 3, 5 and 6 are applied must be chosen distinct from one another and from the variables that already occur in the formulas considered.

Note that 6 follows from 5, in case $\neg p$ is defined as $p \supset . 0 = 1$. We arrive at 5 as follows: we identify (for the special cases at hand) the proposition $(\exists x)H(x) \supset (\exists y)R(y)$ (or $(y)R(y) \supset (x)H(x)$) with the existence of computable functions (defined for all argument sequences of the same type as the variable sequence x) that to each sequence making the antecedent true assign a sequence making the consequent true (or to each sequence making the consequent false assign a sequence making the antecedent false).

Obviously, we do not claim that Definitions 1–6 reproduce the meaning

für das Implicans (bzw. Gegenbeispiel für das Implicatum) ein Beispiel für das Implicatum (bzw. Gegenbeispiel für das Implicans) zuordnen.

Selbstverständlich wird nicht behauptet, dass die Definitionen 1–6 den Sinn der von Brouwer und Heyting eingeführten logischen Partikel wiedergeben. Wieweit sie diese ersetzen können, bedarf einer näheren Untersuchung. Man zeigt leicht, dass, *wenn F im Heytingschen System* Z *der Zahlentheorie beweisbar ist, Funktionen Q in* T *definiert werden können, für welche $A(Q(x), z, x)$ in* T *beweisbar ist.* Es ist nämlich leicht nachprüfbar, dass diese Behauptung für die Axiome von Z gilt und ihre Richtigkeit sich bei Anwendung der Schlussregeln von Z von den Prämissen auf die Konklusion überträgt.

Die Verifikation wird besonders einfach, wenn man folgendes Axiomensystem der intuitionistischen Logik zugrunde legt:[11]

Axiome: Taut, Add, Perm, die zu diesen dualen Axiome für $\wedge$, $0 = 1 . \supset p$ ($\neg p$ wird durch $p \supset . 0 = 1$ definiert).

Schlussregeln: Modus ponens, Einsetzungsregel für freie Zahlvariable, Syll (mit zwei Prämissen), Sum, Exp, Imp, die Regeln über das Hinzufügen und Weglassen eines All- (bzw. Existenz-)Zeichens im Implicatum (bzw. Implicans) einer bewiesenen Implikation.

Für den Widerspruchsfreiheitsbeweis der klassischen Zahlentheorie können die $\vee$ und die $\exists$ enthaltenden Axiome und Schlussregeln weggelassen werden. Bei allen auf Sum folgenden Regeln stellt sich heraus, dass die in T zu beweisende Aussage im wesentlichen dieselbe ist wie die auf Grund der Prämisse bereits bewiesene.

Es ist klar, dass man, von demselben Grundgedanken ausgehend, auch viel stärkere Systeme als T konstruieren kann, zum Beispiel durch Zulassung transfiniter Typen oder der von Brouwer für den Beweis des "Fan-Theorems"[12] benutzten Schlussweise.

287 |

Zusammenfassung

P. Bernays hat darauf hingewiesen, dass man, um die Widerspruchsfreiheit der klassischen Zahlentheorie zu beweisen, den Hilbertschen finiten Standpunkt dadurch erweitern muss, dass man neben den auf Symbole sich beziehenden kombinatorischen Begriffen gewisse abstrakte Begriffe zulässt. Die abstrakten Begriffe, die bisher für diesen Zweck verwendet wurden, sind die der konstruktiven Ordinalzahltheorie und die der intuitionistischen Logik. Es wird gezeigt, dass man statt dessen den Begriff einer berechenbaren Funktion endlichen einfachen Typs über den natürlichen Zahlen benutzen kann, wobei keine anderen Konstruktionsverfahren für solche Funktionen nötig sind, als einfache Rekursion nach einer Zahlvariablen und Einsetzung von Funktionen ineinander (mit trivialen Funktionen als Ausgangspunkt).

[11] Bezüglich der Bezeichnungen, vgl. *Whitehead und Russell 1925*, p. xii. Dieselben Bezeichnungen werden auch für die den Formeln entsprechenden Schlussregeln verwendet.

[12] Vgl. *Heyting 1956*, p. 42.

of the logical particles introduced by Brouwer and Heyting. Further investigation is needed to see how far these can be replaced by our definitions. One can easily show that, *if F is provable in Heyting's system* Z *of number theory, then in* T *functions Q can be defined for which $A(Q(x), z, x)$ is provable in* T. For one can easily verify that this assertion holds for the axioms of Z and that, when we apply any inference rule of Z, it holds for the conclusion whenever it holds for the premises.

The verification is particularly simple if we adopt the following axiom system for intuitionistic logic:[11]

Axioms: Taut, Add, Perm, the axioms dual to these for $\wedge$, and $0 = 1 . \supset p$ ($\neg p$ is defined as $p \supset . 0 = 1$).

Rules of inference: Modus ponens, the rule of substitution for free number variables, Syll (with two premises), Sum, Exp, Imp, and the rules for inserting or deleting a universal quantifier in the consequent (or an existential quantifier in the antecedent) of a proved conditional.

For the consistency proof of classical number theory we can omit the axioms and rules of inference containing $\vee$ or $\exists$. For all the rules following Sum it turns out that the proposition to be proved in T is essentially the same as the one that has already been proved on the basis of the premise.

It is clear that, starting from the same basic idea, one can also construct systems that are much stronger than T, for example by admitting transfinite types or the sort of inference that Brouwer used in proving the 'fan theorem'.[12]

Abstract

P. Bernays has pointed out that, in order to prove the consistency of classical number theory, it is necessary to extend Hilbert's finitary standpoint by admitting certain abstract concepts in addition to the combinatorial concepts referring to symbols. The abstract concepts that so far have been used for this purpose are those of the constructive theory of ordinals and those of intuitionistic logic. It is shown that the concept of a computable function of finite simple type over the integers can be used instead, where no other procedures of constructing such functions are necessary except simple recursion by an integer variable and substitution of functions in each other (starting with trivial functions).

[11]For the notation see *Whitehead and Russell 1925*, p. xii. We use the same notation for the rules of inference that correspond to the formulas.

[12]See *Heyting 1956*, p. 42.

Kurt Gödel and Alfred Tarski, March 1962

Postscript to *Spector 1962* (*1962*)

⟦*Spector 1962* was published posthumously and edited by G. Kreisel.⟧

This important paper ⟦*Spector 1962*⟧ was written by Clifford Spector during his stay at the Institute for Advanced Study in 1960–1961 under a grant from the Office of Naval Research. The discussions P. Bernays and I had with Spector (see footnote 1)[1] took place after the main result (contained in §10 of the paper) had been established already. However, it ought to be mentioned that during the time Spector first established this result he was in close contact with Kreisel. It was Spector's express intention to give to Kreisel a good deal of credit for his work. Originally a joint publication by Spector and Kreisel was envisaged. This plan was dropped because Spector had taken over the elaboration by himself and because the version of the proof which was to be published was due to Spector. Also Spector alone, at that time, was working on an extension of the result in the direction of stricter constructivity which he hoped to include in his paper.

[1]⟦Footnote 1 of *Spector 1962* was written by Kreisel, as were all the footnotes in that paper, and stated in part: "From paragraph 3 of the introduction below, and from conversations with Spector, I know that he valued highly his discussions with P. Bernays and K. Gödel on the subject of the present paper."⟧

What is Cantor's continuum problem? (*1964*)

[This article is a revised and expanded version of *Gödel 1947*. The introductory note to both *1947* and *1964* is found on page 154, immediately preceding *1947*.]

1. The concept of cardinal number

Cantor's continuum problem is simply the question: How many points are there on a straight line in Euclidean space? An equivalent question is: How many different sets of integers do there exist?

This question, of course, could arise only after the concept of "number" had been extended to infinite sets; hence it might be doubted if this extension can be effected in a uniquely determined manner and if, therefore, the statement of the problem in the simple terms used above is justified. Closer examination, however, shows that Cantor's definition of infinite numbers really has this character of uniqueness. For whatever "number" as applied to infinite sets may mean, we certainly want it to have the property that the number of objects belonging to some class does not change if, leaving the objects the same, one changes in any way whatsoever their properties or mutual relations (e.g., their colors or their distribution in space). From this, however, it follows at once that two sets (at least two sets of changeable objects of the space-time world) will have the same cardinal number if their elements can be brought into a one-to-one correspondence, which is Cantor's definition of equality between numbers. For if there exists such a correspondence for two sets A and B it is possible (at least theoretically) to change the properties and relations of each element of A into those of the corresponding element of B, whereby A is transformed into a set completely indistinguishable from B, hence of the same cardinal number. For example, assuming a square and a line segment both completely filled with mass points (so that at each point of them exactly one mass point 259 is situated), it follows, owing to the | demonstrable fact that there exists a one-to-one correspondence between the points of a square and of a line segment and, therefore, also between the corresponding mass points, that the mass points of the square can be so rearranged as exactly to fill out the line segment, and vice versa. Such considerations, it is true, apply directly only to physical objects, but a definition of the concept of "number" which would depend on the kind of objects that are numbered could hardly be considered to be satisfactory.

So there is hardly any choice left but to accept Cantor's definition of equality between numbers, which can easily be extended to a definition of "greater" and "less" for infinite numbers by stipulating that the cardinal number M of a set A is to be called less than the cardinal number N of a set B if M is different from N but equal to the cardinal number of some subset of B. That a cardinal number having a certain property exists is defined to mean that a set of such a cardinal number exists. On the basis of these definitions, it becomes possible to prove that there exist infinitely many different infinite cardinal numbers or "powers", and that, in particular, the number of subsets of a set is always greater than the number of its elements; furthermore, it becomes possible to extend (again without any arbitrariness) the arithmetical operations to infinite numbers (including sums and products with any infinite number of terms or factors) and to prove practically all ordinary rules of computation.

But, even after that, the problem of identifying the cardinal number of an individual set, such as the linear continuum, would not be well-defined if there did not exist some systematic representation of the infinite cardinal numbers, comparable to the decimal notation of the integers. Such a systematic representation, however, does exist, owing to the theorem that for each cardinal number and each set of cardinal numbers[1] there exists exactly one cardinal number immediately succeeding in magnitude and that the cardinal number of every set occurs in the series thus obtained.[2] This theorem makes it possible to denote the cardinal number immediately succeeding the set of finite numbers by $\aleph_0$ (which is the power of the "denumerably infinite" sets), the next one by $\aleph_1$, etc.; the one immediately succeeding all $\aleph_i$ (where i is an integer) by $\aleph_\omega$, the next one by $\aleph_{\omega+1}$, etc. The theory of ordinal numbers provides the means for extending this series further and further.

[1]As to the question of why there does not exist a set of all cardinal numbers, see footnote 15.

[2]The axiom of choice is needed for the proof of this theorem (see *Fraenkel and Bar-Hillel 1958*). But it may be said that this axiom, from almost every possible point of view, is as well-founded today as the other axioms of set theory. It has been proved consistent with the other axioms of set theory which are usually assumed, provided that these other axioms are consistent (see my *1940*). Moreover, it is possible to define in terms of any system of objects satisfying the other axioms a system of objects satisfying those axioms *and* the axiom of choice. Finally, the axiom of choice is just as evident as the other set-theoretical axioms for the "pure" concept of set explained in footnote 14.

260 |

2. The continuum problem, the continuum hypothesis, and the partial results concerning its truth obtained so far

So the analysis of the phrase "how many" unambiguously leads to a definite meaning for the question stated in the second line of this paper: The problem is to find out which one of the $\aleph$'s is the number of points of a straight line or (which is the same) of any other continuum (of any number of dimensions) in a Euclidean space. Cantor, after having proved that this number is greater than $\aleph_0$, conjectured that it is $\aleph_1$. An equivalent proposition is this: Any infinite subset of the continuum has the power either of the set of integers or of the whole continuum. This is Cantor's continuum hypothesis.

But, although Cantor's set theory now has had a development of more than seventy years and the problem evidently is of great importance for it, nothing has been proved so far about the question what the power of the continuum is or whether its subsets satisfy the condition just stated, except (1) that the power of the continuum is not a cardinal number of a certain special kind, namely, not a limit of denumerably many smaller cardinal numbers,[3] and (2) that the proposition just mentioned about the subsets of the continuum is true for a certain infinitesimal fraction of these subsets, the analytic[4] sets.[5] Not even an upper bound, however large, can be assigned for the power of the continuum. Nor is the quality of the cardinal number of the continuum known any better than its quantity. It is undecided whether this number is regular or singular, accessible or inaccessible, and (except for König's negative result) what its character of cofinality (see footnote 4) is. The only thing that is known, in addition to the results just mentioned, is a great number of consequences of, and some propositions equivalent to, Cantor's conjecture.[6]

This pronounced failure becomes still more striking if the problem is considered in its connection with general questions of cardinal arithmetic. It is easily proved that the power of the continuum is equal to $2^{\aleph_0}$. So the continuum problem turns out to be a question from the "multiplication table" of cardinal numbers, namely, the problem of evaluating a certain

[3]See *Hausdorff 1914*, p. 68, or *Bachmann 1955*, p. 167. The discoverer of this theorem, J. König, asserted more than he had actually proved (see his *1905*.)

[4]See the list of definitions on pp. 268–9.

[5]See *Hausdorff 1935*, p. 32. Even for complements of analytic sets the question is undecided at present, and it can be proved only that they either have the power $\aleph_0$ or $\aleph_1$ or that of the continuum or are finite (see *Kuratowski 1933*, p. 246.)

[6]See *Sierpiński 1934* and *1956*.

infinite product (in fact the simplest non-trivial one that can be formed). There is, however, not one infinite product (of factors > 1) for which so much as an upper bound for its value can be assigned. All one knows about the evaluation of infinite products are two lower bounds due to Cantor and König (the latter of which implies the aforementioned negative theorem on the power of | the continuum), and some theorems concerning 261 the reduction of products with different factors to exponentiations and of exponentiations to exponentiations with smaller bases or exponents. These theorems reduce[7] the whole problem of computing infinite products to the evaluation of $\aleph_\alpha^{\mathrm{cf}(\aleph_\alpha)}$ and the performance of certain fundamental operations on ordinal numbers, such as determining the limit of a series of them. All products and powers can easily be computed[8] if the "generalized continuum hypothesis" is assumed, i.e., if it is assumed that $2^{\aleph_\alpha} = \aleph_{\alpha+1}$ for every α, or, in other terms, that the number of subsets of a set of power $\aleph_\alpha$ is $\aleph_{\alpha+1}$. But, without making any undemonstrated assumption, it is not even known whether or not $m < n$ implies $2^m < 2^n$ (although it is trivial that it implies $2^m \leq 2^n$), nor even whether $2^{\aleph_0} < 2^{\aleph_1}$.

3. Restatement of the problem on the basis of an analysis of the foundations of set theory and results obtained along these lines

This scarcity of results, even as to the most fundamental questions in this field, to some extent may be due to purely mathematical difficulties; it seems, however (see Section 4), that there are also deeper reasons involved and that a complete solution of these problems can be obtained only by a more profound analysis (than mathematics is accustomed to giving) of the meanings of the terms occurring in them (such as "set", "one-to-one correspondence", etc.) and of the axioms underlying their use. Several such analyses have already been proposed. Let us see then what they give for our problem.

First of all there is Brouwer's intuitionism, which is utterly destructive in its results. The whole theory of the $\aleph$'s greater than $\aleph_1$ is rejected as meaningless.[9] Cantor's conjecture itself receives several different meanings, all of which, though very interesting in themselves, are quite different from

[7]This reduction can be effected, owing to the results and methods of *Tarski 1925*.

[8]For regular numbers $\aleph_\alpha$, one obtains immediately:

$$\aleph_\alpha^{\mathrm{cf}(\aleph_\alpha)} = \aleph_\alpha^{\aleph_\alpha} = 2^{\aleph_\alpha} = \aleph_{\alpha+1}.$$

[9]See *Brouwer 1909*.

the original problem. They lead partly to affirmative, partly to negative answers.[10] Not everything in this field, however, has been sufficiently clarified. The "semi-intuitionistic" standpoint along the lines of H. Poincaré and H. Weyl[11] would hardly preserve substantially more of set theory.

262 | However, this negative attitude toward Cantor's set theory, and toward classical mathematics, of which it is a natural generalization, is by no means a necessary outcome of a closer examination of their foundations, but only the result of a certain philosophical conception of the nature of mathematics, which admits mathematical objects only to the extent to which they are interpretable as our own constructions or, at least, can be completely given in mathematical intuition. For someone who considers mathematical objects to exist independently of our constructions and of our having an intuition of them individually, and who requires only that the general mathematical concepts must be sufficiently clear for us to be able to recognize their soundness and the truth of the axioms concerning them, there exists, I believe, a satisfactory foundation of Cantor's set theory in its whole original extent and meaning, namely, axiomatics of set theory interpreted in the way sketched below.

It might seem at first that the set-theoretical paradoxes would doom to failure such an undertaking, but closer examination shows that they cause no trouble at all. They are a very serious problem, not for mathematics, however, but rather for logic and epistemology. As far as sets occur in mathematics (at least in the mathematics of today, including all of Cantor's set theory), they are sets of integers, or of rational numbers (i.e., of pairs of integers), or of real numbers (i.e., of sets of rational numbers), or of functions of real numbers (i.e., of sets of pairs of real numbers), etc. When theorems about all sets (or the existence of sets in general) are asserted, they can always be interpreted without any difficulty to mean that they hold for sets of integers as well as for sets of sets of integers, etc. (respectively, that there either exist sets of integers, or sets of sets of integers, or ... etc., which have the asserted property). This concept of set,[12] how-

[10]See *Brouwer 1907*, I, 9; III, 2.

[11]See *Weyl 1932*. If the procedure of construction of sets described there (p. 20) is iterated a sufficiently large (transfinite) number of times, one gets exactly the real numbers of the model for set theory mentioned in Section 4, in which the continuum hypothesis is true. But this iteration is not possible within the limits of the semi-intuitionistic standpoint.

[12]It must be admitted that the spirit of the modern abstract disciplines of mathematics, in particular of the theory of categories, transcends this concept of set, as becomes apparent, e.g., by the self-applicability of categories (see *Mac Lane 1961*). It does not seem, however, that anything is lost from the mathematical content of the theory if categories of different levels are distinguished. If there existed mathematically interesting proofs that would not go through under this interpretation, then the paradoxes of set theory would become a serious problem for mathematics.

ever, according to which a set is something obtainable from the integers (or some other well-defined objects) by iterated application[13] of the operation "set of",[14] not something obtained by | dividing the totality of all existing things into two categories, has never led to any antinomy whatsoever; that is, the perfectly "naïve" and uncritical working with this concept of set has so far proved completely self-consistent.[15] 263

But, furthermore, the axioms underlying the unrestricted use of this concept of set or, at least, a subset of them which suffices for all mathematical proofs devised up to now (except for theorems depending on the existence of extremely large cardinal numbers, see footnote 20), have been formulated so precisely in axiomatic set theory[16] that the question of whether some given proposition follows from them can be transformed, by means of mathematical logic, into a purely combinatorial problem concerning the manipulation of symbols which even the most radical intuitionist must acknowledge as meaningful. So Cantor's continuum problem, no matter what philosophical standpoint is taken, undeniably retains at least this meaning: to find out whether an answer, and if so which answer, can be derived from the axioms of set theory as formulated in the systems cited.

Of course, if it is interpreted in this way, there are (assuming the consistency of the axioms) a priori three possibilities for Cantor's conjecture: It may be demonstrable, disprovable, or undecidable.[17] The third alternative (which is only a precise formulation of the foregoing conjecture, that the difficulties of the problem are probably not purely mathematical) is the most likely. To seek a proof for it is, at present, perhaps the most promising way of attacking the problem. One result along these lines has been

[13]This phrase is meant to include transfinite iteration, i.e., the totality of sets obtained by finite iteration is considered to be itself a set and a basis for further applications of the operation "set of".

[14]The operation "set of x's" (where the variable "x" ranges over some given kind of objects) cannot be defined satisfactorily (at least not in the present state of knowledge), but can only be paraphrased by other expressions involving again the concept of set, such as: "multitude of x's", "combination of any number of x's", "part of the totality of x's", where a "multitude" ("combination", "part") is conceived of as something which exists in itself no matter whether we can define it in a finite number of words (so that random sets are not excluded).

[15]It follows at once from this explanation of the term "set" that a set of all sets or other sets of a similar extension cannot exist, since every set obtained in this way immediately gives rise to further applications of the operation "set of" and, therefore, to the existence of larger sets.

[16]See, e.g., *Bernays 1937, 1941, 1942, 1943, von Neumann 1925*; cf. also *von Neumann 1928a* and *1929, Gödel 1940, Bernays and Fraenkel 1958*. By including very strong axioms of infinity, much more elegant axiomatizations have recently become possible. (See *Bernays 1961*.)

[17]In case the axioms were inconsistent the last one of the four a priori possible alternatives for Cantor's conjecture would occur, namely, it would then be both demonstrable and disprovable by the axioms of set theory.

obtained already, namely, that Cantor's conjecture is not disprovable from the axioms of set theory, provided that these axioms are consistent (see Section 4).

It is to be noted, however, that on the basis of the point of view here adopted, a proof of the undecidability of Cantor's conjecture from the accepted axioms of set theory (in contradistinction, e.g., to the proof of the transcendency of π) would by no means solve the problem. For if the meanings of the primitive terms of set theory as explained on page 262 and in footnote 14 are accepted as sound, it follows that the set-theoretical concepts and theorems describe some well-determined reality, in which Cantor's conjecture | must be either true or false. Hence its undecidability from the axioms being assumed today can only mean that these axioms do not contain a complete description of that reality. Such a belief is by no means chimerical, since it is possible to point out ways in which the decision of a question, which is undecidable from the usual axioms, might nevertheless be obtained.

264

First of all the axioms of set theory by no means form a system closed in itself, but, quite on the contrary, the very concept of set[18] on which they are based suggests their extension by new axioms which assert the existence of still further iterations of the operation "set of". These axioms can be formulated also as propositions asserting the existence of very great cardinal numbers (i.e., of sets having these cardinal numbers). The simplest of these strong "axioms of infinity" asserts the existence of inaccessible numbers (in the weaker or stronger sense) $> \aleph_0$. The latter axiom, roughly speaking, means nothing else but that the totality of sets obtainable by use of the procedures of formation of sets expressed in the other axioms forms again a set (and, therefore, a new basis for further applications of these procedures).[19] Other axioms of infinity have first been formulated by P. Mahlo.[20] These axioms show clearly, not only that the axiomatic

[18]Similarly the concept "property of set" (the second of the primitive terms of set theory) suggests continued extensions of the axioms referring to it. Furthermore, concepts of "property of property of set" etc. can be introduced. The new axioms thus obtained, however, as to their consequences for propositions referring to limited domains of sets (such as the continuum hypothesis) are contained (as far as they are known today) in the axioms about sets.

[19]See *Zermelo 1930*.

[20][*Revised note of September 1966*: See *Mahlo 1911*, pp. 190–200, and *1913*, pp. 269–276. From Mahlo's presentation of the subject, however, it does not appear that the numbers he defines actually exist. In recent years great progress has been made in the area of axioms of infinity. In particular, some propositions have been formulated which, if consistent, are extremely strong axioms of infinity of an entirely new kind (see *Keisler and Tarski 1964* and the material cited there). Dana Scott (*1961*) has proved that one of them implies the existence of non-constructible sets. That these axioms are implied by the general concept of set in the same sense as Mahlo's has not been made clear

system of set theory as used today is incomplete, but also that it can be supplemented without arbitrariness by new axioms which only unfold the content of the concept of set explained above.

It can be proved that these axioms also have consequences far outside the domain of very great transfinite numbers, which is their immediate subject matter: each of them, under the assumption of its consistency, can be shown to increase the number of decidable propositions even in the field of Diophantine equations. As for the continuum problem, there is little hope of | solving it by means of those axioms of infinity which can be set up on the basis of Mahlo's principles (the aforementioned proof for the undisprovability of the continuum hypothesis goes through for all of them without any change). But there exist others based on different principles (see footnote 20); also there may exist, besides the usual axioms, the axioms of infinity, and the axioms mentioned in footnote 18, other (hitherto unknown) axioms of set theory which a more profound understanding of the concepts underlying logic and mathematics would enable us to recognize as implied by these concepts (see, e.g., footnote 23). 265

Secondly, however, even disregarding the intrinsic necessity of some new axiom, and even in case it has no intrinsic necessity at all, a probable decision about its truth is possible also in another way, namely, inductively by studying its "success". Success here means fruitfulness in consequences, in particular in "verifiable" consequences, i.e., consequences demonstrable without the new axiom, whose proofs with the help of the new axiom, however, are considerably simpler and easier to discover, and make it possible to contract into one proof many different proofs. The axioms for the system of real numbers, rejected by the intuitionists, have in this sense been verified to some extent, owing to the fact that analytical number theory frequently allows one to prove number-theoretical theorems which, in a more cumbersome way, can subsequently be verified by elementary methods. A much higher degree of verification than that, however, is conceivable. There might exist axioms so abundant in their verifiable consequences, shedding so much light upon a whole field, and yielding such powerful methods for solving problems (and even solving them constructively, as far as that is possible) that, no matter whether or not they are intrinsically necessary, they would have to be accepted at least in the same sense as any well-established physical theory.

yet (see *Tarski 1962*, p. 134). However, they are supported by strong arguments from analogy, e.g., by the fact that they follow from the existence of generalizations of Stone's representation theorem to Boolean algebras with operations on infinitely many elements. Mahlo's axioms of infinity have been derived from a general principle about the totality of sets which was first introduced by A. Levy (*1960*). It gives rise to a hierarchy of different precise formulations. One, given by P. Bernays (*1961*), implies all of Mahlo's axioms.]

4. Some observations about the question: In what sense and in which direction may a solution of the continuum problem be expected?

But are such considerations appropriate for the continuum problem? Are there really any clear indications for its unsolvability by the accepted axioms? I think there are at least two:

The first results from the fact that there are two quite differently defined classes of objects both of which satisfy all axioms of set theory that have been set up so far. One class consists of the sets definable in a certain manner by properties of their elements;[21] the other of the sets in the sense of arbitrary multitudes, regardless of if, or how, they can be defined. Now, 266 before it has | been settled what objects are to be numbered, and on the basis of what one-to-one correspondences, one can hardly expect to be able to determine their number, except perhaps in the case of some fortunate coincidence. If, however, one believes that it is meaningless to speak of sets except in the sense of extensions of definable properties, then, too, he can hardly expect more than a small fraction of the problems of set theory to be solvable without making use of this, in his opinion essential, characteristic of sets, namely, that they are extensions of definable properties. This characteristic of sets, however, is neither formulated explicitly nor contained implicitly in the accepted axioms of set theory. So from either point of view, if in addition one takes into account what was said in Section 2, it may be conjectured that the continuum problem cannot be solved on the basis of the axioms set up so far, but, on the other hand, may be solvable with the help of some new axiom which would state or imply something about the definability of sets.[22]

The latter half of this conjecture has already been verified; namely, the concept of definability mentioned in footnote 21 (which itself is definable in axiomatic set theory) makes it possible to derive, in axiomatic set theory, the generalized continuum hypothesis from the axiom that every set is definable in this sense.[23] Since this axiom (let us call it "A") turns

[21]Namely, definable by certain procedures, "in terms of ordinal numbers" (i.e., roughly speaking, under the assumption that for each ordinal number a symbol denoting it is given). See my papers *1939a* and *1940*. The paradox of Richard, of course, does not apply to this kind of definability, since the totality of ordinals is certainly not denumerable.

[22]D. Hilbert's program for a solution of the continuum problem (see his *1926*), which, however, has never been carried through, also was based on a consideration of all possible definitions of real numbers.

[23]On the other hand, from an axiom in some sense opposite to this one, the negation of Cantor's conjecture could perhaps be derived. I am thinking of an axiom which (similar to Hilbert's completeness axiom in geometry) would state some maximum

out to be demonstrably consistent with the other axioms, under the assumption of the consistency of these other axioms, this result (regardless of the philosophical position taken toward definability) shows the consistency of the continuum hypothesis with the axioms of set theory, provided that these axioms themselves are consistent.[24] This proof in its structure is similar to the consistency proof of non-Euclidean geometry by means of a model within Euclidean geometry. Namely, it follows from the axioms of set theory that the sets definable in the aforementioned sense form a model of set theory in which the proposition A and, therefore, the generalized continuum hypothesis is true.

A second argument in favor of the unsolvability of the continuum problem on the basis of the usual axioms can be based on certain facts (not known at Cantor's time) which seem to indicate that Cantor's conjecture will turn out | to be wrong,[25] while, on the other hand, a disproof of it is demonstrably impossible on the basis of the axioms being assumed today. 267

One such fact is the existence of certain properties of point sets (asserting an extreme rareness of the sets concerned) for which one has succeeded in proving the existence of non-denumerable sets having these properties, but no way is apparent in which one could expect to prove the existence of examples of the power of the continuum. Properties of this type (of subsets of a straight line) are: (1) being of the first category on every perfect set,[26] (2) being carried into a zero set by every continuous one-to-one mapping of the line onto itself.[27] Another property of a similar nature is that of being coverable by infinitely many intervals of any given lengths. But in this case one has so far not even succeeded in proving the existence of non-denumerable examples. From the continuum hypothesis, however, it follows in all three cases that there exist, not only examples of the power of the continuum,[28] but even such as are carried into themselves (up to denumerably many points) by *every* translation of the straight line.[29]

Other highly implausible consequences of the continuum hypothesis are that there exist: (1) subsets of a straight line of the power of the continuum which are covered (up to denumerably many points) by *every* dense set

property of the system of all sets, whereas axiom A states a minimum property. Note that only a maximum property would seem to harmonize with the concept of set explained in footnote 14.

[24]See my monograph *1940* and my paper *1939a*. For a carrying through of the proof in all details, my *1940* is to be consulted.

[25]Views tending in this direction have been expressed also by N. Luzin in his *1935*, pp. 129 ff. See also *Sierpiński 1935*.

[26]See *Sierpiński 1934a* and *Kuratowski 1933*, pp. 269 ff.

[27]See *Luzin and Sierpiński 1918* and *Sierpiński 1934a*.

[28]For the third case see *Sierpiński 1934*, p. 39, Theorem 1.

[29]See *Sierpiński 1935a*.

of intervals;[30] (2) infinite-dimensional subsets of Hilbert space which contain no non-denumerable finite-dimensional subset (in the sense of Menger-Urysohn);[31] (3) an infinite sequence A^i of decompositions of any set M of the power of the continuum into continuum-many mutually exclusive sets A^i_x such that, in whichever way a set $A^i_{x_i}$ is chosen for each i, $\prod_{i=0}^{\infty}(M - A^i_{x_i})$ is denumerable.[32] (1) and (3) are very implausible even if "power of the continuum" is replaced by "$\aleph_1$".

One may say that many results of point-set theory obtained without using the continuum hypothesis also are highly unexpected and implausible.[33] But, true as that may be, still the situation is different there, in that, in most of those instances (such as, e.g., Peano's curves) the appearance to the contrary can be explained by a lack of agreement between our intuitive geometrical concepts and the set-theoretical ones occurring in the theorems. Also, it is very | suspicious that, as against the numerous plausible propositions which imply the negation of the continuum hypothesis, not one plausible proposition is known which would imply the continuum hypothesis. I believe that adding up all that has been said one has good reason for suspecting that the role of the continuum problem in set theory will be to lead to the discovery of new axioms which will make it possible to disprove Cantor's conjecture.

Definitions of some of the technical terms

Definitions 4–15 refer to subsets of a straight line, but can be literally transferred to subsets of Euclidean spaces of any number of dimensions if "interval" is identified with "interior of a parallelepipedon".

1. I call *the character of cofinality* of a cardinal number m (abbreviated by "cf(m)") the smallest number n such that m is the sum of n numbers $< m$.
2. A cardinal number m is *regular* if $\mathrm{cf}(m) = m$, otherwise singular.
3. An infinite cardinal number m is *inaccessible* if it is regular and has no immediate predecessor (i.e., if, although it is a limit of numbers $< m$, it is not a limit of fewer than m such numbers); it is *strongly inaccessible* if each product (and, therefore, also each sum) of fewer than m numbers $< m$ is $< m$. (See *Sierpiński and Tarski 1930*, *Tarski 1938*.)

[30]See *Luzin 1914*, p. 1259.

[31]See *Hurewicz 1932*.

[32]See *Braun and Sierpiński 1932*, p. 1, proposition (Q). This proposition is equivalent with the continuum hypothesis.

[33]See, e.g., *Blumenthal 1940*.

It follows from the generalized continuum hypothesis that these two concepts are equivalent. $\aleph_0$ is evidently inaccessible, and also strongly inaccessible. As for finite numbers, 0 and 2 and no others are strongly inaccessible. A definition of inaccessibility, applicable to finite numbers, is this: m is inaccessible if (1) any sum of fewer than m numbers $< m$ is $< m$, and (2) the number of numbers $< m$ is m. This definition, for transfinite numbers, agrees with that given above and, for finite numbers, yields 0, 1, 2 as inaccessible. So inaccessibility and strong inaccessibility turn out not to be equivalent for finite numbers. This casts some doubt on their equivalence for transfinite numbers, which follows from the generalized continuum hypothesis.

4. A set of intervals is *dense* if every interval has points in common with some interval of the set. (The endpoints of an interval are not considered as points of the interval.)
5. A *zero set* is a set which can be covered by infinite sets of intervals with arbitrarily small lengths-sum.
6. A *neighborhood* of a point P is an interval containing P.
7. A subset A of B is *dense in* B if every neighborhood of any point of B contains points of A.
8. | A point P is in the *exterior* of A if it has a neighborhood containing 269 no point of A.
9. A subset A of B is *nowhere dense in* B if those points of B which are in the exterior of A are dense in B, or (which is equivalent) if for no interval I the intersection IA is dense in IB.
10. A subset A of B is *of the first category in* B if it is the sum of denumerably many sets nowhere dense in B.
11. A set A is *of the first category on* B if the intersection AB is of the first category in B.
12. A point P is called *a limit point* of a set A if any neighborhood of P contains infinitely many points of A.
13. A set A is called *closed* if it contains all its limit points.
14. A set is *perfect* if it is closed and has no isolated point (i.e., no point with a neighborhood containing no other point of the set).
15. *Borel sets* are defined as the smallest system of sets satisfying the postulates:
    (1) The closed sets are Borel sets.
    (2) The complement of a Borel set is a Borel set.
    (3) The sum of denumerably many Borel sets is a Borel set.
16. A set is *analytic* if it is the orthogonal projection of some Borel set of a space of next higher dimension. (Every Borel set therefore is, of course, analytic.)

Supplement to the second edition

Since the publication of the preceding paper, a number of new results have been obtained; I would like to mention those that are of special interest in connection with the foregoing discussions.

1. A. Hajnal has proved[34] that, if $2^{\aleph_0} \neq \aleph_2$ could be derived from the axioms of set theory, so could $2^{\aleph_0} = \aleph_1$. This surprising result could greatly facilitate the solution of the continuum problem, should Cantor's continuum hypothesis be demonstrable from the axioms of set theory, which, however, probably is not the case.

2. Some new consequences of, and propositions equivalent with, Cantor's hypothesis can be found in the new edition of W. Sierpiński's book.[35] In the first edition, it had been proved that the continuum hypothesis is equivalent with the proposition that the Euclidean plane is the sum of denumerably many "generalized curves" (where a generalized curve is a point 270 set definable | by an equation $y = f(x)$ in some Cartesian coordinate system). In the second edition, it is pointed out[36] that the Euclidean plane can be proved to be the sum of fewer than continuum-many generalized curves under the much weaker assumption that the power of the continuum is not an inaccessible number. A proof of the converse of this theorem would give some plausibility to the hypothesis $2^{\aleph_0}$ = the smallest inaccessible number $> \aleph_0$. However, great caution is called for with regard to this inference,[36a] because the paradoxical appearance in this case (like in Peano's "curves") is due (at least in part) to a transference of our geometrical intuition of curves to something which has only some of the characteristics of curves. Note that nothing of this kind is involved in the counterintuitive consequences of the continuum hypothesis mentioned on page 267.

3. C. Kuratowski has formulated a strengthening of the continuum hypothesis,[37] whose consistency follows from the consistency proof mentioned in Section 4. He then drew various consequences from this new hypothesis.

4. Very interesting new results about the axioms of infinity have been obtained in recent years (see footnotes 20 and 16).

In opposition to the viewpoint advocated in Section 4 it has been suggested[38] that, in case Cantor's continuum problem should turn out to be

[34]See his *1956*.

[35]See *Sierpiński 1956*.

[36]See his *1956*, p. 207 or his *1951*, p. 9. Related results are given by C. Kuratowski (*1951*, p. 15) and R. Sikorski (*1951*).

[36a][*Note added September 1966*: It seems that this warning has since been vindicated by Roy O. Davies (*1963*).]

[37]See his *1948*.

[38]See *Errera 1952*.

undecidable from the accepted axioms of set theory, the question of its truth would lose its meaning, exactly as the question of the truth of Euclid's fifth postulate by the proof of the consistency of non-Euclidean geometry became meaningless for the mathematician. I therefore would like to point out that the situation in set theory is very different from that in geometry, both from the mathematical and from the epistemological point of view.

In the case of the axiom of the existence of inaccessible numbers, e.g., (which can be proved to be undecidable from the von Neumann-Bernays axioms of set theory provided that it is consistent with them) there is a striking asymmetry, mathematically, between the system in which it is asserted and the one in which it is negated.[39]

Namely, the latter (but not the former) has a model which can be defined and proved to be a model in the original (unextended) system. This means that the former is an extension in a much stronger sense. A closely related fact is that the assertion (but not the negation) of the axiom implies new theorems about integers (the individual instances of which can be verified by computation). So the criterion of truth explained on page 264 is satisfied, to some extent, for the assertion, but not for the negation. Briefly speaking, only the assertion | yields a "fruitful" extension, while the negation is sterile outside its own very limited domain. The generalized continuum hypothesis, too, can be shown to be sterile for number theory and to be true in a model constructible in the original system, whereas for some other assumption about the power of $2^{\aleph_\alpha}$ this perhaps is not so. On the other hand, neither one of those asymmetries applies to Euclid's fifth postulate. To be more precise, both it and its negation are extensions in the weak sense. 271

As far as the epistemological situation is concerned, it is to be said that by a proof of undecidability a question loses its meaning only if the system of axioms under consideration is interpreted as a hypothetico-deductive system, i.e., if the meanings of the primitive terms are left undetermined. In geometry, e.g., the question as to whether Euclid's fifth postulate is true retains its meaning if the primitive terms are taken in a definite sense, i.e., as referring to the behavior of rigid bodies, rays of light, etc. The situation in set theory is similar; the difference is only that, in geometry, the meaning usually adopted today refers to physics rather than to mathematical intuition and that, therefore, a decision falls outside the range of mathematics. On the other hand, the objects of transfinite set theory, conceived in the manner explained on page 262 and in footnote 14, clearly do not belong to the physical world, and even their indirect connection with physical experience is very loose (owing primarily to the fact that set-theoretical concepts play only a minor role in the physical theories of today).

[39]The same asymmetry also occurs on the lowest levels of set theory, where the consistency of the axioms in question is less subject to being doubted by skeptics.

But, despite their remoteness from sense experience, we do have something like a perception also of the objects of set theory, as is seen from the fact that the axioms force themselves upon us as being true. I don't see any reason why we should have less confidence in this kind of perception, i.e., in mathematical intuition, than in sense perception, which induces us to build up physical theories and to expect that future sense perceptions will agree with them, and, moreover, to believe that a question not decidable now has meaning and may be decided in the future. The set-theoretical paradoxes are hardly any more troublesome for mathematics than deceptions of the senses are for physics. That new mathematical intuitions leading to a decision of such problems as Cantor's continuum hypothesis are perfectly possible was pointed out earlier (pages 264–265).

It should be noted that mathematical intuition need not be conceived of as a faculty giving an *immediate* knowledge of the objects concerned. Rather it seems that, as in the case of physical experience, we *form* our ideas also of those objects on the basis of something else which *is* immediately given. Only this something else here is *not*, or not primarily, the sensations. That something besides the sensations actually is immediately given follows (independently of mathematics) from the fact that even our ideas referring to physical objects contain constituents qualitatively different from sensations or mere combinations of sensations, e.g., the idea of object itself, whereas, on the other hand, by our thinking we cannot create
272 any qualitatively new elements, but only | reproduce and combine those that are given. Evidently the "given" underlying mathematics is closely related to the abstract elements contained in our empirical ideas.[40] It by no means follows, however, that the data of this second kind, because they cannot be associated with actions of certain things upon our sense organs, are something purely subjective, as Kant asserted. Rather they, too, may represent an aspect of objective reality, but, as opposed to the sensations, their presence in us may be due to another kind of relationship between ourselves and reality.

However, the question of the objective existence of the objects of mathematical intuition (which, incidentally, is an exact replica of the question of the objective existence of the outer world) is not decisive for the problem under discussion here. The mere psychological fact of the existence of an intuition which is sufficiently clear to produce the axioms of set theory and an open series of extensions of them suffices to give meaning to the question of the truth or falsity of propositions like Cantor's continuum hypothesis. What, however, perhaps more than anything else, justifies the

[40]Note that there is a close relationship between the concept of set explained in footnote 14 and the categories of pure understanding in Kant's sense. Namely, the function of both is "synthesis", i.e., the generating of unities out of manifolds (e.g., in Kant, of the idea of *one* object out of its various aspects).

acceptance of this criterion of truth in set theory is the fact that continued appeals to mathematical intuition are necessary not only for obtaining unambiguous answers to the questions of transfinite set theory, but also for the solution of the problems of finitary number theory[41] (of the type of Goldbach's conjecture),[42] where the meaningfulness and unambiguity of the concepts entering into them can hardly be doubted. This follows from the fact that for every axiomatic system there are infinitely many undecidable propositions of this type.

It was pointed out earlier (page 265) that, besides mathematical intuition, there exists another (though only probable) criterion of the truth of mathematical axioms, namely their fruitfulness in mathematics and, one may add, possibly also in physics. This criterion, however, though it may become decisive in the future, cannot yet be applied to the specifically set-theoretical axioms (such as those referring to great cardinal numbers), because very little is known about their consequences in other fields. The simplest case of an application of the criterion under discussion arises when some set-theoretical axiom has number-theoretical consequences verifiable by computation up to any given integer. On the basis of what is known today, however, it is not possible to make the truth of any set-theoretical axiom reasonably probable in this manner.

Postscript

273

[*Revised postscript of September 1966*: Shortly after the completion of the manuscript of the second edition [[*1964*]] of this paper the question of whether Cantor's continuum hypothesis is decidable from the von Neumann–Bernays axioms of set theory (the axiom of choice included) was settled in the negative by Paul J. Cohen. A sketch of the proof has appeared in his *1963* and *1964*. It turns out that for all $\aleph_\tau$ defined by the usual devices and not excluded by König's theorem (see page 260 above) the equality $2^{\aleph_0} = \aleph_\tau$ is consistent and an extension in the weak sense (i.e., it implies no new number-theoretical theorem). Whether, for a satisfactory concept of "standard definition", this is true for *all* definable $\aleph_\tau$ not excluded by König's theorem is an open question. An affirmative answer would require the solution of the difficult problem of making the concept of standard definition, or some wider concept, precise. Cohen's work, which

[41] Unless one is satisfied with inductive (probable) decisions, such as verifying the theorem up to very great numbers, or more indirect inductive procedures (see pp. 265, 272).

[42] I.e., universal propositions about integers which can be decided in each individual instance.

no doubt is the greatest advance in the foundations of set theory since its axiomatization, has been used to settle several other important independence questions. In particular, it seems to follow that the axioms of infinity mentioned in footnote 20, to the extent to which they have so far been precisely formulated, are not sufficient to answer the question of the truth or falsehood of Cantor's continuum hypothesis.]

On an extension of finitary mathematics which has not yet been used[a] (*1972*)

⟦The introductory note to *1972*, as well as to related items, is found on page 217, immediately preceding *1958*.⟧

Abstract

P. Bernays has pointed out that, even in order to prove only the consistency of classical number theory, it is necessary to extend Hilbert's finitary standpoint. He suggested admitting certain abstract concepts in addition to the combinatorial concepts referring to symbols. The abstract concepts that so far have been used for this purpose are those of the constructive theory of ordinals and those of intuitionistic logic. It is shown that a certain concept of computable function of finite simple type over the natural numbers can be used instead, where no other procedures of constructing such functions are necessary except primitive recursion by a number variable and definition of a function by an equality with a term containing only variables and/or previously introduced functions beginning with the function +1.

P. Bernays has pointed out[1] on several occasions that, in view of the fact that the consistency of a formal system cannot be proved by any deduction procedures available in the system itself, it is necessary to go beyond the framework of finitary mathematics in Hilbert's sense in order to prove the consistency of classical mathematics or even of classical number the-

[1]See: *Bernays 1941a*, pp. 144, 147, 150, 152; *Hilbert and Bernays 1939*, pp. 347–349, 357–360; *Bernays 1954*, p. 9; cf. also *Bernays 1935*, pp. 62, 69.

[a]The present paper is not a literal translation of the German original published in *Dialectica* (*1958*). In revising the translation by Leo F. Boron, I have rephrased many passages. But the meaning has nowhere been substantially changed. Some minor inaccuracies have been corrected and a number of notes have been added, to which the letters (**a**)–(**n**) refer. I wish to express my best thanks to Professor Dana Scott for supervising the typing of this and the subsequent paper ⟦*1972a*⟧ while I was ill, and to Professor Paul Bernays for reading the proof sheets and calling my attention to some oversights in the manuscript.

ory. Since finitary mathematics is defined[2] as the mathematics of *concrete intuition*, this seems to imply that *abstract concepts* are needed for the proof of consistency of number theory.[c] An extension of finitism by such concepts was explicitly suggested by Bernays in his *1935*, page 69. By abstract concepts, in this context, are meant concepts which are essentially of the second or higher level, i.e., which do not have as their content properties or relations of *concrete objects* (such as combinations of symbols), but

[2]See Hilbert's explanation in his *1926*, pp. 170–173.[b]

[b]"*Concrete intuition*", "*concretely intuitive*" are used as translations of "Anschauung", "anschaulich". The simple terms "*concrete*" or "*intuitive*" are also used in this sense in the present paper. What Hilbert means by "Anschauung" is substantially Kant's space-time intuition confined, however, to configurations of a finite number of discrete objects. Note that it is Hilbert's insistence on *concrete* knowledge that makes finitary mathematics so surprisingly weak and excludes many things that are just as incontrovertibly evident to everybody as finitary number theory. E.g., while any primitive recursive definition is finitary, the general principle of primitive recursive definition is not a finitary proposition, because it contains the abstract concept of function. There is nothing in the term "finitary" which would suggest a restriction to concrete knowledge. Only Hilbert's special interpretation of it introduces this restriction.

[c]*Accessibility* and some closely related concepts (combined with intuitionistic logic) are those that have been used most of all in consistency proofs (see: 1. *Gentzen 1936*, pp. 555, 558; 2. *Lorenzen 1951*, p. 99, in particular his "induction of the second kind"; 3. *Schütte 1954*, p. 31; 4. *Kreisel 1965*, p. 137, *1967*, p. 246, and *1968*, p. 351, §12; 5. *Takeuti 1957*, *1960*, *1967*.) These concepts create the deceptive impression of being based on a *concrete intuition* of certain *infinite procedures*, such as "counting beyond ω" or "running through" the ordinals smaller than an ordinal α. We do have such an intuition, but it does not reach very far in the series of ordinals, certainly no farther than finitism. In order to make the concept of accessibility *fruitful*, *abstract conceptions* are always necessary, e.g., insights about infinitely many possible insights in Gentzen's original definition, which is somewhat different from that given above (see his *1936*, p. 555, line 7). A closer approximation to Hilbert's finitism can be achieved by using the concept of free choice sequences instead of "accessibility".[d]

[d]This is really an abstract principle about schemes of ramification, which, however, by Brouwer and Heyting is stated and proved only for the case that their elements are integers (although it is not clear that this fact is substantially used in the proof). C. Spector in his *1962* has shown that the abstract principle implies the consistency of classical analysis, while Brouwer's principle yields only the consistency of a certain subsystem of it. Unfortunately, however, no satisfactory constructivistic proof is known for either one of the two principles (except that, according to G. Kreisel (*1965*, p. 143), the weaker principle can be proved relatively consistent with the other accepted axioms of intuitionism). It was G. Kreisel who first suggested using this principle for consistency proofs.

Perhaps the most promising extension of the system T is that obtained by introducing *higher-type computable functions of constructive ordinals*.

rather of *thought structures* or *thought contents* (e.g., proofs, meaningful propositions, and so on), where in the proofs of propositions about these mental objects insights are needed which are not derived from a reflection upon the combinatorial (space-time) properties of the symbols representing them, but rather from a reflection upon the *meanings* involved.[e]

Due to the lack of a precise definition of either concrete or abstract evidence there exists, today, no rigorous proof for the insufficiency (even for the consistency proof of number theory) of finitary mathematics. However, this surprising fact has been made abundantly clear through the examination of induction by ϵ_0 used in Gentzen's consistency proof of number theory. The situation may roughly be described as follows: Recursion for ϵ_0 could be proved finitarily if the consistency of number theory could. On the other hand the validity of this recursion can certainly not be made *immediately* evident, as is possible for example in the case of ω^2. That is to say, one cannot grasp at one glance the various structural possibilities which exist for decreasing sequences, and there exists, therefore, no *immediate* concrete knowledge of the termination of every such sequence. But furthermore such *concrete* knowledge (in Hilbert's sense) cannot be realized either by a stepwise transition from smaller to larger ordinal numbers, because the concretely evident steps, such as $\alpha \to \alpha^2$, are so small that they would have to be repeated ϵ_0 times in order to reach ϵ_0. The same is true of chains of other concretely evident inferences which one may try to use, e.g., Hilbert's ω-rule to the extent to which it is concretely evident. What can be accomplished is only an *abstract* knowledge based on concepts of higher level, e.g., on "accessibility". This concept can be defined by the fact that the validity of induction is constructively demonstrable for the ordinal in question.[3] Whether the necessity of abstract concepts for the proof of induction from a certain point on in the series of constructive ordinals is due solely to the impossibility of grasping intuitively the complicated (though

[3]W. Ackermann in his *1951*, p. 407, says that "accessible" has a concrete meaning if demonstrability is understood as formal provability according to certain rules. However, it is to be noted that from this concrete fact the validity of the rule of transfinite induction applied to a given property follows only with the help of abstract concepts, or with the help of transfinite induction in metamathematics. But it is true that the concept of "accessible", at least for induction up to ϵ_0, can be replaced by weaker abstract concepts (see *Hilbert and Bernays 1939*, p. 363ff.); see footnote **c**.

---

[e]An example is the concept "p implies q" in the sense of: "From a convincing proof of p a convincing proof of q can be obtained".

only *finitely* complicated) combinational relations involved,[4] or arises for some essential reason, cannot be decided off hand.

In the second case it must be possible, after making the concepts in question precise, to give a rigorous proof for the existence of that necessity.

At any rate Bernays' observations in his *1935*, footnote 1, teach us to distinguish two component parts in the concept of finitary mathematics, namely: first, the *constructivistic* element, which consists in admitting reference to mathematical objects or facts only in the sense that they can be exhibited, or obtained by construction or proof; second, the specifically *finitistic* element, which requires in addition that the objects and facts considered should be given in concrete mathematical intuition. This, as far as the objects are concerned, means that they must be finite space-time configurations of elements whose nature is irrelevant except for equality or difference. (In contrast to this, the objects in intuitionistic logic are meaningful propositions and proofs.)

It is the second requirement which must be dropped. Until now this fact was taken into account by adjoining to finitary mathematics parts of intuitionistic logic and of the constructivistic theory of ordinal numbers. It will be shown in the sequel that, instead, one can use, for the proof of consistency of number theory, a certain concept of a *computable function of finite type over the natural numbers* and some very elementary axioms and principles of construction for such functions.

The concept "computable function of type t" is defined as follows: 1. The computable functions of type 0 are the natural numbers. 2. If the

[4]Note that an adequate proof-theoretic characterization of concrete intuition, in case this faculty is *idealized* by abstracting from the practical limitation, will include induction procedures which *for us* are *not* concretely intuitive and which could very well yield a proof of the inductive inference for ϵ_0 or larger ordinals. Another possibility of extending the original finitary viewpoint for which the same comment holds consists in considering as finitary any abstract arguments which only reflect (in a combinatorially finitary manner) on the content of finitary formalisms constructed before, and iterate this reflection transfinitely, using only ordinals constructed in previous stages of this process. A formalism based on this idea was given by G. Kreisel at the International Congress of Mathematicians in Edinburgh, 1958 (*Kreisel 1960*).[f] Note that, if finitism is extended in this manner, the abstract element appears in an *essentially weaker form* than in any other extension mentioned in the present paper.

---

[f]An unobjectionable version is given in *Kreisel 1965*, pp. 168–173, 177–178. Theorem 3.43 on page 172 of these lectures states that ϵ_0 is the limit of this process. Kreisel wants to conclude from this fact that ϵ_0 is the exact limit of idealized concrete intuition. But his arguments would have to be elaborated further in order to be fully convincing. Note that Kreisel's hierarchy can be extended far beyond ϵ_0 by considering as one step any sequence of steps that has been shown to be admissible (e.g., any sequence of ϵ_0 steps). It then provides a means for making the much used concept of accessibility (see footnote **c** above) constructive in a much stricter sense by resolving the general impredicative concept of intuitionistic proof into constructed levels of formal proofs.

concepts "computable function of type t_0", "computable function of type t_1", ..., "computable function of type t_k" (where $k \geq 1$), have already been defined, then a computable function of type $(t_0, t_1, \ldots, t_k)$ is defined to be a well-defined mathematical procedure which can be applied to *any* k-tuple of computable functions of types $t_1, t_2, \ldots, t_k$, and yields a computable function of type t_0 as result; and for which, moreover, this general fact is constructively evident. The phrase "well-defined mathematical procedure" is to be accepted as having a clear meaning without any further explanation.[5] The functions occurring in this hierarchy are called "computable functions of finite type over the natural numbers."[6]

[5]It is well-known that A. M. Turing has given an elaborate definition of the concept of a *mechanically* computable function of natural numbers. This definition most certainly was not superfluous. However, if the term "mechanically computable" had not had a clear, although unanalyzed, meaning before, the question as to whether Turing's definition is adequate would be meaningless, while it undoubtedly has an affirmative answer.[g]

[6]One may doubt that, on the basis of the definition given, we have a sufficiently clear idea of the content of this concept, but not that the axioms of the system T given in the sequel are valid for it. The same apparently paradoxical situation also exists for the concept of an intuitionistically correct proof, which is the basis of intuitionistic logic in Heyting's interpretation. As the subsequent discussion will show, these two concepts can replace each other in building up intuitionistic logic within number theory. Of course, if this replacement is to have any epistemological significance, the concept of computable function used and the insight that these functions satisfy the axioms of T given below must not implicitly involve intuitionistic logic or the concept of proof as used by Heyting. This condition is satisfied for the concept of "computable of finite type" given in the text and footnote **h**.

---

[g]It is easily seen that Turing's functions (where functions as arguments or values of higher-type functions are to be identified with the code numbers of their machines) and certain subclasses of them also satisfy the axioms and rules of the system T given below. As to the meaning of this fact, see footnote **h**.

[h]An elaboration of this idea would lead to the following:

1. A *narrower* concept of proof, which may be called "reductive proof" and which, roughly speaking, is defined by the fact that, up to certain trivial supplementations, the chain of definitions of the concepts occurring in the theorem together with certain axioms about the primitive terms forms by itself a proof, i.e., an unbroken chain of immediate evidences. In special applications (as, e.g., in our case) this concept of proof can be made precise by specifying the supplementations, the axioms, and the evidences to be used.

Note that in this context a definition is to be considered as a theorem stating the existence and unicity of an object satisfying certain conditions and that, in our case, it is convenient that a statement regarding the type character of the function defined should form part of its definition.

2. A more special concept of "computable of type t", obtained by replacing in the definition given above the concept of "constructively evident or demonstrable"

As far as axioms and rules of inference for this concept are concerned, no others are needed except the following: (1) axioms for the two-valued propositional calculus applied to equations between terms of equal type, (2) the axioms of equality,[7] i.e., $x = x$ and $x = y . \supset . t(x) = t(y)$ for variables x, y and terms t of any type, (3) the third and fourth Peano axioms, i.e., $x + 1 \neq 0$, $x + 1 = y + 1 . \supset . x = y$, (4) the rule of substitution of terms of equal type for free variables (bound variables do not occur in the system), (5) rules which permit the definition of a function by an equality with a term constructed from variables and previously defined functions or by primitive recursion with respect to a number variable, (6) the usual version of the inference of complete induction with respect to a number variable.

[7]Equality of functions is to be understood as *intensional equality*. It means that the two functions have the same procedure of computation, i.e., (by our definition of "computable function") that they are identical. This is always decidable for two *given* functions, which justifies the application of the two-valued propositional calculus.

(occurring in it both explicitly, and implicitly through implications of the form: *If* x, y, ... have certain types, *then* ...) by "reductively provable". Note that, because "reductively provable" is a decidable property, the implications occurring may also be interpreted as truth-value functions.

3. The fact that, if the axioms, rules, and primitive concepts of T (note, e.g., that each type is a primitive concept of T) are, by means of the definitions 1, 2, just given, replaced by really primitive concepts and insights (or, at least, by something closer to real primitives), thus obtaining a system T′, only the (in comparison to Heyting's) incomparably narrower concept of *reductive* proof need be used in the propositions and proofs of T′, and that, moreover, because these proofs are uniquely determined by the theorems, quantifications over "any proof" can be avoided. Note that it is *not* claimed that the proofs in T′ *are* reductive. This is true only in certain cases, in particular for the proofs of the axioms of T and of the individual cases of the rules of T (nontrivially for those of groups (4) and (5), trivially for the others). What is claimed is only that no other *concept* of proof than that of reductive proof *occurs in* the propositions and proofs of T′, except, of course, insofar as any theorem P in intuitionism means: A *proof* of P has been given. Substantially the same method for avoiding the use of Heyting's logic or of the general concept of proof should be applicable also if T is interpreted in terms of Turing functions (see note (**g**)).

Item 3 shows that the interpretation of intuitionistic logic, in terms of computable functions, in no way presupposes Heyting's and that, moreover, it is constructive and evident in a higher degree than Heyting's. For it is exactly the elimination of such vast generalities as "any proof" which makes for greater evidence and constructivity.

The higher degree of constructivity also appears in other facts, e.g., that Markov's principle $\neg(x)\phi(x) \supset (\exists x)\neg\phi(x)$ (see *Kleene 1960*, page 157, footnote) is trivially provable for any primitive recursive ϕ and, in a more general setting, for any decidable property ϕ of any objects x. This, incidentally, gives an interest to this interpretation of intuitionistic logic (no matter whether in terms of computable functions of higher types or of Turing functions) even if Heyting's logic *is* presupposed.

Note that the axioms and rules of this system, which will be called T,[i] are formally almost the same as those of primitive recursive number theory, only that the constants and variables (except those to which induction is applied) can have any finite type over the natural numbers.[8] The system T has the same deductive power as a system of recursive number theory in

[8]The only other difference consists in the fact that a function P of higher type can also be defined by a term equation of the form: $[P(x_1, x_2, \ldots, x_n)]\,(y_1, y_2, \ldots, y_m) = t$, where t is a term containing no variables except $x_1, \ldots, x_n, y_1, \ldots, y_m$. This is a combination of two "abstractions", i.e., applications of the λ-operator. Formally, this difference vanishes if functions of several arguments are replaced by functions of one argument by A. Church's method.

---

[i]For a precise description of T the following should be added:

The primitive symbols of T are: 0, +1, =, *variables* and *defined constants* of any finite type, "*application*" of functions to arguments of suitable types (denoted by .(.,.)), and *propositional connectives*. *Terms* are built solely out of constants, variables, and application. *Meaningful formulae* are truth-value functions of equations between terms of equal type.

Regarding the axioms of T note the following:

1. The version of *complete induction* used in the consistency proof is this:

$$A(0, x), A(s, F(s, x)) \supset A(s+1, x) \ \vdash \ A(s, x)$$

where x is a finite sequence of variables of arbitrary types and F a sequence of previously defined functions of suitable types (as to the notation used here, see p. 278 below).

2. For the proofs that the Axioms 1 and 4 of H and the deduction Rule 6 of H hold in the interpretation $'$ defined below, the following principle of *disjunctive definition* is needed:

A function f may be defined by stipulating

$$A \supset f(x) = t_1, \quad \neg A \supset f(x) = t_2,$$

where t_1, t_2 are terms and A is a truth-value function of equations between *number terms*, both containing only previously defined functions and no variables except those of the sequence x.

3. Both the version of complete induction mentioned under 1 and the disjunctive definitions mentioned under 2 can be derived in T, the latter by means of disjunctive functions H defined recursively thus:

$$H(0, f, g) = f, \quad H(n+1, f, g) = g.$$

However, it seems preferable first to formulate axioms from which the consistency proof is immediate, and then reduce them to simpler ones: What adds considerably to the simplicity of the consistency proof also is the fact that we avoid extensional equality, which is an incomparably more intricate concept than logical identity.

4. If no attention is paid to the complexity of the consistency proof, the whole calculus of propositions in T can be dispensed with. For, 1. as applied to number equations it can be replaced by certain purely arithmetical devices, 2. as applied to equations of higher type it can be altogether omitted if (a) the second equality axiom (group (2)) is formulated as a rule of inference (which, incidentally, is used only for substituting the definiens and definiendum for each other) and (b) a disjunctive rule of inference is introduced which says that, if A follows both from $t = 0$ and $N(t) = 0$ (where N is defined by: $N(0) = 1$, $N(x+1) = 0$) by means of the other axioms and rules, except the rule of substitution for variables of t, then A may be asserted.

5. It is a curious fact that axiom group (3) is superfluous due to the recursive definability of a function δ by: $\delta(0) = 0$, $\delta(x+1) = x$, and due to the definability of $\neg p$ by $p \supset . 1 = 0$. For it follows immediately that: $x+1 = y+1 . \supset . \delta(x+1) = \delta(y+1) . \supset . x = y$ and $x+1 = 0 . \supset . \delta(x+1) = \delta(0) . \supset . x = 0 . \supset . 1 = 0$; on the other hand $\neg(1 = 0)$ by definition of $\neg$.

which complete induction is permitted for all ordinal numbers less than ϵ_0.

The reduction of the consistency of classical number theory to that of the system T is achieved by means of the following interpretation of intuitionistic number theory, to which classical number theory is reducible:[9]

With every formula F of the system H of intuitionistic number theory[10] a formula F' of the form $(\exists y)(z)A(y,z,x)$ is associated, where x is the sequence of free variables of F, y and z are finite sequences of variables of finite types, and $A(y,z,x)$ is a formula of T containing exactly the variables occurring in x, y, z. The variables of any one of the three sequences, x, y, z (each of which may also be empty), are always mutually distinct and distinct from those of the other two sequences. We denote by xy the sequence compounded of x and y in this order.[k]

Furthermore, the following notation is used in the formulas 1 to 6 below.

1. v, w are finite sequences of variables which may be of any types; s, t, are number variables; u is a finite sequence of number variables.

2. V is a finite sequence of variables whose number and types are determined by the fact that each of them can be applied to y as an argument sequence, and that the sequence of terms thus obtained (which is denoted by $V(y)$) agrees with the sequence v as regards the number and the types of its members. For the empty sequence Λ we stipulate $x(\Lambda) = x$ and $\Lambda(x) = \Lambda$, so that $y(v)$ is well-determined also in case y or v is empty.

3. The sequence of variables Y (or Z, or $\overline{Z}$, respectively) is determined in the same manner, as regards the number and the types of its members,

[9]See *Gödel 1933e*.

[10]H is supposed to be a system containing no propositional or function variables, but only *number variables*. The axioms and deduction rules of logic (given on p. 280 below) are to be considered as schemata for all possible substitutions of formulas of the system in place of the propositional variables.[j]

---

[j]For a complete description of the system H used in this paper the following should be added: *Number-theoretic functions* are defined only by primitive recursion and by setting the values of a function equal to those of a term composed of variables and previously introduced functions. *Formulas* are what is obtained from equations between terms by (iterated) application of propositional connectives and quantifiers. "*Sequenzen*" in Gentzen's sense, or the *descriptive operator* ι_x, are not used. *Complete induction* is formulated as a rule of inference. *The axioms of equality are*: $x = x$ and $x = y \,.\supset .\, t(x) = t(y)$ for any term $t(x)$. Outside of the axioms mentioned in footnote 10 and in this footnote, only the *third and fourth Peano axioms* are assumed. Evidently the systems T and H overlap.

[k]From here on the reader is asked to pay attention to the fact that the letters and formulas occurring in the subsequent discussions *are not*, but rather *denote*, combinations of symbols of T or H; or like "$(\exists x)$" or "$x(y)$", they denote *operations* to be performed on combinations of symbols yielding other such combinations. This relation of denotation can easily be made perfectly precise. In particular note that in expressions like "$A(\alpha,\beta,\gamma)$" the brackets denote an *operation of substitution*; i.e., "$A(\alpha,\beta,\gamma)$" is to be regarded as an abbreviation of "Subst $\left(A\,{}^{y,z,x}_{\alpha,\beta,\gamma}\right)$". Hence $A(y,z,x) = A$.

by the sequence s (or yw, or y, respectively) and by the sequence y (or z, or z, respectively).

One-member sequences of variables are identified with variables. If x is the empty sequence, $(\exists x)A = (x)A = A$ by definition.

The correspondence of F' to F is defined by induction on the number k of logical operators contained in F. The precautions to be taken in the choice of the symbols for the bound variables and the heuristic grounds for point 5 of the definition are given below the following formulas.

I. Let $F' = F$ for $k = 0$.

II. Suppose

$$F' = (\exists y)(z)A(y, z, x)$$

and

$$G' = (\exists v)(w)B(v, w, u)$$

have already been defined; then, *per definitionem*, we set:

1. $(F \wedge G)' = (\exists yv)(zw)[A(y, z, x) \wedge B(v, w, u)]$.
2. $(F \vee G)' = (\exists yvt)(zw)[t = 0 \wedge A(y, z, x)\ .\vee .\ t = 1 \wedge B(v, w, u)]$.
3. $[(s)F]' = (\exists Y)(sz)A(Y(s), z, x)$.
4. $[(\exists s)F]' = (\exists sy)(z)A(y, z, x)$,
   where s is a number variable contained in x.
5. $(F \supset G)' = (\exists VZ)(yw)[A(y, Z(yw), x) \supset B(V(y), w, u)]$,
   which, by the definition of negation given on page 280 below, implies:
6. $(\neg F)' = (\exists \overline{Z})(y)\neg A(y, \overline{Z}(y), x)$.

Before using Rules 1–6 the bound variables in the formulas F' and G' are, if necessary, to be renamed so that they are all mutually distinct and different from the variables in the sequences x, u. Furthermore, the bound variables of the sequences $t, Y, V, Z, \overline{Z}$, which are newly introduced by the application of the Rules 2, 3, 5, 6, are to be so chosen that they are mutually distinct and different from the variables already occurring in the formulas concerned.

The right-hand side of 5 is obtained by stepwise transforming the formula $F' \supset G'$ according to the rule that propositions of the form $(\exists x)(\ldots) \supset (\exists y)(\ldots)$ (or $(y)(\ldots) \supset (x)(\ldots)$, respectively), where x, y may be sequences of variables of any types, are replaced by propositions stating that there exist computable functions which assign to each example for the implicans (or counterexample for the implicatum, respectively) an example for the implicatum (or a counterexample for the implicans, respectively), taking account of the fact that $\neg B \supset \neg A . \equiv . A \supset B$.[1]

---

[1]The complexity of the definition of $(F \vee G)'$ is necessary in order to ensure the decidability of $\vee$ and, thereby, the validity of the inference $p \supset r . q \supset r \vdash p \vee q . \supset r$.

Of course it is not claimed that the Definitions 1–6 express the meaning of the logical particles introduced by Brouwer and Heyting. The question to what extent they can replace them requires closer investigation. It is easily shown that, *if F is provable in* H, *then the proposition F' is constructively provable in* T; *to be more precise, if $F' = (\exists y)(z)A(y, z, x)$, then a finite (possibly empty) sequence Q of functional constants can be defined in* T *such that $A(Q(x), z, x)$ is provable in* T. The proof consists in verifying that the assertion holds for the axioms of H, and that, if it holds for the premises of a given rule of inference of H, it also holds for the conclusion.

The verification becomes quite simple and straightforward if the following axiom system of intuitionistic logic is used:

Axioms: (1) $p \supset . p \wedge p$, (2) $p \wedge q . \supset p$, (3) $p \wedge q . \supset . q \wedge p$, (4), (5), (6), the axioms for $\vee$ which are dual to these, (7) $0 = 1 . \supset p$.[m] Negation is defined by $\neg p . =: p \supset . 0 = 1$.

Deduction rules: (1) Modus ponens, (2) substitution of terms for free variables, (3) $p \supset q, q \supset r \vdash p \supset r$, (4) $p \wedge q . \supset r \vdash p \supset . q \supset r$ and vice versa, (5) $p \supset q \vdash p \supset (x)q$ and vice versa, provided p does not contain x as a free variable, (6) $p \supset r, q \supset r \vdash p \vee q . \supset r$, (7) $p \supset q \vdash (\exists x)p . \supset q$ and vice versa, provided q does not contain x as a free variable.

It turns out that in the deduction rules 4, 5, and 7 the proposition to be proved in T is substantially the same as that already proved on the basis of the premise.[n] For the consistency proof of classical number theory the axioms and deduction rules containing $\vee$ or $\exists$ can be omitted.

It is clear that, starting with the same basic ideas, much stronger systems than T can be constructed, for example, by admitting transfinite types or the methods of deduction used by Brouwer for the proof of the "fan theorem".[11]

[11]See *Heyting 1956*, p. 42, and footnote **d**.

For quantifier-free F and G, $(F \vee G)'$ may be defined like $(F \wedge G)'$. This has the consequence that $F' = F$ for all quantifier-free F.

[m]Axiom 7 may be omitted without jeopardizing the interpretability of classical in intuitionistic number theory (see *Johansson 1936*). $0 = 1 . \supset \neg p$ follows from the definition of $\neg$ and the other axioms and rules.

[n]Note moreover the following: The proof of the assertion is trivial for all axioms of H and for the deduction rule 2. For the deduction rules 1 and 3 it follows easily from the fact that what the formula $A(Q(x), z, x)$ (corresponding to $(p \supset q)'$) says is exactly *that and how functions Q for q' can be derived from functions Q for p'*. As far as complete induction is concerned, note that the conclusion of this inference specialized to the integer n is obtained from the premises by n-fold application of substitution and modus ponens.

Note also that the deduction theorem can easily be proved for the interpretation $'$. Moreover, as Spector has observed (*1962*, p. 10), the system T enlarged by intuitionistic logic with quantifiers for functions of any finite type (i.e., his system $\Sigma_2 - \{F\}$) can be interpreted in T in exactly the same way as intuitionistic number theory. The proof is carried out in detail in §9, pp. 12–15, of Spector's paper.

Introductory note to *1972a*

The item to which this note serves as introduction consists of three remarks (indicated in the following as Remarks 1, 2 and 3) on the undecidability results, and was found appended to the galley proofs for *1972*. As explained in the introductory note to the latter, Gödel apparently worked on *1972* off and on during the period 1965–1972. We do not know at what point he considered adding these three remarks, but internal evidence suggests that it was later in the period rather than earlier. The remarks deal with preoccupations that Gödel had with both the generality and significance of his incompleteness results ever since the publication of his famous *1931* paper.

In brief, Remark 1 is concerned with improvements in the statement of the second incompleteness theorem which increase its scope. This remark already appeared (in slightly variant form) as a footnote that Gödel wrote in 1966 to accompany the translation of his *1932b* in *van Heijenoort 1967*. Remark 2 promotes ideas concerning the need for axioms of infinity in order to overcome incompleteness, ideas first suggested in footnote 48a of *1931* and expressed more fully in the *1964* supplement to *1947*. Finally, Remark 3 was presented as a footnote to Gödel's 1964 postscript to his *1934* lectures (on the occasion of their reproduction in *Davis 1965*). Gödel was there at pains both to emphasize the generality of his incompleteness results, in consequence of Turing's analysis of the concept of "mechanical procedure", and to reject the idea propounded by Turing and Post that these results establish "bounds for the powers of human reason". Another version of this same remark was communicated to Hao Wang and appeared on pages 325–326 of *Wang 1974* (completed, according to its introduction, in 1972.)

One may speculate that Gödel thought *1972* would be one of his last publications and that it provided a final opportunity to stress certain fundamental points and themes that he felt had been insufficiently appreciated. In addition, though he makes no explicit reference to *1972* in these remarks, there is a more than casual connection. For, by the use in *1958* and *1972* of a new abstract concept (constructive function of finite type) to establish the consistency of elementary number theory, Gödel illustrated his dictum in Remark 3 that incompleteness is to be overcome by the development of human understanding through the use of "more and more abstract terms".

Despite their brevity, these three remarks broach a wealth of matters that must be addressed at length if they are to be dealt with at all adequately. The introductory notes which follow thus discuss each of them separately. The first has been written by me, the second jointly with Robert M. Solovay, and the third by Judson C. Webb. In view of the length of these notes, the reader is advised to study Gödel's remarks first.

Solomon Feferman

Remark 1

As explained above, this remark, entitled "The best and most general version of the unprovability of consistency in the same system",[a] essentially reproduces a footnote which first appeared with the translation of *1932b* in 1967. Gödel excuses its reproduction anew, in 1972, by the statement that "perhaps it has not received sufficient notice".

The remark itself begins with the assertion that the consistency of a system S containing elementary number theory Z (directly or by translation), *may* be provable in S; indeed, the consistency of very strong S may even be provable in a system of primitive recursive number theory. On the face of it, this assertion seems to contradict Gödel's own theorem on the unprovability of consistency in (primitive) recursive consistent extensions of arithmetic (*1931*, *1932b*). However, by 1966 there were several examples in the literature of systems justifying Gödel's apparently contrary statement here. Those examples demonstrated that the applicability of Gödel's theorem on the underivability in a consistent system S of the consistency statement Con_S depends essentially on *how* S is *presented.* That is, they showed that, for suitable S, another presentation S^* of S could be given, with the same set of theorems, for which $S^* \vdash \mathrm{Con}_{S^*}$. In *Takeuti 1955* this was done by changing the set of rules generating the theorems, in *Feferman 1960* by changing the description of the set of axioms, and in *Kreisel 1965* by changing the description of the set of proofs.[b] Of these, the example by Takeuti is perhaps the most natural, since it deals with systems that have established significance in the literature, namely Gentzen-style sequential systems with or

[a]This remark, dated 18 May 1966, was added as a footnote to *1932b* in *van Heijenoort 1967*, p. 616. Gödel erroneously refers to it as appearing in the translation of *1931* rather than *1932b*.

[b]This glosses over some essential points of difference as to just what is demonstrated by the examples of Takeuti and Feferman. In the case of the former, only a weak form of the consistency of S^* is demonstrated in S^*, namely a formula $\mathrm{Con}^0_{S^*}$ which expresses the non-provability of $\mathit{0} = \mathit{1}$; the general form, which expresses that for each A not both A and $\neg A$ are provable, is not provable in the system S^* used by Takeuti. The example provided by *Feferman 1960* is not on its face effective, though S^* happens to be presented by a formula which binumerates (numeralwise defines) in S the same set of axioms as S.

Note also that the word 'presentation' in the text is not used in any specific technical sense.

without the so-called *cut rule* (a form of *modus ponens*, or *detachment*, appropriate to Gentzen-style systems). The *cut-elimination theorem* for suitable such S shows that the cut rule is dispensable; the system S^* is then S without the cut rule. So presented, S^* can prove its own consistency.[c] The examples due to Feferman and Kreisel have only more limited technical significance.[d] Gödel was undoubtedly familiar with all these examples, but we do not know whether he had any of them specifically in mind when he referred here to the possibility of a system's proving its own consistency.

While Gödel had sketched a proof of the underivability of consistency in *1931*, a detailed proof was first given in *Hilbert and Bernays 1939* (pages 285–328) for a system Z of elementary number theory (and a related system Z_μ). The work there was broken into two parts. First, three "derivability conditions" D1–3 were set down on the arithmetical formula $\text{Prov}_S(x)$ expressing in S that x is the Gödel number of a formula provable from S, and it was shown that, for the sentence Con_S expressing the consistency of S and defined by

$$\text{Con}_S = \forall x \neg[\text{Prov}_S(x) \wedge \text{Prov}_S(\text{neg}(x))],$$

we have $S \nvdash \text{Con}_S$ whenever S is consistent and satisfies those conditions (*ibid.*, pages 285–288). Second, the derivability conditions were verified for Z and Z_μ.[e]

The first derivability condition, D1, states that if B follows from A in S then $S \vdash \text{Prov}_S(\ulcorner A \urcorner) \rightarrow \text{Prov}_S(\ulcorner B \urcorner)$.[f] Condition D2 expresses a special case of closure of the provable formulas under a rule of numerical substitution. Finally, the third condition, D3, expresses a form of the adequacy of S for primitive recursive arithmetic, namely that for each primitive recursive function f (and corresponding function symbol in S) we can prove in S the formalization of

$$\text{if } f(m) = 0 \text{ then } S \vdash f(\boldsymbol{m}) = \boldsymbol{0},$$

where $\boldsymbol{m}$ is the numeral for m. Here S must contain primitive recursive arithmetic, either directly or by translation. The derivation of Gödel's second incompleteness theorem for consistent S also requires that the

[c]In the weak sense explained in footnote b.

[d]However, Jeroslow (*1975*) has shown that Feferman's result applies in a natural way to a class of non-effective systems called "experimental logics".

[e]*Hilbert and Bernays 1939*, pp. 289–328. According to G. Kreisel, this procedure followed a plan Gödel outlined to Bernays on a transatlantic voyage to the U.S.A. in 1935.

[f]We use $\ulcorner A \urcorner$ for the numeral in S corresponding to the Gödel number of A.

relation $\mathrm{Proof}_S(x, y)$, which holds when y is the number of a proof in S of the formula with number x, be primitive recursive, and that $\mathrm{Prov}_S(x) = \exists y\, \mathrm{Proof}_S(x, y)$.

By the work of Hilbert and Bernays, at least one of the derivability conditions D1–3 must fail for each of the three examples above of (presentations of) systems which prove their own consistency. The obvious candidate in the case of Takeuti's cut-free system is D1. However, it was later shown by Jeroslow (*1973*) that the derivability condition D1 is actually dispensable, in other words that D2 and D3 suffice for Gödel's second incompleteness theorem. Moreover, of these it is D3 which is crucial, since D2 can generally be trivially verified; indeed, all three examples above fail to satisfy D3. It happens that the examples due to Kreisel and Takeuti both also fail to satisfy D1, while that of Feferman does satisfy D1.

The general result stated informally by Gödel in Remark 1 is that a certain instance of what is now called the Π^0_1-*reflection principle for* S (denoted Π^0_1-RP_S in the following) is underivable in S, provided only that (i) S contains primitive recursive arithmetic (*PRA*) and is provably closed under the rules of the equational calculus and (ii) that Π^0_1-RP_S is correct for S. Here Π^0_1 statements are those of the form $\forall x f(x) = 0$, where f is primitive recursive, and Π^0_1-RP_S is the scheme

$$(\Pi^0_1\text{-RP}_S) \qquad \mathrm{Prov}_S(\ulcorner A \urcorner) \rightarrow A, \quad \text{for } A \text{ in } \Pi^0_1.$$

This scheme is correct for S if, whenever $S \vdash \forall x f(x) = 0$ with f primitive recursive, then for each natural number m, $f(m) = 0$. Actually, Gödel takes a slightly variant form Π^0_1-RP'_S of this principle, one which expresses that every equation proved in S using only the rules of the equational calculus is correct for each numerical instance. It is this form that Gödel calls the "outer consistency" of S.[g] Thus Gödel's result can be restated as:

(∗) If S contains *PRA* and is outer consistent, then an instance of the outer consistency of S is not provable in S.

Gödel does not indicate a proof of (∗) here, but such a proof can be reconstructed following standard lines for his second incompleteness theorem, with a small but essential technical change at one point. First, assume again that S has a primitive recursive presentation. Then there is a primitive recursive relation $\mathrm{Proof}_S(n, m)$ which holds just in case m is the number of a proof in S of the formula with number n. If we let p

[g]This has not become established terminology.

be the characteristic function of this relation, then from the assumption that *PRA* is contained in S we have

(1) $$\text{Proof}(n,m) \text{ implies } S \vdash p(\boldsymbol{n},\boldsymbol{m}) = \boldsymbol{1}, \text{ and}$$
$$\neg\text{Proof}(n,m) \text{ implies } S \vdash p(\boldsymbol{n},\boldsymbol{m}) = \boldsymbol{0}.$$

Let $\text{Prov}_S(x)$ be the formula $\exists y\, p(x,y) = \boldsymbol{1}$. Then

(2) $$S \vdash A \text{ implies } S \vdash \text{Prov}_S(\ulcorner A \urcorner).$$

In the usual line of argument for the second incompleteness theorem, a sentence G_0 is formed (by diagonalization) in such a way that

(3) $$S \vdash [G_0 \leftrightarrow \neg\text{Prov}_S(\ulcorner G_0 \urcorner)].$$

Then one shows (first part of the first incompleteness theorem) that if S is consistent, then $S \nvdash G_0$. By formalizing this argument one obtains $S \vdash (\text{Con}_S \rightarrow G_0)$, whence $S \nvdash \text{Con}_S$. In the first part, one proceeds by assuming $S \vdash G_0$ and applies (2) to conclude $S \vdash [\text{Prov}_S(\ulcorner G_0 \urcorner)]$; then from $S \vdash [G_0 \rightarrow \neg\text{Prov}_S(\ulcorner G_0 \urcorner)]$ and $S \vdash G_0$ we conclude that S is inconsistent. In formalizing this, one must apply first D3 and then D1.

For the new argument here the main technical point is that, without change in the basic diagonal technique used to obtain (3), one can construct a sentence G and a primitive recursive term t such that

(4) $$\text{(i) } G \text{ is } \forall y[p(t,y) = \boldsymbol{0}], \text{ and}$$
$$\text{(ii) } S \vdash t = \ulcorner G \urcorner.$$

(In *Kreisel and Takeuti 1974* such statements are called *literal Gödel sentences*. The first published use of statements of this kind appears to have been in *Jeroslow 1973*.) Now, by analogy with the standard argument, one first shows the following:

(5) $$\text{If } S \text{ is outer consistent, then } S \nvdash G.$$

The proof of (5) proceeds simply, as follows: If $S \vdash G$ and n is the number of G, then for some m, $p(n,m) = 1$. Also, $S \vdash \forall y[p(t,y) = \boldsymbol{0}]$, so $S \vdash p(t,\boldsymbol{m}) = \boldsymbol{0}$ and by (4)(ii), $S \vdash p(\boldsymbol{n},\boldsymbol{m}) = \boldsymbol{0}$. Hence if S is outer consistent we have $p(n,m) = 0$, in contradiction to $p(n,m) = 1$. Now, by formalizing this argument for (5), one obtains:

(6) $$\text{For a certain instance } A \text{ of } \Pi^0_1\text{-RP}'_S,\ S \vdash A \rightarrow G.$$

Hence, under the hypotheses of $(*)$, $S \nvdash A$. The use of literal Gödel sentences thus permits one to deal entirely with assumptions about derivability via the equational calculus in S, in place of conditions D1 and D3.

For the usual systems, as Gödel points out, outer consistency is equivalent to consistency; the argument, which is quite simple, goes back to Hilbert.[h] Moreover, the formal equivalence of Con_S with $\Pi_1^0\text{-RP}'_S$ (and at the same time with $\Pi_1^0\text{-RP}_S$) is provable in a system satisfying the Hilbert–Bernays (or similar) derivability conditions.[i]

In fact, as Gödel stresses, the question of establishing outer consistency (by finitary means) is the central one for Hilbert's program (as formulated, for example, in *Hilbert 1926*). Hilbert had divided the statements of a language into *ideal* ones and *real* ones. By the latter he meant the purely universal statements, each numerical instance of which was subject to a finitary check. Hilbert's program aimed to show that for various systems S encompassing mathematical practice the "ideal" statements can be eliminated from derivations of the "real" statements, in other words, that the reflection principle holds for the latter class of statements. Thus the program requires the outer consistency of S, for which, as Hilbert observed, it would be sufficient to establish the ordinary consistency of S, at least for the usual systems S. But Gödel's second incompleteness theorem showed that for these systems one cannot hope to prove the consistency of S within S. What Gödel accomplishes in the present remark is to show, even more generally, that one cannot hope to prove the outer consistency of S within S, if indeed outer consistency holds for S. Thus, with respect to Hilbert's program, Gödel can fairly claim to have established "the best and most general version" of his second incompleteness theorem.

However, Gödel ignores generalizations of his incompleteness theorems to other situations, for example to various non-constructive systems in *Rosser 1937*, *Mostowski 1952*, and elsewhere. Nor does he concern himself with generalizations of the reflection principles, such as were dealt with in *Feferman 1962* and in *Kreisel and Levy 1968* and which have a variety of important applications outside of Hilbert's program. Moreover, systems encompassing ordinary mathematical practice must include *modus ponens* (or the cut rule), so in this respect the kind of generalization obtained by Gödel is of marginal interest. This is not to deny that cut-free systems have been of fundamental importance in proof theory (see, for example, *Takeuti 1975*) or that they provide a useful context in which to illustrate various technical aspects of the use of self-referential statements, as shown for example in *Kreisel and Takeuti*

[h]See, for example, page 474 of the translation of *Hilbert 1928* in *van Heijenoort 1967*.

[i]For details, see *Kreisel and Levy 1968*, p. 105, or *Smoryński 1977*, p. 846.

1974. But for the usual systems and the various non-constructive extensions that have been considered, it is both much more natural and of greater generality to follow the lead of *Löb 1955*, in which quite elegant abstract derivability conditions (modifying those of Hilbert and Bernays) proved to be the appropriate means for settling the status of various self-referential statements and reflection principles in such systems. Löb's results have been put in an even more general logical context through the work of Solovay (*1976*) on the completeness of certain modal logics under the provability interpretation of the necessity operator.[j] Still, to study the question of applicability of Löb's derivability conditions, one must consider how formal systems may be presented within themselves. Here, as Kreisel has often stressed (see for example his *1965*, page 154), dealing with the question of what constitutes a *canonical presentation* of a formal system becomes the central concern. One solution has been provided in *Feferman 1982*.

One final technical point concerns incompleteness theorems for systems (much) weaker than arithmetic, for example those such as *PRA* which are quantifier-free. Gödel points out that his "most general" version of the second incompleteness theorem can be extended to apply to such systems. For the technical tools needed to deal with related versions of the theorem, see *Jeroslow 1973*.

Solomon Feferman

Remark 2

This remark begins with what Gödel terms "another version of the first undecidability theorem", which concerns the degree of complexity (or "complication", in Gödel's words) of axioms needed to settle problems of "Goldbach type" of high complexity. Gödel had also referred to problems of this type in *1964*, and he explained there (in footnote 42) that by such he meant "universal propositions about integers which can be decided in each individual instance".[k] Most generally, then, such propositions are statements of the form $\forall x R(x)$ with R general recursive (or effectively decidable, by Church's thesis). It is shown in recursion theory that every such statement is equivalent to one of the same form with R primitive recursive, and by definition these comprise the class of Π^0_1 statements. In fact, it is known through the work of

[j]See also *Boolos 1979*.

[k]See p. 269 above. Goldbach's own statement, dating from his 1742 letter to Euler, is the still unsettled conjecture that every even integer is the sum of two primes. (For Goldbach and Euler, 1 was a prime.)

Matiyasevich that every Π_1^0 statement is equivalent to one of the form $\forall x_1 \ldots \forall x_n[p(x_1, \ldots, x_n) \neq q(x_1, \ldots, x_n)]$, where p and q are polynomials with integer coefficients and $n \leq 13$.[l]

Gödel here takes the degree of complexity $d(A)$ of a formula A (in a given language) to be the number of basic symbols occurring in it. In other words, if, for a given basic stock of symbols $s_1, \ldots, s_m$, the formula is written as a concatenation $A = s_{i_1} \ldots s_{i_k}$, then $d(A)$ is defined to be k. For S a finite set of (distinct) formulas $A_1, \ldots, A_n$, considered as a system of non-logical axioms, the degree $d(S)$ is defined to be $d(A_1) + \cdots + d(A_n) + (n-1)$. The theorem stated informally by Gödel is that in order to solve all problems A of Goldbach type of a "certain" degree k, one needs a system of axioms S with degree $d(S) \geq k$, "up to a minor correction". It is not clear what kind of minor correction Gödel intended here, so we do not know just how he would have stated this as a precise result. After examining this question more closely, the authors have arrived at some results of the same character as Gödel's, but not quite as strong as what would be suggested by a first reading of his assertion; we have not, however, been able to establish the latter itself. These various statements and their status are explained as follows.

Let $\mathcal{L}$ be a language with a finite stock $s_1, \ldots, s_m$ of basic symbols, including logical symbols such as '$\neg$', '$\wedge$', '$\forall$', a constant symbol '$\boldsymbol{0}$', the successor symbol '$'$', a means for systematically forming variable symbols 'v_i' for $i = 0, 1, 2, \ldots$ from the basic symbols,[m] the equality symbol '$=$', and parentheses '(', ')'. $\mathcal{L}$ should also contain symbols, either directly or by definition, for a certain number of primitive recursive functions $f_0, \ldots, f_j$, where f_0 and f_1 are $+$ and $\cdot$, respectively. It is assumed that we have a consistent finite axiom system S_0 in $\mathcal{L}$ which contains (or proves) defining equations for $f_0, \ldots, f_j$, and enough of the axiom system of primitive recursive arithmetic for these functions in order to carry out Gödel's first incompleteness theorem. In particular, S_0 should be consistent and complete for Σ_1^0 sentences (and hence correct for Π_1^0 sentences). For the assertion of Gödel's being examined here, only those systems S are considered which are consistent and contain S_0. Then the following theorem can be proved:

[l]See *Davis, Matiyasevich and Robinson 1976*. Matiyasevich later showed that one could take $n \leq 9$; see his *1977*.

[m]One obvious way to do this is to identify v_i with $v\,\underbrace{\boldsymbol{0}'' \ldots '}_{i}$ where 'v' is a new basic symbol; this makes $d(v_i) = i + 2$. However, there are somewhat more efficient ways of building v_i from basic symbols, so that $d(v_i) = \log_2 i + O(\log_2 \log_2 i)$; we shall assume that such an encoding is being used in the discussion that follows.

(∗) There are positive integers c_1, c_2 such that for all $k > c_2$ and $k_1 = (k - c_2)/c_1$, no finite consistent extension S of S_0 with $d(S) \leq k_1$ proves all true Π_1^0 statements A having $d(A) \leq k$.

The proof of (∗) rests on an examination of Gödel's construction in his *1931* (for the first incompleteness theorem) of a true Π_1^0 statement G_S which is not provable from S for any finite consistent S extending S_0. G_S can be regarded as a statement which expresses that $\text{Conj}(S) \rightarrow G_S$ is not logically provable, where $\text{Conj}(S) = A_1 \wedge \cdots \wedge A_n$ for $S = \{A_1, \ldots, A_n\}$. This construction is uniform in S; that is, for a suitable Π_1^0 formula $B(v_0)$ with at most v_0 free, we have G_S equivalent to $B(\ulcorner\text{Conj}(S)\urcorner)$, where $\ulcorner\text{Conj}(S)\urcorner$ is the numeral in $\mathcal{L}$ for a Gödel number of $\text{Conj}(S)$. Using this, it may be shown that G_S can be chosen with $d(G_S) \leq c_1 d(S) + c_2$, where c_1 is a constant depending on the efficiency of the Gödel numbering of expressions. It turns out that one can take $c_1 = [\log_2 m] + 1$, where m is the number of basic symbols in $\mathcal{L}$.[n] For the usual logical systems m is between 8 and 16, hence $c_1 = 4$. But a first reading of Gödel's assertion under consideration would put $c_1 = 1$ in (∗); call that assertion (‡). (If (‡) holds, Gödel's "minor correction" would simply be c_2.)

The remainder of Gödel's remark does not depend essentially on whether one can obtain (‡) or not, but only that we at least have (∗). For Gödel's way of measuring complexity, the crucial thing is that the degree of complexity of axiom systems needed to establish true Π_1^0 sentences A increases roughly in direct proportion c_1 to the complexity of A, where c_1 is small.

We now pass from these technical questions to Gödel's discussion of their significance. This shifts, in effect, to systems of set theory. The reason is that all of present-day mathematics can be formalized in a relatively simple finite system S_1 of set theory (for example, the Bernays–Gödel system of sets and classes). According to Gödel, it follows from the result (‡), or (∗), that in order to solve problems of Goldbach type which can be formulated in a few pages, the axioms of S_1 "will have to be supplemented by a great number of new ones or by axioms of great complication." Naturally, one would be led to accept as axioms only those statements that are recognized to be evident, though not necessarily immediately so for the intended interpretation (that being, in the case of BG, sets in the cumulative hierarchy together with arbitrary classes of sets). Thus Gödel says that one may be led to doubt "whether evident

[n] Gödel's own numbering of expressions in *1931* is rather inefficient and gives a comparatively large value for c_1. The proof that $c_1 = [\log_2 m] + 1$ suffices relies particularly on the more efficient coding of variables mentioned in footnote m.

axioms in such great numbers (or of such great complexity) can exist at all, and therefore the theorem mentioned might be taken as an indication for the existence of mathematical yes or no questions undecidable for the human mind."

In response to such doubts, Gödel points out "the fact that there *do* exist unexplored series of axioms which are analytic in the sense that they only explicate the content of the concepts occuring in them". As his main example, he cites the axioms of infinity in set theory, "which assert the existence of sets of greater and greater cardinality or of higher and higher transfinite types" and "which only explicate the content of the general concept of set." Here Gödel repeats ideas broached in *1947* and more fully in its revised version *1964*.[o] There he said that the axioms for set theory "can be supplemented without arbitrariness by new axioms which only unfold the concept of set ..." (*1964*, page 264). Moreover, the axioms of set theory are recognized to be correct by a faculty of mathematical intuition, which Gödel says is analogous to that of sense perception of physical objects: "... we do have something like a perception also of the objects of set theory, as is seen from the fact that the axioms force themselves upon us as being true" (*1964*, page 271). He goes on to note there that "mathematical intuition need not be conceived of as a faculty giving an *immediate* knowledge of the objects concerned." In *1964* that point is elaborated by reference to Kantian philosophy. But at the end of the present remark, Gödel puts the matter in a way that is supported by the working experience of set theorists who have been led to accept axioms of infinity, namely: "These principles show that ever more (and ever more complicated) axioms appear during the development of mathematics. For, in order only to understand the axioms of infinity, one must first have developed set theory to a considerable extent." The implicit but unstated conclusion of all this is that such axioms of increasing complexity can be used to settle more and more complicated problems of "Goldbach type". In other words, *despite* results such as (∗) (or even (‡), if true) "mathematical yes or no questions undecidable for the human mind" need *not* exist, in principle.

There is one essential difference of aim in the discussions of *1964* and of the present remark, concerning the possible utility of axioms of infinity. In *1964*, Gödel thought that such axioms could be used to decide *CH*, whereas here he aims to use them to solve number-theoretic problems. The study of the so-called axioms of infinity goes back to Hausdorff (*1908*), followed by several publications by Mahlo (*1911, 1912, 1913*). After that, there was only scattered work in the subject until the late

[o]See particularly *1964*, pp. 264–265 and 271–272. Gödel first touched on axioms of infinity in footnote 48a of his *1931* and in *1932b*.

1950s, when it began to undergo intensive development that continues to this day. Contrary to Gödel's views, there is no universal agreement among those who have studied set theory deeply as to the acceptability of these statements as axioms. For very favorable views, see *Reinhardt 1974* or *Kanamori and Magidor 1978* (the latter also being a very useful survey paper); for a completely opposite (negative) view, see *Cohen 1971*. While it is certainly true that one must do considerable work in the subject in order to understand these statements and thus (perhaps) be led to accept them, it is not the case that the complexity (in Gödel's sense) of the additional axioms has grown enormously, since new and stronger axioms simply displace old ones. Thus a few pages suffice to formulate the strongest such statements that have been considered, and the complexity of the additional axioms is still relatively low. It is true that, for each new axiom of infinity A which has been considered and which goes beyond a previously accepted S, A "solves" the above G_S, simply because $S + A$ establishes the existence of a model for S and thus proves the consistency of S (a statement equivalent to G_S). On the other hand, Gödel's hope in *1964* that use of axioms of infinity might settle *CH* has simply not been realized. As Martin explains in his report on Cantor's continuum problem, *CH* has been shown to be undecidable relative to any remotely plausible extension of the usual axioms ZFC of set theory by axioms of infinity (*Martin 1976*, page 86).[p] The situation has not changed at the time of this writing.

Axioms of infinity are offered by Gödel as an example of further axioms that might help solve previously unsettled problems. In *1964*, page 265, he suggested that there might be "other (hitherto unknown) axioms of set theory which a more profound understanding of the concepts underlying logic and mathematics would enable us to recognize as implied by these concepts". But his indication (*1964*, footnote 23) of the nature of such (as stating some kind of "maximum property") is rather vague. Since then a number of other specific axioms have in fact been proposed, some of which have been studied intensively by set theorists—in particular, the so-called "axiom of determinacy" (see, for example, *Martin 1976*). With respect to the present discussion, all these share with the axioms of infinity the following characteristics: (i) though they have received some degree of acceptance among set theorists, none of these axioms is widely accepted by the general community of mathematicians, (ii) their complexity is relatively low, (iii) they serve

[p]The "axiom of constructibility" $V = L$ does prove *GCH*, as we know from *Gödel 1938–1940*, but is incompatible with strong axioms of infinity. It is also seen to be intuitively false in the intended interpretation, since it says that all sets are definable in a certain way. See the introductory note to *1938–1940* in this volume.

to establish previously undecided propositions G_S, and (iv) they do not settle *CH*.

While the two authors of this note disagree about the foundational status of all such proposed new axioms, they agree that—contrary to Gödel's view expressed here—there are simple "mathematical yes or no problems" which will probably never be settled by the human mind because they are beyond all remotely feasible computational power and provide absolutely no conceptual foothold. For example, we can ask the following question: Is it true that if w is the sequence of the first $2^{2^{2^{100}}}$ terms in the binary expression of π–3, then the last term of w is 0?

To conclude, it is our view that whatever technical interest there may be in such measures of complexity as those offered here by Gödel (and related ones by Kolmogorov and Chaitin[q]), they are irrelevant to the experience of working mathematicians. It is not the complexity of the *axioms* needed for solving problems which is at issue in practice, but rather the complexity of the *proofs* required, and here there is no simple relationship between results and proofs. Relatively complicated problems can have relatively simple proofs once the right key is found, while relatively simple problems may require amassing an enormous amount of (conceptually) complicated machinery in order finally to settle them. Moreover, complexity is a shifting matter in the eyes of mathematicians: As mathematics develops, previously complicated notions and results are assimilated and become everyday tools for the attack on yet more difficult problems. A realistic mathematical theory of this common psychological experience has yet to be provided.

Solomon Feferman and Robert M. Solovay

Remark 3

Gödel's 1964 Postscriptum to *1934* began by stressing that the "precise and unquestionably adequate definition of the general concept of formal system" made possible by Turing's work allows his incompleteness theorems to be "proved rigorously for *every* consistent formal system containing a certain amount of finitary number theory" (Gödel in *Davis 1965*, page 71 = *Gödel 1986*, page 369). He insisted, however, that such generalized undecidability results "do not establish any bounds for the powers of human reason, but rather for the potentialities of pure formalism in mathematics" (*1986*, page 370). Gödel was no doubt responding

[q]See *Chaitin 1974*.

here to the claim in *Post 1936* that the generality of the incompleteness and undecidability theorems for "all symbolic logics and all methods of solvability" required that Church's thesis be seen as "a natural law", for "to mask this identification under a definition hides the fact that a fundamental discovery in the limitations of the mathematicizing power of Homo sapiens has been made and blinds us to the need of its continual verification" (Post in *Davis 1965*, page 291). But on reflection Gödel realized that, insofar as they tried to show that "mental procedures cannot go beyond mechanical procedures" effectively, Turing's *arguments* for his "unquestionably adequate definition" of computability would imply the same kind of limitation on human reason as claimed by Post, and so he wrote this Remark 3 of *1972a* on Turing's "philosophical error" as a footnote elaborating his disclaimer quoted above. Our problem is to understand how Gödel could enjoy the generality conferred on his results by Turing's work, despite the error of its ways. Since this clearly involves not only his interpretation of Turing's work but also of Church's thesis generally, a brief review of Gödel's role in the emergence of this thesis may help to put some aspects of our problem in perspective.

In *1934* Gödel claimed that the primitive recursive functions used in his arithmetization of syntax can all be "computed by a finite procedure". In a footnote he said that the converse "seems to be true" if we allow "recursions of other forms". Herbrand (*1931*) admitted arbitrary recursion equations (axioms of Group C) into his formalism for arithmetic, provided that, "considered intuitionistically, they make the actual computation of the $f_i(x_1, \ldots, x_n)$ possible for every given set of numbers, and it is possible to prove intuitionistically that we obtain a well-determined result" (Herbrand in *van Heijenoort 1967*, page 624). By this he meant that the computation be carried out informally in "ordinary language" and shown constructively to terminate. He also claimed that it was impossible to "describe outright" *all* these f_i, since otherwise $f_x(x)+1$ would be an "intuitionistically defined function" not in the list generated by such a description. He concluded that Gödel's incompleteness theorems did not hold for his arithmetic, for "to carry out Gödel's argument, we have to number all objects occurring in proofs" (*ibid.*, page 627); but to number the objects of his formalism one would have to focus on "a definite group of schemata" for his recursions, and since the diagonal function $f_x(x)+1$ "cannot be among these functions" (*ibid.*), it could not have any Gödel number. This raised the question of the generality of his recursions as well as that of Gödel's theorem itself. In a letter to Gödel he formulated a general but precise notion of recursion equations and called "recursive" those functions which are the unique solutions of such equations.

Gödel then realized, after rejecting as "thoroughly unsatisfactory" Church's proposal to identify effectiveness with λ-definability, that

Herbrand's definition could be modified in the direction of effectiveness, and he proposed to call "general recursive" those functions whose values could be deduced from his equations by two explicitly stated rules. This made his aforementioned footnote suggesting that 'finite computability' could be identified with a wide enough class of recursions sound like Church's thesis after all. But Gödel had qualified this as only a "heuristic principle", and later wrote to Davis that he was *not* proposing Church's thesis, but rather

(GT) The functions "computed by a finite procedure" are those definable by "recursions of the most general kind",

explaining that in 1934 he was "not at all convinced that my concept of recursion comprises all possible recursions" (Gödel in *Davis 1982*, page 8) and that the equivalence of his concept, based on Herbrand's equations, with Kleene's, based on minimalization, "is not quite trivial" (*ibid.*). Indeed, if the proof of the general recursiveness of all the μ-recursive functions by *Kleene 1936* were trivial, Gödel would presumably not have been in such doubt about the generality of his recursions. In fact, this result is behind the "kind of miracle" that Gödel (*1946*) saw in the closure of recursiveness under diagonalization, which allowed it to provide an "absolute definition" for the "epistemological notion" of computability: given any recursive sequence of general recursive functions, Kleene defined by minimalization a new diagonal function $\phi_x(x) + 1$, which is nevertheless still general recursive by Kleene's result. On the other hand, Kleene's normal form theorem shows by explicit arithmetization of the Herbrand–Gödel formalism that all the general recursive functions will have Gödel numbers, and hence refutes Herbrand's claim that the diagonalization of any "definite group of schemata" for his recursions must yield a new recursive function with no Gödel number. Adding to this Kleene's general result on "the undecidability, in general, which systems of equations define recursive functions" (*Kleene 1936* = *Davis 1965*, page 248), we cannot but wonder why Kleene's analysis did not eliminate all Gödel's qualms about replacing the right side of (GT) by general recursiveness.[r] Certainly it showed that one could not reasonably expect

[r]Davis (*1982*) remarks that Kleene's normal form theorem "must have gone a considerable distance towards convincing Gödel that his 'concept of recursion' indeed 'comprises all possible recursions'" (p. 11). I have tried to spell out here why Kleene's results *should* perhaps have done so, but in fact we have no direct evidence that they *did*. Kleene's paper would even seem to be the plausible source for Gödel's remark in *1946* about the "miracle" of diagonalization, but it may well be that he never really looked at it until after he read and was convinced by Turing's very different analysis of computability.

to diagonalize out of this class by any "finite procedure": For (i) clearly effective diagonalizations miraculously do not lead outside this class, and (ii) those which do are clearly not effective, since they would depend on knowing effectively which sets of Herbrand's equations lead to his "well-determined result", by Gödel's result.[s] In fact, Kleene shows that in any of "certain formal logics" the arithmetization of infinitely many true claims of this form will be unprovable, bringing out the importance of formally undecidable sentences in justifying (ii). It would thus seem that Kleene's analysis should not only have convinced Gödel that his recursions were wide enough to comprise the "finite procedures", but also provided him with answers to Herbrand's argument against the generality of his incompleteness theorems. In translating classical into intuitionistic arithmetic, Gödel (*1933e*) had used the system of *Herbrand 1931* to represent the former, where his formulation spoke of "the denumerable set of function signs f_i" introduced by Herbrand's Group C of axioms.[t] But in *1934*, where he formulated the "conditions that a formal system must satisfy" for his incompleteness theorems to apply by asking that it be so Gödel numbered that its axioms and relation of immediate consequence be primitive recursive, he was conspicuously silent about Herbrand's claim that his *functions* themselves could not be numbered. Why then, in view of his own explanation for not having advanced Church's thesis in 1934, did Gödel never cite *Kleene 1936*, either as having helped to persuade him of it or to settle his score with Herbrand?

The answer to the latter question seems to be that Gödel was never able to satisfy himself with any answer to Herbrand. Indeed, in his 1965 elaboration of footnote 34 of *1934* he still claimed that the equivalence of Herbrand's notion of a constructively provable recursive function with general recursiveness was "a largely epistemological question which has not yet been answered". The question was whether or not the concept of a computable function depends on that of an intuitionistic or materially correct proof.[u] Gödel evidently thought that Turing's analysis showed

[s] Since Herbrand had already stressed in an unsigned note that "all the functions introduced must actually be calculable for all values of their arguments by means of operations described wholly beforehand" (*1971*, p. 273), it seems fair to say that the *concept* of general recursiveness was already his (cf. *van Heijenoort 1982*). The real importance of actually writing down such rules as Gödel did was that it allowed the arithmetization of the entire theory of recursive functions: only then could Kleene's plan to "treat the defining equations formally, as sequences of symbols" (*1936*, p. 729 = *Davis 1965*, p. 239) lead to substantial results.

[t] As is pointed out in the introductory note to *1933e*, in volume I of these *Works*, Gödel's use of these axioms plays no special role in his proofs.

[u] One should consult the introductory note to *1958* and *1972*, where the considerable amount of trouble and hard thinking this issue caused Gödel are analyzed at length.

the notion of "mechanical computability" to be independent of such a proof concept, and went on to stress in his Postscriptum to *1934*, as we have seen, that Turing's analysis also provided an unquestionably adequate definition of a formal system. For the "essence" of such a system, says Gödel, is to completely replace reasoning by "mechanical operations on formulas", and Turing had shown this to be "equivalent" to a Turing machine (*1965*, page 72). But Gödel never uses the word "effective" to describe the explicanda of Turing's analysis, and says explicitly that general recursiveness can take its place in (GT) only "if 'finite procedure' is understood to mean 'mechanical procedure'" (*ibid.*, page 73). So it is unclear whether *any* analysis ever convinced him that any of these equivalent mathematical concepts comprised the "effective" functions. He stressed that Turing's analysis is independent of "the question of whether there exist finite *non-mechanical* procedures ... such as involve the use of abstract terms on the basis of their meaning" (*ibid.*), referring to his own *1958* where his abstract concept of a "computable function of type *t*" is used to prove the consistency of arithmetic. Yet he also emphasized, as we have seen, that the undecidability results made possible by Turing placed no limitation on human reason, but only on pure formalism—a point he tries to explain in his Remark 3 of *1972a* on Turing's "philosophical error" of assuming that a human computer would be capable of only finitely many distinguishable mental states. To understand how Gödel could allow an analysis based on such an error to stand as "unquestionably adequate" to establish the generality of his theorems, we turn to Turing's work.

In a personal communication to the author, Feferman has plausibly conjectured that the basic new feature of Turing's machines which convinced Gödel of their adequacy for defining a general concept of "mechanical procedure" was the *deterministic* character of their computations, since this automatically ensured the *consistency* of Turing's definition of computability. In Church's λ-calculus the calculation of a normal form for a term representing a function may take many different courses, and the Church–Rosser theorem on the uniqueness of existing normal forms, which ensures the consistency of Church's definition, was indeed, as Feferman (*1984a*) points out, "exceptionally difficult" to follow—to say nothing of the ontological obscurity surrounding the λ-calculus. This may also explain why *Kleene 1936* failed to completely persuade Gödel.[v] While this feature alone of Turing's analysis may

[v]See *Kleene 1943*, where the analogous problems of consistency for various formalisms for recursive functions not mentioned in *1936* are discussed fully. In particular, Kleene points out that the consistency proof for one formalism of partial recursive functions "seems to require the type of argument used in the Church–Rosser consistency proof for λ-conversion."

explain why Gödel found it more convincing than Church's and Kleene's, we have to look more closely at Turing's arguments to see how Gödel might have acquiesced in its advantages despite their error.

In *Turing 1937* (reproduced in *Davis 1965*, pages 116–149), Turing's arguments for the thesis that his machines could compute any function "calculable by finite means" are divided into three types. Type I presents his analysis of the operations an ideal "human computer" could perform and depends on the assumption, questioned by Gödel, that he is capable of only finitely many "states of mind". Type II shows that the entire deductive apparatus of the predicate calculus can be simulated by one of his machines. Type III is a "modification" of the type I argument that replaces the problematic notion of a state of mind by "a more physical and definite counterpart of it" (Turing in *Davis 1965*, page 139), namely, a "note of instructions" enabling his computer "to break off from his work" and later resume it. Since each such note "must enable him to carry out one step and write the next note" (*ibid.*), it follows that each stage of his computation is "completely determined" by such notes. Turing argues that, since the instantaneous "state of the system" comprised of a note of instructions and tape symbols can be represented by single expressions, the entire computation of his computer could be formalized in the predicate calculus, and therefore, by the type II argument, carried out by one of his machines. Thus Feferman suggests that Gödel rejected only Turing's type I argument, while accepting his "more physical" type III argument. Indeed, Wang (*1974*, page 326) reports that in discussions with Gödel about this remark, Gödel admitted the validity of Turing's argument under two additional assumptions: (1) "There is no mind separate from matter," and (2) "The brain functions basically like a digital computer," or (2′) "The physical laws, in their observable consequences, have a finite limit of precision". Although Gödel accepted both (2) and (2′), he rejected (1) as "a prejudice of our time" which would eventually be scientifically refuted, possibly by showing that "there aren't enough nerve cells to perform the observable operations of the mind" (*ibid.*). Gödel believed that Turing's argument depended on some form of physicalism, and indeed Turing says of the *elementary* operations of his human computer that "every such operation consists of some change of the physical system consisting of the computer and his tape" (Turing in *Davis 1965*, page 136). But since there is no doubt that such "observable" mental operations as reading symbols can be performed by one's nerve cells, it is not yet clear what is lost in Turing's physicalist analysis of computability, or even that it really depends on assumption (1). Moreover, since the physicalist claim just quoted occurs in the type I argument, it is clear that in analyzing Gödel's critique of Turing we cannot simply ask which of the three types of argument he may have found valid or invalid. Indeed, a closer look at

Turing's paper will show, I believe, that he basically presented just one argument, one which is still plausible without any physicalist premises, but also one that Gödel could accept only under his own interpretation on the *conclusion* of Turing's argument, that is, on Turing's thesis. Let us examine this more closely.

For Turing the question of what can be done "effectively" concerns *memory*: though it can only scan one symbol at a time, by altering its state "the machine can effectively remember some of the symbols which it has 'seen' (scanned) previously" (*ibid.*, page 117). The restriction to a finite number of states thus limits its memory, and Turing justifies its comparison with a human computer by reference to "the fact that the human memory is necessarily limited" (*ibid.*). How else is one to represent this limitation except in terms of the number of states? Since Gödel accepted such a limitation on the brain, it seems that he may have envisaged, as did Leibniz, some kind of purely mental memory, "separate" from that of the brain. But this would still not undermine Turing's type I argument, for it does not deny the *existence* of an infinity of mental states, much less assume (1), but argues rather that "if we admitted an infinity of states of mind, some of them will be 'arbitrarily close' and will be confused" (*ibid.*, page 136). Turing's point is that only an *effectively distinguishable* set of states could be used to "effectively remember" symbols, and hence to effectively compute. Gödel admits that this set is finite even for the mind in its *current* stage of development, but envisaged the possibility of "systematic methods" for so actualizing the development of our understanding of abstract terms that it would "converge" to infinity.[w]

[w]That Turing's finiteness hypothesis is perfectly compatible with a "dynamic" view of mind, however, emerges clearly when we consider the suggestion that it was Turing's type III argument that convinced Gödel (that "mechanically computable" functions are Turing-computable). This argument actually contains a sketch for a rather different but still direct formulation of the computability idea, indeed the very one simultaneously and independently worked out in *Post 1936*. (See *Hodges 1983* for some interesting remarks on this.) Here the "memorial" role of states, that they depend on previous states and scanned symbols, is played by instructions so numbered that they can refer to each other. The equivalence of Turing's computability and Post's is thus essentially contained in Turing's type III argument itself: it can be regarded as essentially a sketch for a proof that his machines could simulate any of Post's finite 1-processes. Turing's restriction of his human computer to finitely many states is thus equivalent to Post's restriction of his "worker", who executes the same atomic operations, to finitely many instructions. Post (*1941*) saw his restriction as evident from the fact that "the system of symbolizations ... is essentially to be a human product and each symbolization a human way of describing the original mathematical state" (p. 427). He emphasized the dualism between "the static outer symbol-space" and our "dynamic mental world" which nevertheless has "its obvious limitations", ones that are fully emphasized by Turing's "finite number of mental states" hypothesis (p. 431).

But does it follow, as Gödel claimed, that from the convergence to infinity of the number and precision of our abstract terms the number of *distinguishable* states of mind must so converge? Turing might say: If we admit an infinity of abstract terms some of them will be "confused" with each other. Gödel would say: Not if each of them is understood "precisely". This assumption, that coming to understand such a term precisely always creates a distinct new mental state for using it, not only depends on the special significance Gödel attaches to "abstract" terms, but also shows that he has in mind a notion of "state" different from Turing's. This seems clear from the superficial similarity of Gödel's argument to the familiar fallacious one that, since the mind can in principle "think" of any natural number, it must be capable of an infinity of distinct states and hence not be a Turing machine. This rests on the false assumption that we need to be in distinct single states to think of distinct numbers: Since "thinking of a number" in general involves symbolic calculation, it is best analyzed precisely as Turing does as a process of passing through a finite *sequence* of states, each carrying a bounded record of the history of the calculation. Indeed, in a penetrating analysis of Turing's arguments, Gandy replaces his finiteness condition on states by a "principle of local causation" excluding any instantaneous action at a distance in the causal relations between successive states, and conjectures that Gödel's "non-mechanical intelligence would, so to speak, see the state x as a *Gestalt*, and by abstract thought make global determinations which could not be got at by local methods" (*Gandy 1980*, page 146). By Turing's own lights, Gödel's intelligence would have an essentially *omniscient memory.* But, as Gandy's formulation reminds us, it is the *abstractness* of the terms more precisely understood by Gödel's developing mind that is supposed to make its states different from those relevant to Turing's human computer and which also distinguishes his argument from the familiar fallacious one for an infinity of mental states. When we note, furthermore, that it is the *complexity* of abstract terms that distinguishes them from concrete ones by Gödel's lights, his argument can be seen as more than a mere formulation of the possibility that we might in the unforseeable future develop an "infinite mind".[x]

We see this in fact from Gödel's Remark 2 of *1972a* on "another version" of his first theorem, where he takes it as showing that to solve even relatively simple problems we shall need ever new axioms of "great complication". His favorite examples are the stronger axioms of infinity

[x]This phrase was used by Gödel in *1944* to criticize the attempt by Ramsey (*1926*) to reduce classes to infinitely long propositions. What else could an infinite truth-function be, asks Gödel, but another infinite structure more complicated than classes, "endowed in addition with a hypothetical meaning, which can be understood only by an infinite mind"? (*1944*, p. 142).

which are "evident" even though "ever more complicated" because they assert the existence of increasingly abstract and complicated objects. Indeed, Gödel's argument against Turing must hinge on this assumption about abstract terms: that by understanding them more precisely we become capable of *states which are themselves more and more complicated.* In fact, immediately after his claim quoted above that we could effectively distinguish only finitely many of a supposed infinity of states, Turing had admitted that this restriction "is not one which seriously affects computation, since the use of more complicated states of mind can be avoided by writing more symbols on the tape" (*Davis 1965*, page 136). So even in Turing's own mind the issue was more the complexity of the states than their actual infinity, and indeed his construction of the universal machine comprised the most striking confirmation of this kind of 'compensation' for the lack of complicated states. Here the 'cybernetical aspect' of the enumeration theorems comes to the fore: to simulate machines with arbitrarily more and more states which are more and more complicated, the universal machine, with its fixed finite number of states, has only to be given their Gödel number on its tape, and then, as Turing later put it, "the complexity of the machine to be imitated is concentrated in the tape and does not appear in the universal machine proper in any way" (*Hodges 1983*, page 320). Gödel was presumably not convinced, the universal machine notwithstanding, that all the states entered by a human computer using "finite *non*-mechanical procedures" could always be compensated for in Turing's purely symbolic manner, for in such states it just might exploit the meanings of ever more abstract concepts of proof and infinity to grasp *infinitely* complicated combinatorial relations. It is really this kind of possibility more than any convergence to an infinity of states that could undermine Turing's arguments, but, far from having disregarded it completely, it seems that Turing himself must have initially thought such an objection plausible; yet once he discovered the universal machine he saw that it could indeed compensate symbolically for a surprisingly wide class of increasingly complicated *machine* states.[y] Otherwise he would never have claimed that "a man

[y]The critique of Turing's analysis of computability in *Kreisel 1972* is closely related to Gödel's, with whom Kreisel agrees that "Turing's error" of assuming that a human computer can enter only finitely many distinguishable states "does not invalidate his analysis of *mechanical* instructions" (p. 318). His second error, "not unrelated" to "the *petitio principii* concerning the finiteness of our thinking" already opposed by Cantor, is "assuming that the basic relations between (finite) codes of mental states must themselves be mechanical" (p. 319). That is, the succession of mental states in human computation cannot be described by "rudimentary" functions as can those of Turing machines. This suggests to Kreisel that "the sequence of steps needed to execute an h-effective definition" cannot be reproduced by such machines. In the case of number-theoretic functions, "we can say loosely that the human

provided with paper, pencil, and rubber, and subject to strict discipline, is in effect a universal machine" (*Turing 1970*, page 9).

That Turing's mechanical model of the human computer is by no means unduly simple is brought out by another version of Gödel's theorem due to Kleene: In every consistent formal system F in which the universal machine U can be described, there are infinitely many expressible facts about U's halting behavior which cannot be proved in F. Since F may have as axioms, for example, any effectively specifiable set of strong axioms of infinity, it would seem that U must be rather complicated after all, despite the seeming simplicity of its states. That this point was not lost on Gödel is clear from a letter he wrote Arthur Burks, who had queried him about von Neumann's notion of an automaton so complex that its behavior was "asymptotically [?] infinitely longer to describe" than the automaton itself. Gödel replied (as quoted in *von Neumann 1966*, page 56) that

> what von Neumann perhaps had in mind appears more clearly from the universal Turing machine. There it might be said that the complete description of its behavior is infinite because, in view of the non-existence of a decision procedure predicting its behavior, the complete description could be given only by an enumeration of all instances ... The universal Turing machine, where the ratio of the two complexities is infinity, might then be considered to be a limiting case of other finite mechanisms. This immediately leads to von Neumann's conjecture.

He added, however, that this presupposes "the finitistic way of thinking" about descriptions, so he was presumably still not willing to concede that U poses an absolutely unsolvable decision problem, even by finite non-mechanical procedures. But his Remark 3 of *1972a* admits that

computations are more 'complicated' or, better, more abstract than the objects on which they operate—our thoughts may be more complicated than the objects thought about" (*ibid.*, p. 320). This is the same connection between abstractness and complexity that we find in Gödel's argument. Kreisel errs, however, when he claims "in the case of (Turing) machines whose *states* are finite spatio-temporal configurations it is quite clear how to code states by natural numbers" (p. 319), as if only their "finite spatio-temporal" character had allowed Turing to code them. But his coding of machine states depends only on the fact that they are finite in *number* and has nothing to do with such properties. We agree with Kreisel that the coding of purely mental states should be "a more delicate matter". But the only property of such states he uses is that a human computer may think about proofs in Heyting's formal arithmetic, and his examples of prima facie non-mechanical rules depending on such states have turned out to be Turing-computable. Still, Kreisel's discussion does engage more explicitly than Gödel's argument the problem Turing himself saw at issue in this hypothesis, namely, the 'complexity' of states, and this is its virtue.

our development of such procedures is still eons away from being able to "actually carry out" the computations needed to predict the behavior of Turing's model of the human computer.

Finally, we note that the interpretation of Gödel's argument against Turing as depending on the complexity of the states entered by a human computer using "finite non-mechanical procedures" does not conflict with his claim that the adequacy of the latter's analysis of "mechanical procedure" has nothing to do with the existence of such procedures. Still, this claim does imply a certain interpretation of his own of Turing's work, namely, that all Turing was really *analyzing* was the concept of "mechanical procedure", but that in his *arguments* for the adequacy of his analysis he overstepped himself by dragging in the mental life of a human computer. As Wang's authorized formulation of Gödel's view puts it, "we had not perceived the sharp concept of mechanical procedures sharply before Turing, who brought us to the right perspective" (*1974*, page 85). This sounds plausible enough until we look at what Turing actually said. As we saw, Turing offers (in Section 9 of his paper) three "types of argument" for the thesis that his machines can compute the decimal expansion of any real number which is "calculable by finite means". Gödel took issue with his type I argument, though I believe that none of them can hold by his lights (see footnote w above). Perhaps in reflecting on this argument in 1972, Gödel forgot that "the word 'computer' here meant only what that word meant in 1936: a person doing calculations" (*Hodges 1983*, page 105). Where then did he find Turing's "analysis" of mechanical operations shorn of human aspect? Really, nowhere: as Turing himself says, his type I argument is "only an elaboration of the ideas" presented in Section 1 of his paper, and his type III argument "may be regarded as a modification of I or as a corollary of II". In fact, Turing has *one* basic argument, which is presented in Section 1 and discussed above, and whose central premise is "the fact that the *human* memory is necessarily limited". Turing refers to *this* "fact" as the "justification" of his definition of the computable real numbers in terms of his machines. The heart of his argument was a novel abstract logical analysis of what it means to "effectively remember" things relevant to computation, such as symbols or how many times one has executed a subroutine: to do so one must be able to change from one distinguishable state to another, *whether you are human or a machine.* We presume indeed that "states of mind" may also carry memories beyond the wildest dreams of machines, but the only ones relevant to effective computation are those you are put into by symbols and processes arising in the course of the computation. But our memory is just as "necessarily limited" as a machine's—in either case, to a finite number of recognizable state changes. We saw, however, that a crucial problem does surface in Turing's "elaboration" of this limitation, namely, the apparent need to represent "more complicated" states as their number increases. Here a prima facie difference between a human computer and a Turing machine presented itself to Turing as soon as he wrote down the definition of the latter: for then Turing himself could number his machines like Gödel and define a universal partial function which was obviously effective for him, whereas it could not have been obvious from his definition that one of his machines could compute it. Perhaps he could compute it only because he entered more complicated states enabling himself to simulate machines with ever more complicated states. In any case, his discovery of the universal machine, compensating for an internal memory of bounded complexity by an external one of unbounded size, is really his main contribution to science, having already made possible a better grip on a basic concept which we are still far from seeing "sharply", namely, that of 'complexity'.

We now try to focus the questions raised by Gödel's view of the significance of Turing's analysis and his own results. The mere existence of undecidable sentences in formalisms like *Principia mathematica*, while interesting, would not of itself force one to reexamine the scope of formalization in mathematics. There would simply be axioms one had overlooked. The full force of the results of Gödel and Church is only made explicit in Rosser's *1936* extension:

(T) The set of unprovable sentences of any adequate formalism is not r.e.

Gödel saw this as no limitation on human reason but only "pure formalism in mathematics"—in opposition to Post, who regarded Church's thesis as a *hypothesis about the significance of* (T), namely, that it implied that the decision problem for any formalism containing such sentences was absolutely unsolvable by any effective method. Gödel had been sure only of (GT) until Turing discovered the first completely deterministic, and hence obviously consistent, formulation of computability. But even when Turing computability proved equivalent to both λ-definability and general recursiveness (and the latter to μ-recursiveness), he still insisted that the latter comprised the "most general recursions" of (GT) *only if "finite procedure" was interpreted as "mechanical procedure"*—but not if it referred to what a human computer could "effect". For this, one needed, according to Gödel, the assumption that such a computer was a completely physical system, an assumption he could not accept. He saw clearly that Turing's analysis, if allowed to stand for humanly effective processes generally, would imply that human minds are not more effective than machines in dealing with (T). Hence he looked for "systematic methods" for developing our use of abstract notions of infinity and proof to the point of being able to handle such problems in constructive but non-mechanical ways. But his example of looking for stronger axioms of infinity, while it may well actualize and sharpen our ability to use abstract concepts, is not necessarily a process that would actually increase the number of our distinguishable mental states. It might be argued that when such axioms "force themselves on us as being true", we enter meaning-using states which, in some as yet undefined sense, are more complex than those of any machine; but this would seem to depend on Gödel's Platonism. His other example of "the process of systematically constructing" all the recursive ordinals suggests that he may have contemplated the study of Turing's ordinal logics in search of non-mechanical but effective ways of overcoming (T). Clearly he was under no illusions about the prospects that either approach could lead to the construction of a sufficiently "well-defined procedure" to refute Turing's analysis. I conclude that Gödel saw the difficulty of interpreting (T) in a way that both preserved the generality of his incompleteness theorems and avoided a mechanistic hypothesis to the effect that humanly effective processes are mechanizable. The degree of this difficulty can perhaps be seen by how far into the future of human development he felt he had to look for a way out of it.

Judson C. Webb[z]

[z]I would like to thank Solomon Feferman for innumerable and invaluable suggestions, which have helped me find my way through Gödel's thought.

Some remarks on the undecidability results (*1972a*)

1. *The best and most general version of the unprovability of consistency in the same system.*[1] Under the sole hypothesis that Z (number theory) is recursively one-to-one translatable into S, with demonstrability preserved in this direction, the consistency (in the sense of non-demonstrability of both a proposition and its negation), even of very strong systems S, *may* be provable in S, and even in primitive recursive number theory. However, what can be shown to be unprovable in S is the fact that the rules of the equational calculus applied to equations demonstrable in S between primitive recursive terms yield only correct numerical equations (provided that S possesses the property which is asserted to be unprovable). Note that it is necessary to prove this "outer" consistency of S (which for the usual systems is trivially equivalent with consistency) in order to "justify" the transfinite axioms of a system S in the sense of Hilbert's program. ("Rules of the equational calculus" in the foregoing means the two rules of substituting primitive recursive terms for variables and of substituting one such term for another one to which it has been proved equal.)

This theorem remains valid for much weaker systems than Z. With insignificant changes in the wording it even holds for any recursive translation of the primitive recursive equations into S.

2. *Another version of the first undecidability theorem.* The situation may be characterized by the following theorem: In order to solve all problems of Goldbach type of a certain degree of complication k one needs a system of axioms whose degree of complication, up to a minor correction, is $\geq$ k (where the degree of complication is measured by the number of symbols necessary to formulate the problem (or the system of axioms), of course with inclusion of the symbols occurring in the definitions of the non-primitive terms used). Now all of present day mathematics can be derived from a handful of rather simple axioms about a very few primitive terms. Therefore, even if only those problems are to be solvable which can be formulated in a few pages, the few simple axioms being used today will have to be supplemented by a great number of new ones or by axioms of great complication. It may be doubted whether evident axioms in such great numbers (or of such great complexity) can exist at all, and therefore the theorem mentioned might be taken as an indication for the existence of mathematical yes or no questions undecidable for the human mind. But

[1]This has already been published as a remark to footnote 1 of the translation (*1967*, p. 616) of my *1931*, but perhaps it has not received sufficient notice.

what weighs against this interpretation is the fact that there *do* exist unexplored series of axioms which are analytic in the sense that they only explicate the content of the concepts occurring in them, e.g., the axioms of infinity in set theory, which assert the existence of sets of greater and greater cardinality or of higher and higher transfinite types and which only explicate the content of the general concept of set. These principles show that ever more (and ever more complicated) axioms appear during the development of mathematics. For, in order only to understand the axioms of infinity, one must first have developed set theory to a considerable extent.

3. *A philosophical error in Turing's work.*[2] Turing in his *1937*, page 250 (*1965*, page 136), gives an argument which is supposed to show that mental procedures cannot go beyond mechanical procedures. However, this argument is inconclusive. What Turing disregards completely is the fact that *mind, in its use, is not static, but constantly developing*, i.e., that we understand abstract terms more and more precisely as we go on using them, and that more and more abstract terms enter the sphere of our understanding. There may exist systematic methods of actualizing this development, which could form part of the procedure. Therefore, although at each stage the number and precision of the abstract terms at our disposal may be *finite*, both (and, therefore, also Turing's number of *distinguishable states of mind*) may *converge toward infinity* in the course of the application of the procedure. Note that something like this indeed seems to happen in the process of forming stronger and stronger axioms of infinity in set theory. This process, however, today is far from being sufficiently understood to form a well-defined procedure. It must be admitted that the construction of a well-defined procedure which could actually be carried out (and would yield a non-recursive number-theoretic function) would require a substantial advance in our understanding of the basic concepts of mathematics. Another example illustrating the situation is the process of systematically constructing, by their distinguished sequences $\alpha_n \to \alpha$, all recursive ordinals α of the second number-class.

[2]This remark may be regarded as a footnote to the word "mathematics" on page 73, line 3, of my 1964 postscript to *Gödel 1965*.

Introductory note to *1974*

"In this test, however, the infinitely small has completely failed."[a] The test was the foundation of the differential and integral calculus; the author was Abraham Fraenkel (*1928*, page 116); the view expressed was the canonical one at the time.

But infinitesimals refused to go away; by a rather circuitous route they have re-emerged as a part of a viable foundation for mathematical analysis. This was achieved by Abraham Robinson, using the construction of non-standard models by logical means that had originally been introduced by Skolem (*1933a*, *1934*). In Robinson's own words (*1966*, page vii):

> In the fall of 1960 it occurred to me that the concepts and methods of contemporary Mathematical Logic are capable of providing a suitable framework for the development of the Differential and Integral Calculus by means of infinitely small and infinitely large numbers. I first reported my ideas in a seminar talk at Princeton University (November 1960) ....

The influence of Skolem's work was explicitly acknowledged by Robinson. But whereas Skolem's aim had been deeply negative, namely to show the limitation of axiomatic foundations, Robinson was able to turn the non-standard method to positive advantage by providing a new, efficient and rigorous technique for the use of infinitesimals in mathematical analysis.

There had been previous attempts in the same direction since Leibniz' time, but none achieved satisfactory levels of rigor.[b] Robinson succeeded by bringing modern logical notions and results to bear on the problem; by these means one can explain exactly which properties transfer from the standard structure of reals to the non-standard structure with infinitely large and infinitely small numbers. The non-standard extension he constructed is, first of all, an elementary extension in the sense of model theory and, secondly, one rich in points in the sense that any (internal) family of sets with the finite intersection property has a common point (that is, a point belonging to all the sets of the family).

The first of these properties is called the *transfer principle*, and it

[a] "Bei dieser Probe hat aber das Unendlichkleine restlos versagt."

[b] The history of attempts at a theory of infinitesimals from Leibniz to Robinson cannot be dealt with in this brief note; for some reviews of it, see *Robinson 1966* and *Laugwitz 1978*.

guarantees that a non-standard extension has the same "algebra" as the reals. The second property is called the *concurrence principle* or *saturation*; it is a very important uniformity principle, which lies behind many mathematical arguments, and has the form of a transition from a statement with quantifier structure ∀∃ (to express the "local" property of finite intersection) to one of the form ∃∀ (to express the "global", or uniform, property). This is a principle which is at the heart of many finiteness, compactness and uniform-boundedness arguments.

Gödel forcefully expressed his views on the importance of Robinson's work in some remarks following a talk by Robinson at the Institute for Advanced Study in March 1973. Gödel's statement on that occasion was reproduced in *Robinson 1974*, page x, with Gödel's permission, and has been extracted here as *Gödel 1974*. In these remarks Gödel noted that non-standard analysis "frequently simplifies substantially the proofs", that it is not a "fad of mathematical logicians", and further that "there are good reasons to believe that non-standard analysis, in some version or other, will be the analysis of the future".

Gödel further remarked that it is "a great oddity" that the "natural step after the reals, namely the introduction of infinitesimals, has simply been omitted". He linked this to another "oddity" of modern mathematics, "namely the fact that such problems as Fermat's, which can be written down in ten symbols of elementary arithmetic, are still unsolved 300 years after they have been posed". He saw a reason for this failure in the enormous concentration on the development of abstract mathematics, while work on concrete numerical problems was neglected.

At first sight one may indeed wonder why "the next quite natural step" after the reals was not taken sooner. From any one of several points of view, the status of the non-standard system of (hyper)reals may be considered to be on a par with that of the standard system of real numbers; the case for that was put as follows in *Robinson 1966* (page 282):

> Whatever our outlook and in spite of Leibniz' position, it appears to us today that the infinitely small and infinitely large numbers of a non-standard model of Analysis are neither more nor less real than, for example, the standard irrational numbers. This is obvious if we introduce such numbers axiomatically; while in the genetic approach both standard irrational numbers and non-standard numbers are introduced by certain infinitary processes. This remark is equally true if we approach the problem from the point of view of the empirical scientist. For all measurements are recorded in terms of integers or rational numbers, and if our theoretical framework goes beyond these then there is no compelling reason why we should stay within an Archimedean number system.

Gödel's remarks suggest that he would not have disagreed with this statement of Robinson's. But their views would probably have diverged on the question of the ontological status of the "new" numbers. For, Robinson—who on several occasions expressed a strong formalist conviction (see also his *1965* and *1975*)—goes on to say in *1966*: "From a formalist point of view we may look at our theory syntactically and may consider that what we have done is to introduce *new deductive procedures* rather than new mathematical entities."

In contrast, Gödel's remarks here and his general Platonist position would lead him to hold that there is no ontological difference between the integers, the rationals, the standard irrationals, and the infinitesimals: by a series of "quite natural steps" we may become familiar with and gain insight into what already exists.

Beyond these obvious points of agreement and disagreement, Gödel's remarks contain some provocative statements and phrases, which deserve to be singled out for special comment.

"*The next quite natural step*": The use of the definite article suggests that adding infinitesimals results in a *unique* extended number system. This is not so in Robinson's non-standard analysis. One may enforce uniqueness by somewhat arbitrary restrictions, for example, by requiring the extension to be $\aleph_1$-saturated and of power $\aleph_1$, but few will be satisfied by such a move. In fact arguments can be made that non-uniqueness is not a feature to be criticized but an opportunity to be exploited; see *Fenstad 1985*.

"*A great oddity in the history of mathematics*": Could the "next step" really have been taken earlier? It seems to me that the success of Robinson's non-standard analysis presupposes in an essential way an understanding of the notions of *elementary extension* and *concurrence*. This is why Skolem's work was such an important influence; perhaps Robinson should have acknowledged at the same time the importance of the works of A. Maltsev (*1936* and *1941*), in which compactness arguments (giving concurrence) are used for the first time.

The reals and the infinitesimals were used in the early development of the integral and differential calculus and in its applications to the physical sciences. The reals were then tamed in the latter part of the last century (by Weierstrass, Dedekind and Cantor). Infinitesimals were more troublesome and were at first banned ("restlos versagt"). An insight in mathematical logic was necessary for their taming; so, contrary to Gödel, it seems that the next step really had to wait its time.

"*Another oddity*": The intended meaning of Gödel's remarks is problematic; for example, the recent (partial) success on the Fermat problem due to G. Faltings (*1983*) proceeds via an "enormous development of abstract mathematics". Perhaps Gödel meant to hint at some incompleteness phenomenon and the necessity of new axioms to solve concrete problems of mathematics. Such examples are known; but one should not forget that, when presented axiomatically, Robinson's non-standard analysis is a *conservative extension* of the standard theory (*Kreisel 1969*).

"*The analysis of the future*": Here one can add the following point of view in support of Gödel's "good reasons". If we take seriously the idea that the informally understood geometric line can support point sets richer than the standard reals and that *one* non-standard extension is but one way of "constructing" points in this extended continuum, then we have in hand a framework for a geometric analysis of physical phenomena on many, even infinitesimal, scales, including physical phenomena that are too singular to fit in a direct way into the standard frame. In such a geometric analysis, infinitesimals appear not merely as a convenience in "simplifying proofs" but as an essential notion in the very description of the phenomena in question. There is a growing body of new results in the Robinson non-standard analysis exploiting this point of view; see *Cutland 1983* and *Albeverio et alii 1986* for some representative samples.

Jens Erik Fenstad

⟦ Remark on non-standard analysis ⟧
(*1974*)

I would like to point out a fact that was not explicitly mentioned by Professor Robinson, but seems quite important to me; namely that non-standard analysis frequently simplifies substantially the proofs, not only of elementary theorems, but also of deep results. This is true, e.g., also for the proof of the existence of invariant subspaces for compact operators, disregarding the improvement of the result; and it is true in an even higher degree in other cases. This state of affairs should prevent a rather common misinterpretation of non-standard analysis, namely the idea that it is some kind of extravagance or fad of mathematical logicians. Nothing could be farther from the truth. Rather, there are good reasons to believe that non-standard analysis, in some version or other, will be the analysis of the future.

One reason is the just mentioned simplification of proofs, since simplification facilitates discovery. Another, even more convincing reason, is the following: Arithmetic starts with the integers and proceeds by successively enlarging the number system by rational and negative numbers, irrational numbers, etc. But the next quite natural step after the reals, namely the introduction of infinitesimals, has simply been omitted. I think in coming centuries it will be considered a great oddity in the history of mathematics that the first exact theory of infinitesimals was developed 300 years after the invention of the differential calculus. I am inclined to believe that this oddity has something to do with another oddity relating to the same span of time, namely the fact that such problems as Fermat's, which can be written down in ten symbols of elementary arithmetic, are still unsolved 300 years after they have been posed. Perhaps the omission mentioned is largely responsible for the fact that, compared to the enormous development of abstract mathematics, the solution of concrete numerical problems was left far behind.

Textual notes

All of Gödel's articles printed here were previously published, except for *1972* and *1972a*. The copy-text of each work, i.e., the version printed in this volume, is the first published version of the text, except in the case of *1940* where the 1970 printing has been used. The articles *1972* and *1972a* occur as galley proofs in Gödel's *Nachlass*. The copy-text of various papers has been emended to incorporate his later alterations, and these are indicated either in the textual notes below or by single square brackets [ ] in the text. Likewise, editorial additions or corrections are indicated either by textual notes or by double square brackets ⟦ ⟧ in the text. (Minor editorial changes in punctuation have not been recorded in the textual notes.) In these notes, the pairs of numbers on the left indicate page and line number in the present volume.

All articles in this volume were written in English except for *1958*, which is printed here in German and in an English translation; see the textual notes under *1958* for a discussion of the translation. Abbreviations in English (such as Cont. Hyp., Cor., def., langu., prop. funct., math., resp., Th.) and in Latin (such as ad inf.) have been silently expanded throughout, except for p. and pp. in footnotes and for standard abbreviations (such as cf., e.g., etc., i.e.).

Gödel's occasional British spelling has been changed to American spelling in those papers where British spelling occurs, namely *1940*, *1949a* and *1972*.

The original pagination for all previously published texts is indicated by a page number in the margin, with a vertical bar in the text indicating where the page begins. The first page number is always omitted.

Gödel 1938

An offprint of this article in Gödel's *Nachlass* has the following correction.

	Original	Replaced by
26, 5–6	Axiom III3*	i.e., replacing Axiom III3* by Axiom III3

Gödel 1939

	Original	Replaced by
27, 19	is a set	is the set

Gödel 1939a

Gödel noted corrections in the use of α in *1939a* when he published *1947*, and these have been incorporated in the text; see the textual notes below to *1947*. The two lines following Theorem 4 were mistakenly set in italics in the original printing, and are now set in roman.

	Original	Replaced by
29, 9ff	α	μ
29, 28	follows	This follows
32, 14	it follows	there follows

Gödel 1940

The copy-text is the 1970 printing of *1940*, which includes notes that Gödel added in 1951 and in 1965. To improve the readability of proofs, a comma or semi-colon has been introduced from time to time. Likewise, commas are inserted between different members of an ordered pair. Axiom, definition, lemma and theorem have been capitalized whenever they had a number in *1940*.

	Original	Replaced by
35, 24	$(x)(y)(\exists z)$	$(x)(y)(\exists z)(u)$
38, 8	$(x)(\exists y)$	$(x)(\exists y)(u)$
39, 16	the axiom E	Axiom E
39, 16	by a $*$	by $*$
45, 31	only fits to	only applies to
48, 13	function $(x_1, \ldots, x_n)$	function $\phi(x_1, \ldots, x_n)$
49, 19	$zSy)]$	$zSy)$
53, 4–5	hence a set	hence $\mathfrak{D}(F \upharpoonright x)$ is a set
56, 32	by 6.51,	by 6.51, we have
60, 20	$\mathfrak{D}(F)$ is an ordinal	Now $\mathfrak{D}(F)$ is an ordinal
62, 30	$P`\langle\gamma, o\rangle$	$P`\langle\gamma, 0\rangle$
62, 31	$P`\langle\gamma, o\rangle$	$P`\langle\gamma, 0\rangle$
62, 32	$P`\langle\gamma, o\rangle$	$P`\langle\gamma, 0\rangle$
63, 19–20	X is	$\overline{\overline{X}}$ is
64, 23	$\alpha \dotplus 1 \leq \overline{\overline{\alpha^2}}$	$\overline{\overline{\alpha \dotplus 1}} \leq \overline{\overline{\alpha^2}}$
65, 4	Dfn	8.42 Dfn
67, 28	closure	*closure*
68, 27	fundamental operations	*fundamental operations*
71, 21	occuring	occurring
75, 15	$\mathfrak{L}[\mathfrak{Cnv}_k(\overline{A})]$,	$\mathfrak{L}[\mathfrak{Cnv}_k(\overline{A})]$ for $k = 1, 2, 3$,
76, 27	notions, operations	notions and operations

76, 28	notions, operations	notions and operations
76, 29	argument	arguments
77, 5	cf. introduction p. 1.	cf. page 1.
77, 28	$\overline{u} \epsilon \overline{X}$	$\overline{u} \epsilon \overline{Z}$
79, 14	$\equiv . \overline{x} \epsilon \overline{A}$	$\equiv . \overline{x} \epsilon \overline{y}$
79, 16	$\equiv . \sim \overline{x} \epsilon \overline{y}$	$\equiv . \sim \overline{x} \epsilon \overline{A}$
79, 39	$A] . \equiv$	$\overline{A}] . \equiv$
80, 7	$\mathfrak{Cnv}(A)$	$\mathfrak{Cnv}(\overline{A})$
81, 20	applies	applied
82, 17	$x \epsilon \mathfrak{D}(A)$	$x \epsilon \mathfrak{D}(\overline{A})$
84, 6	$\overline{A}^{\text{`}l}\overline{Z}$	$(\overline{A}\text{`})_l\overline{X}$
84, 10	$\mathfrak{Comp}(X)$	$\mathfrak{Comp}(\overline{X})$
84, 11	$\mathfrak{Comp}_l(X)$	$\mathfrak{Comp}_l(\overline{X})$
88, 17	different, since	different. For
89, 3	m is	Furthermore, m is
89, 17	$F\text{`}\delta' \cdot F\text{`}\omega_\alpha$	$F\text{`}\delta' \cdot F\text{``}\omega_\alpha$
89, 37	m^2.	m^2. Now
91, 26	$F\text{`}G\text{`}\alpha \epsilon F\text{`}G\text{`}\alpha$	$F\text{`}G\text{`}\alpha \epsilon F\text{`}G\text{`}\beta$
93, 20	$\alpha \epsilon m$, by (2), $Od\, y$	Now $\alpha \epsilon m$, by (2); $Od\text{`}\, y$
94, 11	$u \epsilon z$	So $u \epsilon z$
94, 11	$u \epsilon z$	Then $u \epsilon z$
95, 13	hypothesis II	hypothesis I
95, 19	symmetry reasons	symmetry
95, 22–23	$\alpha \epsilon m \cdot \eta$.	$\alpha \epsilon m \cdot \eta$. So
95, 27	symmetry reasons	symmetry
96, 19	$x \epsilon r$	Now $x \epsilon r$
96, 24–25	$F\text{`}\eta = J\text{`}_5\langle\beta,\gamma\rangle$	$\eta = J\text{`}_5\langle\beta,\gamma\rangle$
96, 25	$F\text{`}\eta' = J\text{`}_5\langle\beta',\gamma'\rangle$	$\eta' = J\text{`}_5\langle\beta',\gamma'\rangle$
100, 42	$\overline{X}, \overline{Y}$	$\overline{x}, \overline{y}$

Gödel *1944*

The first change below was made in *1972b*, a reprinting of *1944*, and footnote 50 was then omitted (but is retained here).

	Original	Replaced by
122, 43–44	latest book	latest book, *An inquiry into meaning and truth.*
126, 16	XI and XII	xl and xli
135, 12	arbitrary	arbitrarily

Gödel's *Nachlass* contains four annotated offprints of *1944*, catalogued there as items 040265–040268, and designated herein by the letters A–D, respectively. In addition, there is a single annotated page from a fifth offprint, *Nachlass* item 040269, designated here by the letter E; its annotations refer not to *1944* itself, but to Bernays' review of it (*Bernays 1946*). The annotations are variously in English, German, and Gabelsberger shorthand.

In the list below, annotations to *1944* are cited by the page and line in this volume to which they refer. Where an annotation alters the printed text by insertion, deletion, or replacement of material, the original textual passage is reproduced, followed by a slash, the letter designation for the offprint bearing the annotation, a colon, and the text as altered. Other types of annotations are described within editorial brackets following the letter designation and colon. German annotations are followed by an English translation enclosed within parentheses; in the German text itself, words transcribed from shorthand are set in *slanted roman* type.

119, 6	the/D: the most general
120, 7	thorough-going/C: thoroughgoing
120, 21	rule/C: rules
120, 23	symbols/C: symbols,
120, 33	ideas/A: ideas,
121, 37	An/D: Another
122, fn 7	D: ⟦At the end of the footnote Gödel wrote in the margin "in the foll⟦owing⟧ paper"⟧
124, 11	assuming/D: assuming either 1.
124, 12	or/D: or 2.
124, 15	primarily given/D: arrived at first in the construction of language starting with the primitive terms of the language *oder* ⟦(or)⟧ initially
124, 30	"simplicity"/D: "simplicity", or perhaps one should rather say "nonselfreflexivity" ⟦This insertion is preceded and followed by a question mark.⟧
125, 9	appear/C: appear in v⟦on⟧ Neum⟦ann's⟧ syst⟦em⟧ of ax⟦ioms⟧
125, 10	replace/D: replace only
125, 20	terms/D: terms or terms denoting special classes or concepts
125, 23	principles/D: principles concerning the solution of the paradoxes which were
125, 24	principles/C: principles nec⟦essary⟧ for avoiding the paradoxes
125, 33–35	C: ⟦Quotation marks to be deleted and the passage enclosed to be italicized.⟧
125, fn 11	dealt with/D: dealt with in axiomatic set ⟦theory⟧
126, 3	quantifications/D: quantifications or classes
126, 21	axiom/D: axiom (or rather with the decision to restrict oneself to such functions)

126, fn 15	refer/D: refer and which is assumed when they are used
127, 2	nothing to do/C: ⟦These words are underlined and Gödel has written in the right margin "? Zermelo hat jedenfalls etwas damit zu tun" (? In any case, Zermelo has something to do with that).⟧
127, 4	the/C: the merits ⟦of the⟧
127, 15–16	axiom of reducibility/D: axioms of reducibility and of infinity
127, 18	defining outside/C: defining definitions not representable (expressible) in
127, 18	defining/D: defining the real nu⟦mber⟧s mentioned
127, 18	outside/C: in
127, 20	involve/D: involve the same or
127, fn 18	such classes u/C: such classes
128, 13–15	for translating ... contain it, ... fiction. [23]/D: for its use, [23](translating ... contain it) ... fiction.
128, fn 23	One ... impossible/D: E.g. a rule for translating this latter conception of notions is, it is true, impossible for all notions,
128, fn 23	maintaining ... notions/D: maintaining this conception for all abstract notions
128, fn 23	or in fact/D: or
129, 3–4	rules ... containing/D: rules of use for
129, 5	thing/D: concept
129, 8	following/D: following tentative
129, 9	There/D: This definition is impossible because there
129, 14	classes or propositions/D: classes of a given type or propositions containing some entity a
129, 23	would/C: would likewise
130, 4	properties/D: properties of a given type
130, 5–7	the ... propositions/C: ⟦This clause to be italicized.⟧
130, 7	There is no doubt/A: ⟦These words are underlined and "falsch" (false) is written in the right margin.⟧
130, 10	contain themselves/A: ⟦These words are underlined and "falsch" (false) is written in the right margin, followed slightly below by "wegen (x) *bedeutet nicht* wie die Konjunktion" (since (x) *does not mean* the same as the conjunction.)⟧
131, 28	element/A: elements
132, 18	in/D: in some
133, 28	, or combinations of such,/C: ⟦This matter deleted.⟧
136, 17	function/D: function of integers
136, 24–25	this ... consistency of/A: it can be shown that

136, 25	of Cantor's/A: Cantor's
136, 26	of the generalized/A: the generalized
136, 27	set/D: infinite set ⟦The preceding word, 'arbitrary', is enclosed in brackets (to indicate deletion?)⟧
136, 28	subsets)/A: subsets) hold in the system of sets of all transf⟦inite⟧ order⟦s⟧ & that there these prop⟦ositions⟧ are compatible
136, 31–32	the former is, ... quite/A: the former, ... is quite
137, 8	clearly/D: in some sense ⟦to which Gödel added the footnote: propositions prior to their constituents (*Wir erkennen zuerst Zahlen und dann erst verstehen wir Sätze.*) (First we recognize numbers and only then do we understand propositions.)⟧
137, 36	concept/D: well-defined concept
137, 36–37	for any ... arguments/D: of any object as argument
139, 24	140/C: 141
139, fn 47	reduced to/A: ⟦These words are underlined and "falsch" (false) is written in the right margin.⟧
140, 33	axioms/D: true axioms

In addition to the foregoing annotations, there are a number of more general remarks not tied to specific textual passages. They are grouped here according to the reprint on which they appear.

Reprint B:

The title page is covered with pencilled notes, mostly in shorthand, as follows:

At the top: *Die Stellen wo konstruktiv vorkommt sind hier angestrichen.* (The places where 'constructive' occurs are marked herein.) ⟦And in this reprint, all occurrences of the words 'construct', 'constructive', 'constructions' and 'constructivistic' are indeed underlined.⟧

Next below: *Meine* ph. *Meinungen ausgesprochen auf*: p. 127–128, 131, 135, 137, 138–139, 140, 150–151, 152. (My philosophical views expressed on pages 127–128, 131, 135, 137, 138–139, 140, 150–151, 152.)

Above the title: $4'$ p. 134 *unten* weniger *wichtig: selbst in der konstruktierbaren Mathematik kann ein Begriff angewendet werden auf etwas daraus definiert.*

$4''$ p. 136 *Ein gewisses* vic. circle principle *gilt für Konstruktionen.*

(4′, page 134 below, less important: even in constructive mathematics a notion can be applied to something defined by means of it.

4″, page 136, a certain vicious-circle principle holds for constructions.)

Below the title (as an index):

1. Sense perc *und* Theorie *in der Mathematik* p. 127–8
   (Sense perception and theory in mathematics)
2. *Die beiden verschiedenen Interpretationen von* Df (*Kants* "analytisch" *kann so interpretiert werden*) p. 131
   (The two different interpretations of definition (Kant's "analytic" can be so interpreted))
3. Log⟦ical⟧ intuition self contradictory p. 131
4. Vicious Circle principle false p. 135
5. Classes and some real objects p. 137
6. *Man kann von ihnen allen sprechen und das* vic. circle princ. *anwenden.* p. 137–140
   (One can speak of all of them and apply the vicious-circle principle.)
7. Church typenfrei Th. p. 150
   (Church's type-free theory)
8. Th. *der natürlichen Zahlen nachweislich nicht analytisch im Kantschen Sinn.* p. 150
   (The theory of the natural numbers ⟦is⟧ demonstrably not analytic in Kant's sense.)
9. Princ. Math. analytisch im allgemeinen Sinn p. 151
   (*Principia mathematica* ⟦is⟧ analytic in the general sense)
10. *Unser unvollkommenes Verständnis der Grundbegriffe ist der Grund daß die Logik bisher unfruchtbar.* p. 152
    (Our incomplete understanding of the fundamental notions is the reason that logic has so far not been fruitful.)

On a loose slip of paper inserted between pages 132–133 of this reprint, Gödel wrote: "*bis* p. 131 *gelesen* (*wegen* ph *Inhalt und ob es gut ist*) 14./XI.68" (Read to page 131 (on account of its philosophical content and whether it is good) 14 November 1968).

In the left margin of page 140, opposite the underlined words "where it does not apply in the second form either", Gödel wrote "*müße das Ganze lesen*" (would have to read all of it).

Reprint C:

The following remarks are written at the top of page 125:

"p. 140–141 *Das* vic. circ. pr. II for mere pluralities" *sehr plausibel aus das eingeschränkte Aussonderungsax. implied by the conc. of set as plur.* (*über die anderen* Ax. *wird nichts gesagt*) *Unterschied zwischen* 2 *Mengenbegriffe wird deutlich gemacht.* ("pages 140–141: The vicious-circle principle II for mere pluralities" ⟦follows⟧ very plausibly from ⟦the fact that⟧ the restricted separation axiom ⟦is⟧ *implied by the concept of set as plurality* (nothing is said about the other axioms). ⟦The⟧ difference between two notions of set is made clear.)

p. 132 Zermelo *kann als* "elaboration" *der Idee von* limited size *betrachtet werden.* (⟦The work of⟧ Zermelo can be regarded as ⟦an⟧ "elaboration" of the idea of limited size.)

Reprint D:

The following notes appear on the title page:

gelesen bis p. 135 *oben* (read to top of page 135)

p. 136 "constructive" *definiert* (page 136: 'constructive' defined)

The top of page 125 bears the notation:

Bedeutung des terms "constructivistic" (meaning of the term "constructivistic")

A loose sheet was inserted between pages 130–131 of this reprint. The column on the left half of the sheet contains the following fragmentary remarks:

Nach ⟦(according to) an⟧ antireal⟦istic⟧ kind of constr⟦uctivism⟧; i.e., the starting point and means of the constr⟦uction⟧ are to be exclusively sensual & material (e.g. symbols, their perc⟦eptual⟧ prop⟦erties⟧ & rel⟦ations⟧ and the actual or imagined handling of them), not the element⟦ary⟧ operations and int⟦uitions⟧ of a new & irreducible entity called mind. The meaning of the term in question therefore is not ...

I.e., the first alternative of footnote 23 applies while the second leaves room for irred⟦ucible⟧ abstr⟦act⟧ elements.

The column on the right half of the same sheet contains these comments:

Warum nichts von Weyl? (Why nothing by Weyl?)

Warum kein Index? (*Sachregister*) *Macht jedes Buch doppelt so wertvoll.* (Why no (subject) index? ⟦It⟧ makes every book twice as valuable.)

p. 211 *Fußnote* (footnote): strictly antirealistic (i.e., nominalistic)

The extension of this concept is equal to that of predicativity in a rather narrow sense, but admitting quantification in the def⟦initions⟧, & therefore ...

On the back of the inserted sheet is the additional remark:

since even Hilbert's much more restricted "Finitism" does not start with symbols as such objects, but rather with ⟦illegible word inserted above the line⟧ a priori intuition of an idealized space & time ⟦In this passage, the words "more restricted" are somewhat illegible; further down the page Gödel wrote "the much narrower Finitism of Hilbert himself".⟧

The following appears at the bottom of page 135:

Das stimmt nicht (*rekursive* Df. *sind* Df. *außerhalb des Systems, die als solche nicht im System sind*).

(That is not correct (recursive definitions are definitions outside the system, which as such are not in the system).)

In the right margin near the bottom of page 137, is the remark:

welche die einzige interessante Mathematik sind" (which are the only interesting mathematics).

In the left margin of the middle of page 140 is the annotation:

? Brouwers selfreflex. *und meine Beispiele oben*

(? Brouwer's selfreflexivity and my examples above)

Reprint E:

This single page bears the heading "Bernays Rev. *meiner Arbeit über* Russell" (Bernays' review of my paper on Russell), and contains the following remarks:

1.) *Misverständnis meiner* Interpret. *der Typentheorie für* concepts *an zwei Stellen* (Misunderstanding, in two places, of my interpretation of type theory for concepts)
2.) "The whole of math." *muß vorausgesetzt werden.* ("The whole of mathematics" must be presupposed.)
3.) *Vermöge des* meaning *kann man alle math. Sätze auf* $a = a$ *reduzieren.* (By virtue of meaning, one can reduce all mathematical propositions to $a = a$.)
4.) *Das* Probl. *der Beschreibung ist durch "Sinn" und "Bedeutung" in befriedigender Weise gelöst.* (The problem of description is solved in a satisfactory way by "sense" and "denotation".)

5.) *Das* Extens. axiom *gilt nicht für Begriffe.* (The axiom of extensionality does not hold for concepts.)

6.) *Das Meiste, woran* L⟦eibniz⟧ *dachte, ist bereits in der heutigen math. Logik enthalten.*

? *und er deutet an: Man kann aus seinen Worten ersehen, daß was er sagt Unsinn ist?* (5 *Jahre nötig um es zu entwickeln zu einem* powerf. inst. of reas.)

(Most of what Leibniz thought of is already contained in today's mathematical logic.

? and he suggests: one can see from his words that what he says is nonsense? (5 years ⟦will be⟧ necessary in order to develop it into a powerful instrument of reason.) ⟦The referents for 'he', in the passage enclosed within question marks, are unclear.⟧

Toward the bottom of the page Gödel wrote "*Andere* Rev. *zitiert in* vol. XI *von* J.S.L., p. 75". (Other reviews cited in volume XI, page 75, of the *Journal of symbolic logic.*)

Gödel 1946

Items marked with a single asterisk were changed to the new version in the *Davis 1965* printing, while those marked with a pair of asterisks were changed to the new version in the *Klibansky 1968* reprinting.

	Original	Replaced by
*150, 21	different and by	different. By
*151, 11	proposition	propositions
*151, 12	non constructivistic	non-constructive
*151, 14	infinity and it	infinity. It
*151, 20	set	set-theoretic
*151, 29	"mathematical definability"	mathematical definability
*151, 31	it and again	it. Again
*151, 37	e.g.	i.e.
*151, 39	"definability in terms of ordinals"	definability in terms of ordinals
*151, 43	property, i.e. by	property, by
**151, 43	property, by	property: By
**152, 7	sets, namely	sets. Namely
**152, 11	assumed) and for	assumed). For
*152, 20	question	question of
152, 29	which	who
**152, 40	Of course, you will	You may
**152, 42	sets as described	sets as conceived

**152, 44	give nothing	give, or are to give, nothing
**153, 3	any other	any others
**153, 5	But, irrespective	In conclusion I would like to say that, irrespective
**153, 5–6	this concept of definability	the concept of definability suggested in this lecture
**153, 7	I think it has	it has
**153, 7–8	are questions	are two questions
**153, 11–12	It can be proved that	It follows from the axiom of replacement that
**153, 13	can be at all defined	can at all be defined

Gödel 1947

In *1947*, footnote 23, Gödel made corrections to *1939a* in the following sentence, which is omitted here since these corrections are now incorporated in the text of *1939a* printed in the present volume: "I take this opportunity to correct a mistake in the notation and a misprint which occurred in the latter paper: in the lines 25 to 29 of page 221, 4 to 6 and 10 of page 222, 11 to 19 of page 223, the letter α should be replaced (in all places where it occurs) by μ. Also, in Theorem 6 on page 222 the symbol '$\equiv$' should be inserted between $\phi_{\alpha}(x)$ and $\phi_{\overline{\alpha}}(x')$." See also the textual notes under *Gödel 1964* below.

	Original	Replaced by
177, 22	the partila results	the partial results
178, 14	confinality	cofinality
178, 39	or continuum	or that of the continuum
180, 27	suffice	suffices
186, 25	confinality	cofinality

Gödel 1949

Gödel used ch, sh and tg for cosh, sinh and tanh, respectively.

	Original	Replaced by
195, 1	$+x$	$+x_0$
196, 38	so-called	so called
197, 3	$u' = p \cdot u \cdot 9$	$u' = p \cdot u \cdot q$
197, 3	and 9	and q
197, 9	$u' = \sigma^{\beta} \cdot u \cdot 9$	$u' = \sigma^{\beta} \cdot u \cdot q$
197, 9	and g	and q
197, 42	transformation	transformations

Gödel 1949a

Gödel's additions to the *1955* German translation of *1949a* have been inserted in square brackets in the text.

	Original	Replaced by
203, 13–14	destroy	destroys
203, 14	distinguish	distinguishes
204, 1–2	notion the mean	motion the mean
204, 46	my paper forthcoming in	my forthcoming
205, 24	passed according	past according
206, 18	that, whether	that whether
207, 1	exists),	exists)

Gödel 1952

Throughout this article "Newtonean" has been changed to "Newtonian", "Hamiltonean" to "Hamiltonian", and "Lagrangean" to "Lagrangian".

	Original	Replaced by
212, 12	assymmetry	asymmetry
214, 2	connect	connects

Gödel 1958

Since in *1972* above we produce a rather free translation of *1958*, as revised by Gödel, here Stefan Bauer-Mengelberg and Jean van Heijenoort have endeavored to give as literal a translation as possible. They have eschewed paraphrases and have rendered one word by one word (for example, *Anschauung* by *intuition*, and *anschaulich* by *intuitive*).

In footnote 1 Gödel refers to page 2 of *Bernays 1954*. He was apparently using an offprint with its own pagination, beginning with page 1. The proper reference is to page 10, the second page of the article.

There is reason to believe that the "Zusammenfassung" was written by Gödel, but no positive evidence has been found. The "Abstract", reproduced here from *Dialectica*, was probably translated from the "Zusammenfassung" by an editor of *Dialectica* rather than by Gödel, but here again no direct evidence exists.

	Original	Replaced by
251, 32	integral	integer

Gödel 1964

During September 1966 Gödel prepared two typed sheets of changes to *1964* in anticipation of a third edition of *1947* (the paper *1964* constituting the second edition); he added a third sheet in October 1967. These sheets were found in his *Nachlass* and are incorporated in our text of *1964*. The major changes in the text are dated and enclosed there in square brackets, while the minor changes are indicated below by a single asterisk. On the other hand, a pair of asterisks below indicate errors introduced inadvertently in the printing of *1964* but not found in *1947*. The term "euclidean" has been changed throughout to "Euclidean" to agree with the usage in *1947*.

Two of the major changes introduced on those sheets replaced passages in *1964*. The first of them substituted a new version of footnote 20, found in our text, for the version in *1964* as given here:

"[20]See *Mahlo 1911*, pp. 190–200, and *1913*, pp. 269–276. From Mahlo's presentation of the subject, however, it does not appear that the numbers he defines actually exist. In recent years considerable progress has been made as to the axioms of infinity. In particular, some have been formulated that are based on principles entirely different from those of Mahlo, and Dana Scott has proved that one of them implies the negation of proposition A (mentioned on p. 266). So the consistency proof for the continuum hypothesis explained on p. 266 does *not* go through if this axiom is added. However, that these axioms are implied by the general concept of set in the same sense as Mahlo's has not been made clear yet. See *Tarski 1962*, *Scott 1961*, *Hanf and Scott 1961*. Mahlo's axioms have been derived by Azriel Levy from a general principle about the system of all sets. See his *1960*. See also *Bernays 1961*, where almost all set-theoretical axioms are derived from Levy's principle."

The second major change consisted of substituting a new version of the postscript, printed in our text, for the version in *1964* as given below:

"Shortly after the completion of the manuscript of this paper the question of whether Cantor's continuum hypothesis is provable from the von Neumann–Bernays axioms of set theory (the axiom of choice included) was settled in the negative by Paul J. Cohen. A sketch of the proof will appear shortly in the *Proceedings of the National Academy of Sciences*. It turns out that for a wide range of $\aleph_\tau$, the equality $2^{\aleph_0} = \aleph_\tau$ is consistent and an extension in the weak sense (that is, it implies no new number-theoretical theorems). Whether for a suitable concept of "standard" definition there exist definable $\aleph_\tau$ not excluded by König's theorem (see p. 260 above) for which this is not so is still an open question (of course, it must be assumed that the existence of the $\aleph_\tau$ in question is either demonstrable or has been postulated)."

Finally, there is omitted from footnote 24 of *1964* the same sentence (of corrections to *1939a*) that was omitted from footnote 23 of *1947*. See the textual notes above to *1947*.

	Original	Replaced by
256, 26	confinality	cofinality
**256, 31–32	So the continum	So the continuum
**256, 36	definitions pp.	definitions on pp.
256, 39	or continuum or	or that of the continuum or
*259, 9	for theorems about	for theorems depending on
**261, 8	continum problem	continuum problem
262, fn 21	*in terms of ordinal numbers*	in terms of ordinal numbers
264, 25	confinality	cofinality
*267, 20	Cantor's	The generalized
*267, 23	the continuum	$2^{\aleph_\alpha}$

Gödel 1972

	Original	Replaced by
271, 27	proofsheets	proof sheets
273, 28	ordinals, is	ordinals is
275, 29	higher type functions	higher-type functions
278, 30	*Number theoretic*	*Number-theoretic*
278, 36	Outside the	Outside of the
279, fn 1	insure	ensure
280, 29	F, G	F and G

Gödel 1972a

In Gödel's galley proofs, from which *1972a* is printed, he replaced, at 305, 2, "weaker hypotheses" in his earlier version of Remark 1 (*1967*, 616) by "sole hypothesis".

	Original	Replaced by
306, 2	this	the
306, 27	non recursive	non-recursive

References

Aanderaa, Stål

See Dreben, Burton, Peter Andrews and Stål Aanderaa.

Aanderaa, Stål, and Warren D. Goldfarb

1974 The finite controllability of the Maslov case, *The journal of symbolic logic 39*, 509–518.

Ackermann, Wilhelm

1924 Begründung des "tertium non datur" mittels der Hilbertschen Theorie der Widerspruchsfreiheit, *Mathematische Annalen 93*, 1–36.

1928 Zum Hilbertschen Aufbau der reellen Zahlen, *ibid. 99*, 118–133; English translation by Stefan Bauer-Mengelberg in *van Heijenoort 1967*, 493–507.

1928a Über die Erfüllbarkeit gewisser Zählausdrücke, *Mathematische Annalen 100*, 638–649.

1937 Die Widerspruchsfreiheit der allgemeinen Mengenlehre, *ibid. 114*, 305–315.

1940 Zur Widerspruchsfreiheit der Zahlentheorie, *ibid. 117*, 162–194.

1951 Konstruktiver Aufbau eines Abschnitts der zweiten Cantorschen Zahlenklasse, *Mathematische Zeitschrift 53*, 403–413.

1954 *Solvable cases of the decision problem* (Amsterdam: North-Holland).

See also Hilbert, David, and Wilhelm Ackermann.

Addison, John W.

1958 Separation principles in the hierarchies of classical and effective descriptive set theory, *Fundamenta mathematicae 46*, 123–135.

1959 Some consequences of the axiom of constructibility, *ibid.*, 337–357.

See also Henkin et alii.

Addison, John W., Leon Henkin and Alfred Tarski

1965 (eds.) *The theory of models. Proceedings of the 1963 International Symposium at Berkeley* (Amsterdam: North-Holland).

Albeverio, Sergio, Jens E. Fenstad, Raphael Hoegh-Krøhn and Tom Lindstrøm

1986 *Nonstandard methods in stochastic analysis and mathematical physics* (Orlando: Academic Press).

Aleksandrov, Pavel Sergeyevich (Alexandroff, Paul; Александров, Павел Сергеевич)

1916 Sur la puissance des ensembles mesurables B, *Comptes rendus hebdomadaires des séances de l'Académie des Sciences, Paris 162*, 323–325.

Alt, Franz

1933 Zur Theorie der Krümmung, *Ergebnisse eines mathematischen Kolloquiums 4*, 4.

See also *Menger 1936*.

Andrews, Peter

See Dreben, Burton, Peter Andrews and Stål Aanderaa.

Asquith, Peter D., and Philip Kitcher

1985 (eds.) *PSA 1984: Proceedings of the 1984 biennial meeting of the Philosophy of Science Association* (East Lansing, MI: Philosophy of Science Association), vol. 2.

Ax, James, and Simon Kochen

1965 Diophantine problems over local fields II: A complete set of axioms for p-adic number theory, *American journal of mathematics 87*, 631–648.

Bachmann, Heinz

1955 *Transfinite Zahlen*, Ergebnisse der Mathematik und ihrer Grenzgebiete, vol. 1 (Berlin: Springer).

Barendregt, Hendrick P.

1981 *The lambda calculus. Its syntax and semantics* (Amsterdam: North-Holland).

1984 Second edition of *Barendregt 1981*, with revisions (mainly in Part V) and addenda.

Bar-Hillel, Yehoshua

1965 (ed.) *Logic, methodology, and philosophy of science. Proceedings of the 1964 International Congress* (Amsterdam: North-Holland).

See also Fraenkel, Abraham A., and Yehoshua Bar-Hillel.

Bar-Hillel, Yehoshua, E. I. J. Poznanski, Michael O. Rabin and Abraham Robinson

1961 (eds.) *Essays on the foundations of mathematics, dedicated to A. A. Fraenkel on his seventieth anniversary* (Jerusalem: Magnes Press; Amsterdam: North-Holland).

Barwise, Jon
1977 (ed.) *Handbook of mathematical logic* (Amsterdam: North-Holland).

Barwise, Jon, H. Jerome Keisler and Kenneth Kunen
1980 (eds.) *The Kleene symposium. Proceedings of the symposium held at the University of Wisconsin, Madison, Wisconsin, June 18–24, 1978* (Amsterdam: North-Holland).

Barzin, Marcel
1940 Sur la portée du théorème de M. Gödel, *Académie royale de Belgique, Bulletin de la classe des sciences* (*5*) *26*, 230–239.

Baumgartner, James E., and Fred Galvin
1978 Generalized Erdös cardinals and $0^{\#}$, *Annals of mathematical logic 15*, 289–313.

Becker, Oskar
1930 Zur Logik der Modalitäten, *Jahrbuch für Philosophie und phänomenologische Forschung 11*, 497–548.

Beeson, Michael
1978 A type-free Gödel interpretation, *The journal of symbolic logic 43*, 213–227.

Benacerraf, Paul, and Hilary Putnam
1964 (eds.) *Philosophy of mathematics: selected readings* (Englewood Cliffs, N. J.: Prentice-Hall; Oxford: Blackwell).
1983 Second edition of *Benacerraf and Putnam 1964*.

Bergmann, Gustav
1931 Zur Axiomatik der Elementargeometrie, *Ergebnisse eines mathematischen Kolloquiums 1*, 28–30.

Bernays, Paul
1923 Erwiderung auf die Note von Herrn Aloys Müller: "Über Zahlen als Zeichen", *Mathematische Annalen 90*, 159–163; reprinted in *Annalen der Philosophie und philosophischen Kritik 4* (1924), 492–497.
1926 Axiomatische Untersuchung des Aussagen-Kalkuls der *Principia mathematica*, *Mathematische Zeitschrift 25*, 305–320.
1927 Probleme der theoretischen Logik, *Unterrichtsblätter für Mathematik und Naturwissenschaften 33*, 369–377; reprinted in *Bernays 1976*, 1–16.

1935 Sur le platonisme dans les mathématiques, *L'enseignement mathématique 34*, 52–69; English translation by Charles D. Parsons in *Benacerraf and Putnam 1964*, 274–286.

1935a Hilberts Untersuchungen über die Grundlagen der Arithmetik, in *Hilbert 1935*, 196–216.

1937 A system of axiomatic set theory. Part I, *The journal of symbolic logic 2*, 65–77; reprinted in *Bernays 1976a*, 1–13.

1941 A system of axiomatic set theory. Part II, *The journal of symbolic logic 6*, 1–17; reprinted in *Bernays 1976a*, 14–30.

1941a Sur les questions méthodologiques actuelles de la théorie hilbertienne de la démonstration, in *Gonseth 1941*, 144–152.

1942 A system of axiomatic set theory. Part III. Infinity and enumerability. Analysis, *The journal of symbolic logic 7*, 65–89; reprinted in *Bernays 1976a*, 31–55.

1942a A system of axiomatic set theory. Part IV. General set theory, *The journal of symbolic logic 7*, 133–145; reprinted in *Bernays 1976a*, 56–68.

1943 A system of axiomatic set theory. Part V. General set theory (continued), *The journal of symbolic logic 8*, 89–106; reprinted in *Bernays 1976a*, 69–86.

1946 Review of *Gödel 1944*, *The journal of symbolic logic 11*, 75–79.

1954 Zur Beurteilung der Situation in der beweistheoretischen Forschung, *Revue internationale de philosophie 8*, 9–13.

1961 Zur Frage der Unendlichkeitsschemata in der axiomatischen Mengenlehre, in *Bar-Hillel et alii 1961*, 3–49; English translation by John L. Bell and M. Plänitz in *Bernays 1976a*, 121–172.

1967 Hilbert, David, in *Edwards 1967*, vol. 3, 496–504.

1976 *Abhandlungen zur Philosophie der Mathematik* (Darmstadt: Wissenschaftliche Buchgesellschaft).

1976a *Sets and classes: On the work by Paul Bernays*, edited by Gert H. Müller (Amsterdam: North-Holland).

1976b Kurze Biographie, in *Bernays 1976a*, xiv–xvi; English translation in *Bernays 1976a*, xi–xiii.

See also Hilbert, David, and Paul Bernays.

Bernays, Paul, and Abraham A. Fraenkel

1958 *Axiomatic set theory* (Amsterdam: North-Holland).

Bernays, Paul, and Moses Schönfinkel

1928 Zum Entscheidungsproblem der mathematischen Logik, *Mathematische Annalen 99*, 342–372.

Bernstein, Benjamin A.

1931 Whitehead and Russell's theory of deduction as a mathematical science, *Bulletin of the American Mathematical Society 37*, 480–488.

Bernstein, Felix
1901 *Untersuchungen aus der Mengenlehre* (doctoral dissertation, Göttingen; printed at Halle); reprinted with several alterations in *Mathematische Annalen 61* (1905), 117–155.

Betsch, Christian
1926 *Fiktionen in der Mathematik* (Stuttgart: Frommanns).

Bianchi, Luigi
1918 *Lezioni sulla teoria dei gruppi continui finiti di transformazioni* (Pisa: E. Spoerri).

Birkhoff, Garrett
1933 On the combination of subalgebras, *Proceedings of the Cambridge Philosophical Society 29*, 441–464.
1935 Combinatorial relations in projective geometries, *Annals of mathematics (2) 36*, 743–748.
1938 Lattices and their applications, *Bulletin of the American Mathematical Society 44*, 793–800.
1940 *Lattice theory*, Colloquium publications, vol. 25 (New York: American Mathematical Society).

Blackwell, Kenneth
1976 A non-existent revision of *Introduction to mathematical philosophy*, *Russell: The journal of the Bertrand Russell Archives* no. 20, 16–18.

Blum, Manuel
1967 A machine-independent theory of the complexity of recursive functions, *Journal of the Association for Computing Machinery 14*, 322–336.

Blumenthal, Leonard M.
1940 "A paradox, a paradox, a most ingenious paradox", *American mathematical monthly 47*, 346–353.
See also Menger, Karl, and Leonard M. Blumenthal.

Boffa, Maurice, Dirk van Dalen and Kenneth McAloon
1979 (eds.) *Logic Colloquium '78. Proceedings of the colloquium held in Mons, August 1978* (Amsterdam: North-Holland).

Boolos, George
1979 *The unprovability of consistency. An essay in modal logic* (Cambridge, U.K.: Cambridge University Press).

Borel, Emile
1898 *Leçons sur la théorie des fonctions* (Paris: Gauthier-Villars).

Braun, Stefania, and Wacław Sierpiński
1932 Sur quelques propositions équivalentes à l'hypothèse du continu, *Fundamenta mathematicae 19*, 1–7.

Brouwer, Luitzen E. J.
1907 *Over de grondslagen der wiskunde* (Amsterdam: Maas & van Suchtelen); English translation by Arend Heyting and Dr. Gibson in *Brouwer 1975*, 11–101.
1909 Die möglichen Mächtigkeiten, *Atti dei IV Congresso Internazionale dei Matematici, Roma, 6–11 Aprile 1908* (Rome: Accademia dei Lincei), III, 569–571.
1929 Mathematik, Wissenschaft und Sprache, *Monatshefte für Mathematik und Physik 36*, 153–164; reprinted in *Brouwer 1975*, 417–428.
1930 *Die Struktur des Kontinuums* (Vienna: Gistel); reprinted in *Brouwer 1975*, 429–440.
1975 *Collected works*, edited by Arend Heyting, vol. 1 (Amsterdam: North Holland).

Browder, Felix E.
1976 (ed.) *Mathematical developments arising from Hilbert problems*, Proceedings of symposia in pure mathematics, vol. 28 (Providence, R.I.: American Mathematical Society).

Buchholz, Wilfried, Solomon Feferman, Wolfram Pohlers and Wilfried Sieg
1981 *Iterated inductive definitions and subsystems of analysis: recent proof-theoretical studies*, Springer lecture notes in mathematics, no. 897 (Berlin: Springer).

Bukovský, Lev
1965 The continuum problem and the powers of alephs, *Commentationes mathematicae Universitatis Carolinae 6*, 181–197.

Bulloff, Jack J., Thomas C. Holyoke and S. W. Hahn
1969 (eds.) *Foundations of mathematics. Symposium papers commemorating the sixtieth birthday of Kurt Gödel* (New York: Springer).

Cantor, Georg
1874 Über eine Eigenschaft des Inbegriffes aller reellen algebraischen Zahlen, *Journal für die reine und angewandte Mathematik 77*, 258–262; reprinted in *Cantor 1932*, 115–118.
1878 Ein Beitrag zur Mannigfaltigkeitslehre, *Journal für die reine und angewandte Mathematik 84*, 242–258; reprinted in *Cantor 1932*, 119–133.

1883 Ueber unendliche, lineare Punktmannichfaltigkeiten. V, *Mathematische Annalen 21*, 545–591; reprinted in *Cantor 1932*, 165–209.

1884 De la puissance des ensembles parfaits de points, *Acta mathematica 4*, 381–392; reprinted in *Cantor 1932*, 252–260.

1891 Über eine elementare Frage der Mannigfaltigkeitslehre, *Jahresbericht der Deutschen Mathematiker-Vereinigung 1*, 75–78; reprinted in *Cantor 1932*, 278–281.

1895 Beiträge zur Begründung der transfiniten Mengenlehre. I, *Mathematische Annalen 46*, 481–512.

1932 *Gesammelte Abhandlungen mathematischen und philosphischen Inhalts. Mit erläuternden Anmerkungen sowie mit Ergänzungen aus dem Briefwechsel Cantor–Dedekind*, edited by Ernst Zermelo (Berlin: Springer); reprinted in 1962 (Hildesheim: Olms).

Capelli, Alfredo

1897 Saggio sulla introduzione dei numeri irrazionali col metodo delle classi contigue, *Giornale di matematiche di Battaglini 35*, 209–234.

Carnap, Rudolf

1931 Die logizistische Grundlegung der Mathematik, *Erkenntnis 2*, 91–105; English translation by Erna Putnam and Gerald J. Massey in *Benacerraf and Putnam 1964*, 31–41.

1934 Die Antinomien und die Unvollständigkeit der Mathematik, *Monatshefte für Mathematik und Physik 41*, 263–284.

1934a *Logische Syntax der Sprache* (Vienna: Springer); translated into English as *Carnap 1937*.

1935 Ein Gültigkeitskriterium für die Sätze der klassischen Mathematik, *Monatshefte für Mathematik und Physik 42*, 163–190.

1937 *The logical syntax of language* (London: Paul, Trench, Trubner; New York: Harcourt, Brace, and Co.); English translation by Amethe Smeaton of *Carnap 1934a*, with revisions.

See also Hahn et alii.

Cavaillès, Jean

See Noether, Emmy, and Jean Cavaillès.

Chaitin, Gregory J.

1974 Information-theoretic limitations of formal systems, *Journal of the Association for Computing Machinery 21*, 403–424.

Chen, Kien-Kwong

1933 Axioms for real numbers, *Tôhoku mathematical journal 37*, 94–99.

Chihara, Charles S.

1973 *Ontology and the vicious-circle principle* (Ithaca, N.Y.: Cornell University Press).

1982 A Gödelian thesis regarding mathematical objects: Do they exist? And can we perceive them?, *The philosophical review 91*, 211–227.

Christian, Curt

1980 Leben und Wirken Kurt Gödels, *Monatshefte für Mathematik 89*, 261–273.

Church, Alonzo

1932 A set of postulates for the foundation of logic, *Annals of mathematics (2) 33*, 346–366.

1933 A set of postulates for the foundation of logic (second paper), *ibid. (2) 34*, 839–864.

1935 A proof of freedom from contradiction, *Proceedings of the National Academy of Sciences, U.S.A. 21*, 275–281.

1936 An unsolvable problem of elementary number theory, *American journal of mathematics 58*, 345–363; reprinted in *Davis 1965*, 88–107.

1936a A note on the Entscheidungsproblem, *The journal of symbolic logic 1*, 40–41; correction, *ibid.* 101–102; reprinted in *Davis 1965*, 108–115, with the correction incorporated.

1941 *The calculi of lambda-conversion*, Annals of mathematics studies, vol. 6 (Princeton: Princeton University Press); second printing, 1951.

1942 Review of *Quine 1941*, *The journal of symbolic logic 7*, 100–101.

1943 Carnap's introduction to semantics, *Philosophical review 52*, 298–304.

1968 Paul J. Cohen and the continuum problem, *Proceedings of the International Congress of Mathematicians (Moscow–1966)*, 15–20.

1976 Comparison of Russell's resolution of the semantical antinomies with that of Tarski, *The journal of symbolic logic 41*, 747–760.

Church, Alonzo, and J. Barkley Rosser

1936 Some properties of conversion, *Transactions of the American Mathematical Society 39*, 472–482.

Chwistek, Leon
1933 Die nominalistische Grundlegung der Mathematik, *Erkenntnis 3*, 367–388.

Cohen, Paul E.
1979 Partition generation of scales, *Fundamenta mathematicae 103*, 77–82.

Cohen, Paul J.
1963 The independence of the continuum hypothesis. I, *Proceedings of the National Academy of Sciences, U.S.A. 50*, 1143–1148.
1963a A minimal model for set theory, *Bulletin of the American Mathematical Society 69*, 537–540.
1964 The independence of the continuum hypothesis. II, *Proceedings of the National Academy of Sciences, U.S.A. 51*, 105–110.
1966 *Set theory and the continuum hypothesis* (New York: Benjamin).
1971 Comments on the foundations of set theory, in *Scott 1971*, 9–15.

Crossley, John N., and Michael A. E. Dummett
1965 (eds.) *Formal systems and recursive functions* (Amsterdam: North-Holland).

Cutland, Nigel J.
1983 Nonstandard measure theory and its applications, *Bulletin of the London Mathematical Society 15*, 529–589.

Dauben, Joseph W.
1982 Peirce's place in mathematics, *Historia mathematica 9*, 311–325.

Davies, Roy O.
1963 Covering the plane with denumerably many curves, *The journal of the London Mathematical Society 38*, 433–438.

Davis, Martin
1965 (ed.) *The undecidable: Basic papers on undecidable propositions, unsolvable problems and computable functions* (Hewlett, N.Y.: Raven Press).
1982 Why Gödel didn't have Church's thesis, *Information and control 54*, 3–24.

Davis, Martin, Yuri Matiyasevich and Julia Robinson
1976 Hilbert's tenth problem. Diophantine equations: positive aspects of a negative solution, in *Browder 1976*, 323–378.

Dawson, John W., Jr.

1983 The published work of Kurt Gödel: An annotated bibliography, *Notre Dame journal of formal logic 24*, 255–284; addenda and corrigenda, *ibid. 25*, 283–287.

1984 Discussion on the foundation of mathematics, *History and philosophy of logic 5*, 111–129.

1984a Kurt Gödel in sharper focus, *The mathematical intelligencer 6*, no. 4, 9–17.

1985 The reception of Gödel's incompleteness theorems, in *Asquith and Kitcher 1985*, 253–271.

1985a Completing the Gödel–Zermelo correspondence, *Historia mathematica 12*, 66–70.

Dedekind, Richard

1872 *Stetigkeit und irrationale Zahlen* (Braunschweig: Vieweg); English translation in *Dedekind 1901*, 1–27.

1888 *Was sind und was sollen die Zahlen?* (Braunschweig: Vieweg); English translation in *Dedekind 1901*, 31–115.

1901 *Essays on the theory of numbers: Continuity and irrational numbers. The nature and meaning of numbers*, English translation of *1872* and *1888* by Wooster W. Beman (Chicago: Open Court); reprinted in 1963 (New York: Dover).

Dehn, Max

1926 Die Grundlegung der Geometrie in historischer Entwicklung, in *Pasch 1926*, 185–271.

Dekker, Jacob C. E.

1962 (ed.) *Recursive function theory*, Proceedings of symposia in pure mathematics, vol. 5 (Providence R.I.: American Mathematical Society).

Denton, John

See Dreben, Burton, and John Denton.

Devlin, Keith J.

1973 *Aspects of constructibility*, Springer lecture notes in mathematics, no. 354 (Berlin: Springer).

1984 *Constructibility* (Berlin: Springer).

Devlin, Keith J., and Ronald B. Jensen

1975 Marginalia to a theorem of Silver, in *Müller et alii 1975*, 115–142.

Diller, Justus

1968 Zur Berechenbarkeit primitiv-rekursiver Funktionale endlicher Typen, in *Schmidt et alii 1968*, 109–120.

1979 Functional interpretations of Heyting's arithmetic in all finite types, *Nieuw archief voor wiskunde (3) 27*, 70–97.

Diller, Justus, and Gert H. Müller

1975 (eds.) $\models$ *ISILC Proof theory symposion. Dedicated to Kurt Schütte on the occasion of his 65th birthday. Proceedings of the International Summer Institute and Logic Colloquium, Kiel 1974*, Springer lecture notes in mathematics, no. 500 (Berlin: Springer).

Diller, Justus, and Werner Nahm

1974 Eine Variante zur Dialectica-Interpretation der Heyting-Arithmetik endlicher Typen, *Archiv für mathematische Logik und Grundlagenforschung 16*, 49–66.

Diller, Justus, and Kurt Schütte

1971 Simultane Rekursionen in der Theorie der Funktionale endlicher Typen, *Archiv für mathematische Logik und Grundlagenforschung 14*, 69–74.

Diller, Justus, and Helmut Vogel

1975 Intensionale Funktionalinterpretation der Analysis, in *Diller and Müller 1975*, 56–72.

Dingler, Hugo

1931 *Philosophie der Logik und Arithmetik* (Munich: Reinhardt).

Dodd, Anthony J.

198? Strong cardinals (handwritten notes).

Dodd, Anthony J., and Ronald B. Jensen

1981 The core model, *Annals of mathematical logic 20*, 43–75.

Dragalin, Albert G. (Драгалин, Альберт Г.)

1968 The computability of primitive recursive terms of finite type and primitive recursive realizability (Russian), *Zapiski nauchnyk seminarov Leningradskogo otdeleniya Matematicheskogo Instituta im. V. A. Steklova, Akademii nauk S.S.S.R. (Leningrad) 8*, 32–45; English translation in *Slisenko 1970*, 13–18.

1980 New forms of realizability and Markov's rule, *Soviet mathematics doklady 21*, 461–464.

Dreben, Burton

1952 On the completeness of quantification theory, *Proceedings of the National Academy of Sciences, U.S.A. 38*, 1047–1052.

1962 Solvable Surányi subclasses: An introduction to the Herbrand theory, *Proceedings of a Harvard symposium on digital computers and their applications, 3–6 April 1961* (*The annals of the Computation Laboratory of Harvard University 31*) (Cambridge, Mass.: Harvard University Press), 32–47.

Dreben, Burton, Peter Andrews and Stål Aanderaa

1963 False lemmas in Herbrand, *Bulletin of the American Mathematical Society 69*, 699–706.

Dreben, Burton, and John Denton

1970 Herbrand-style consistency proofs, in *Myhill et alii 1970*, 419–433.

Dreben, Burton, and Warren D. Goldfarb

1979 *The decision problem: Solvable classes of quantificational formulas* (Reading, Mass.: Addison–Wesley).

Dummett, Michael A. E.

1959 A propositional calculus with denumerable matrix, *The journal of symbolic logic 24*, 97–106.

1978 *Truth and other enigmas* (London: Duckworth).

See also Crossley, John N., and Michael A. E. Dummett.

Dyson, Freeman

1983 Unfashionable pursuits, *The mathematical intelligencer 5*, no. 3, 47–54.

Easton, William B.

1964 *Powers of regular cardinals* (doctoral dissertation, Princeton University); reprinted in part as *Easton 1970.*

1964a Proper classes of generic sets, *Notices of the American Mathematical Society 11*, 205.

1970 Powers of regular cardinals, *Annals of mathematical logic 1*, 139–178.

Edwards, Paul

1967 (ed.) *The encyclopedia of philosophy* (New York: Macmillan and the Free Press).

Ehrenfeucht, Andrzej, and Andrzej Mostowski

1956 Models of axiomatic theories admitting automorphisms, *Fundamenta mathematicae 43*, 50–68.

Ehrenfeucht, Andrzej, and Jan Mycielski
1971 Abbreviating proofs by adding new axioms, *Bulletin of the American Mathematical Society 77*, 366–367.

Eklof, Paul C.
1976 Whitehead's problem is undecidable, *American mathematical monthly 83*, 775–788.

Ellentuck, Erik
1975 Gödel's square axioms for the continuum, *Mathematische Annalen 216*, 29–33.

Ellis, George F. R.
See Hawking, Stephen W. and George F. R. Ellis.

Erdös, Paul
1963 On a problem in graph theory, *Mathematical gazette 47*, 220–223.

Errera, Alfred
1952 Le problème du continu, *Atti della Accademia Ligure di Scienze e Lettere (Roma) 9*, 176–183.

Faltings, Gerd
1983 Endlichkeitssätze für abelsche Varietäten über Zahlkörpern, *Inventiones mathematicae 73*, 349–366.

Feferman, Solomon
1955 Review of *Wang 1951*, *The journal of symbolic logic 20*, 76–77.
1960 Arithmetization of metamathematics in a general setting, *Fundamenta mathematicae 49*, 35–92.
1962 Transfinite recursive progressions of axiomatic theories, *The journal of symbolic logic 27*, 259–316.
1964 Systems of predicative analysis, *ibid. 29*, 1–30.
1965 Some applications of the notions of forcing and generic sets, *Fundamenta mathematicae 56*, 325–345.
1966 Predicative provability in set theory, *Bulletin of the American Mathematical Society 72*, 486–489.
1971 Ordinals and functionals in proof theory, *Actes du Congrès international des mathématiciens, 1–10 septembre 1970, Nice, France* (Paris: Gauthier-Villars), vol. I, 229–233.
1974 Predicatively reducible systems of set theory, in *Jech 1974*, 11–32.
1977 Theories of finite type related to mathematical practice, in *Barwise 1977*, 913–971.

1982 Inductively presented systems and the formalization of metamathematics, in *van Dalen et alii 1982*.

1984 Toward useful type-free theories. I., *The journal of symbolic logic 49*, 75–111.

1984a Kurt Gödel: Conviction and caution, *Philosophia naturalis 21*, 546–562.

1985 Working foundations, *Synthese 62*, 229–254.

See also Buchholz, Wilfried, et alii.

Feferman, Solomon, and Clifford Spector

1962 Incompleteness along paths in progressions of theories, *The journal of symbolic logic 27*, 383–390.

Feferman, Solomon, and Alfred Tarski

1953 Review of *Rasiowa and Sikorski 1951*, *The journal of symbolic logic 18*, 339–340.

Feigl, Herbert

1969 The Wiener Kreis in America, in *Fleming and Bailyn 1969*, 630–673.

Felgner, Ulrich

1971 Comparison of the axioms of local and universal choice, *Fundamenta mathematicae 71*, 43–62.

Fenstad, Jens E.

1971 (ed.) *Proceedings of the second Scandinavian logic symposium* (Amsterdam: North-Holland).

1985 Is nonstandard analysis relevant for the philosophy of mathematics?, *Synthese 62*, 289–301.

See also Albeverio et alii.

Fleming, Donald, and Bernard Bailyn

1969 (eds.) *The intellectual migration: Europe and America, 1930–1960* (Cambridge, Mass.: Harvard University Press).

Foreman, Matthew

1986 Potent axioms, *Transactions of the American Mathematical Society 294*, 1–28.

Foreman, Matthew, Menachem Magidor and Saharon Shelah

198? Martin's maximum, saturated ideals, and non-regular ultrafilters. Part I, *Annals of mathematics*, to appear.

Foreman, Matthew, and W. Hugh Woodin

198? The G.C.H. can fail everywhere (to appear).

Fourman, Michael P., Christopher J. Mulvey and Dana S. Scott
1979 (eds.) *Applications of sheaves*, Springer lecture notes in mathematics, no. 753 (Berlin: Springer).

Fourman, Michael P., and Dana S. Scott
1979 Sheaves and logic, in *Fourman et alii 1979*, 302–401.

Fraenkel, Abraham A.
1919 *Einleitung in die Mengenlehre* (Berlin: Springer).
1922 Der Begriff 'definit' und die Unabhängigkeit des Auswahlaxioms, *Sitzungsberichte der Preussischen Akademie der Wissenschaften, Physikalisch-mathematische Klasse*, 253–257; English translation by Beverly Woodward in *van Heijenoort 1967*, 284–289.
1922a Zu den Grundlagen der Cantor–Zermeloschen Mengenlehre, *Mathematische Annalen 86*, 230–237.
1925 Untersuchungen über die Grundlagen der Mengenlehre, *Mathematische Zeitschrift 22*, 250–273.
1927 *Zehn Vorlesungen über die Grundlegung der Mengenlehre* (Leipzig: Teubner).
1928 Third, revised edition of *Fraenkel 1919* (Berlin: Springer).
See also Bernays, Paul, and Abraham A. Fraenkel.

Fraenkel, Abraham A., and Yehoshua Bar-Hillel
1958 *Foundations of set theory* (Amsterdam: North-Holland).

Frayne, Thomas, Anne Morel and Dana S. Scott
1962 Reduced direct products, *Fundamenta mathematicae 51*, 195–228.

Frege, Gottlob
1879 *Begriffsschrift, eine der arithmetischen nachgebildete Formelsprache des reinen Denkens* (Halle: Nebert); reprinted in *Frege 1964*; English translation by Stefan Bauer-Mengelberg in *van Heijenoort 1967*, 1–82, and by Terrell W. Bynum in *Frege 1972*, 101–203.
1892 Über Sinn und Bedeutung, *Zeitschrift für Philosophie und philosophische Kritik* (n. s.) *100*, 25–50; English translation by Max Black in *Frege 1952*, 56–78.
1903 *Grundgesetze der Arithmetik, begriffsschriftlich abgeleitet* (Jena: Pohle), vol. 2.
1952 *Translations from the philosophical writings of Gottlob Frege*, edited by Peter Geach and Max Black (Oxford: Blackwell); third edition, 1980.

1964 *Begriffsschrift und andere Aufsätze*, edited by Ignacio Angelelli (Hildesheim: Olms).
1972 *Conceptual notation and related articles*, translated and edited by Terrell W. Bynum (Oxford: Clarendon Press).

Freiling, Chris
1986 Axioms of symmetry: Throwing darts at the real number line, *The journal of symbolic logic 51*, 190–200.

Friedman, Harvey
1973 The consistency of classical set theory relative to a set theory with intuitionistic logic, *The journal of symbolic logic 38*, 315–319.
1978 Classically and intuitionistically provably recursive functions, in *Müller and Scott 1978*, 21–27.

Friedman, Joel I.
1971 The generalized continuum hypothesis is equivalent to the generalized maximization principle, *The journal of symbolic logic 36*, 39–54.

Friedrich, Wolfgang
1984 Spielquantorinterpretation unstetiger Funktionale der höheren Analysis, *Archiv für mathematische Logik und Grundlagenforschung 24*, 73–99.
1985 Gödelsche Funktionalinterpretation für eine Erweiterung der klassischen Analysis, *Zeitschrift für mathematische Logik und Grundlagen der Mathematik 31*, 3–29.

Gaifman, Haim
1964 Measurable cardinals and constructible sets, *Notices of the American Mathematical Society 11*, 771.
1974 Elementary embeddings of models of set-theory and certain subtheories, in *Jech 1974*, 33–101.
1976 Models and types of Peano's arithmetic, *Annals of mathematical logic 9*, 223–306.

Galvin, Fred
See Baumgartner, James E., and Fred Galvin.

Galvin, Fred, and András Hajnal
1975 Inequalities for cardinal powers, *Annals of mathematics (2) 101*, 491–498.

Gandy, Robin O.

1980 Church's thesis and principles for mechanisms, in *Barwise et alii 1980*, 123–148.

Gentzen, Gerhard

1935 Untersuchungen über das logische Schließen, *Mathematische Zeitschrift 39*, 176–210, 405–431; English translation by M. E. Szabo in *Gentzen 1969*, 68–131.

1936 Die Widerspruchsfreiheit der reinen Zahlentheorie, *Mathematische Annalen 112*, 493–565; English translation by M. E. Szabo in *Gentzen 1969*, 132–213.

1969 *The collected papers of Gerhard Gentzen*, edited and translated into English by M. E. Szabo (Amsterdam: North-Holland).

Girard, Jean-Yves

1971 Une extension de l'interprétation de Gödel à l'analyse, et son application à l'élimination des coupures dans l'analyse et la théorie des types, in *Fenstad 1971*, 63–92.

1972 *Interprétation fonctionelle et élimination des coupures de l'arithmétique d'ordre supérieur* (doctoral dissertation, Université de Paris VII).

1982 Herbrand's theorem and proof theory, in *Stern 1982*, 29–38.

Glivenko, Valerii Ivanovich (Гливенко, Валерий Иванович)

1929 Sur quelques points de la logique de M. Brouwer, *Académie royale de Belgique, Bulletin de la classe des sciences (5) 15*, 183–188.

Gödel, Kurt

1929 *Über die Vollständigkeit des Logikkalküls* (doctoral dissertation, University of Vienna).

1930 Die Vollständigkeit der Axiome des logischen Funktionenkalküls, *Monatshefte für Mathematik und Physik 37*, 349–360.

1930a Über die Vollständigkeit des Logikkalküls, *Die Naturwissenschaften 18*, 1068.

1930b Einige metamathematische Resultate über Entscheidungsdefinitheit und Widerspruchsfreiheit, *Anzeiger der Akademie der Wissenschaften in Wien 67*, 214–215.

1931 Über formal unentscheidbare Sätze der *Principia mathematica* und verwandter Systeme I, *Monatshefte für Mathematik und Physik 38*, 173–198.

1931a Diskussion zur Grundlegung der Mathematik (Gödel's remarks in *Hahn et alii 1931*), *Erkenntnis 2*, 147–151.

1931b Review of *Neder 1931*, *Zentralblatt für Mathematik und ihre Grenzgebiete 1*, 5–6.

1931c Review of *Hilbert 1931*, *ibid. 1*, 260.

1931d Review of *Betsch 1926*, *Monatshefte für Mathematik und Physik* (*Literaturberichte*) *38*, 5.

1931e Review of *Becker 1930*, *ibid. 38*, 5–6.

1931f Review of *Hasse and Scholz 1928*, *ibid. 38*, 37.

1931g Review of *von Juhos 1930*, *ibid. 38*, 39.

1932 Zum intuitionistischen Aussagenkalkül, *Anzeiger der Akademie der Wissenschaften in Wien 69*, 65–66; reprinted, with additional comment, as *1933n*.

1932a Ein Spezialfall des Entscheidungsproblems der theoretischen Logik, *Ergebnisse eines mathematischen Kolloquiums 2*, 27–28.

1932b Über Vollständigkeit und Widerspruchsfreiheit, *ibid. 3*, 12–13.

1932c Eine Eigenschaft der Realisierungen des Aussagenkalküls, *ibid. 3*, 20–21.

1932d Review of *Skolem 1931*, *Zentralblatt für Mathematik und ihre Grenzgebiete 2*, 3.

1932e Review of *Carnap 1931*, *ibid. 2*, 321.

1932f Review of *Heyting 1931*, *ibid. 2*, 321–322.

1932g Review of *von Neumann 1931*, *ibid. 2*, 322.

1932h Review of *Klein 1931*, *ibid. 2*, 323.

1932i Review of *Hoensbroech 1931*, *ibid. 3*, 289.

1932j Review of *Klein 1932*, *ibid. 3*, 291.

1932k Review of *Church 1932*, *ibid. 4*, 145–146.

1932l Review of *Kalmár 1932*, *ibid. 4*, 146.

1932m Review of *Huntington 1932*, *ibid. 4*, 146.

1932n Review of *Skolem 1932*, *ibid. 4*, 385.

1932o Review of *Dingler 1931*, *Monatshefte für Mathematik und Physik* (*Literaturberichte*) *39*, 3.

1933 Untitled remark following *Parry 1933*, *Ergebnisse eines mathematischen Kolloquiums 4*, 6.

1933a Über Unabhängigkeitsbeweise im Aussagenkalkül, *ibid. 4*, 9–10.

1933b Über die metrische Einbettbarkeit der Quadrupel des R_3 in Kugelflächen, *ibid. 4*, 16–17.

1933c Über die Waldsche Axiomatik des Zwischenbegriffes, *ibid. 4*, 17–18.

1933d Zur Axiomatik der elementargeometrischen Verknüpfungsrelationen, *ibid. 4*, 34.

1933e Zur intuitionistischen Arithmetik und Zahlentheorie, *ibid. 4*, 34–38.

1933f Eine Interpretation des intuitionistischen Aussagenkalküls, *ibid.* *4*, 39–40.

1933g Bemerkung über projektive Abbildungen, *ibid.* *5*, 1.

1933h (with K. Menger and A. Wald) Diskussion über koordinatenlose Differentialgeometrie, *ibid.* *5*, 25–26.

1933i Zum Entscheidungsproblem des logischen Funktionenkalküls, *Monatshefte für Mathematik und Physik* *40*, 433–443.

1933j Review of *Kaczmarz 1932*, *Zentralblatt für Mathematik und ihre Grenzgebiete* *5*, 146.

1933k Review of *Lewis 1932*, *ibid.* *5*, 337–338.

1933l Review of *Kalmár 1933*, *ibid.* *6*, 385–386.

1933m Review of *Hahn 1932*, *Monatshefte für Mathematik und Physik* (*Literaturberichte*) *40*, 20–22.

1933n Reprint of *Gödel 1932*, with additional comment, *Ergebnisse eines mathematischen Kolloquiums* *4*, 40.

1934 *On undecidable propositions of formal mathematical systems* (mimeographed lecture notes, taken by Stephen C. Kleene and J. Barkley Rosser); reprinted with revisions in *Davis 1965*, 39–74.

1934a Review of *Skolem 1933*, *Zentralblatt für Mathematik und ihre Grenzgebiete* *7*, 97–98.

1934b Review of *Quine 1933*, *ibid.* *7*, 98.

1934c Review of *Skolem 1933a*, *ibid.* *7*, 193–194.

1934d Review of *Chen 1933*, *ibid.* *7*, 385.

1934e Review of *Church 1933*, *ibid.* *8*, 289.

1934f Review of *Notcutt 1934*, *ibid.* *9*, 3.

1935 Review of *Skolem 1934*, *ibid.* *10*, 49.

1935a Review of *Huntington 1934*, *ibid.* *10*, 49.

1935b Review of *Carnap 1934*, *ibid.* *11*, 1.

1935c Review of *Kalmár 1934*, *ibid.* *11*, 3–4.

1936 Untitled remark following *Wald 1936*, *Ergebnisse eines mathematischen Kolloquiums* *7*, 6.

1936a Über die Länge von Beweisen, *ibid.* *7*, 23–24.

1936b Review of *Church 1935*, *Zentralblatt für Mathematik und ihre Grenzgebiete* *12*, 241–242.

1938 The consistency of the axiom of choice and of the generalized continuum hypothesis, *Proceedings of the National Academy of Sciences, U.S.A.* *24*, 556–557.

1939 The consistency of the generalized continuum hypothesis, *Bulletin of the American Mathematical Society* *45*, 93.

1939a Consistency proof for the generalized continuum hypothesis, *Proceedings of the National Academy of Sciences, U.S.A.* *25*, 220–224; errata in *1947*, footnote 23.

1940 *The consistency of the axiom of choice and of the generalized continuum hypothesis with the axioms of set theory*, Annals of mathematics studies, vol. 3 (Princeton: Princeton University Press), lecture notes taken by George W. Brown; reprinted with additional notes in 1951 and with further notes in 1966.

1944 Russell's mathematical logic, in *Schilpp 1944*, 123–153.

1946 Remarks before the Princeton bicentennial conference on problems in mathematics, 1–4; first published in *Davis 1965*, 84–88.

1947 What is Cantor's continuum problem?, *American mathematical monthly 54*, 515–525; errata, *55*, 151.

1949 An example of a new type of cosmological solutions of Einstein's field equations of gravitation, *Reviews of modern physics 21*, 447–450.

1949a A remark about the relationship between relativity theory and idealistic philosophy, in *Schilpp 1949*, 555–562.

1952 Rotating universes in general relativity theory, *Proceedings of the International Congress of Mathematicians; Cambridge, Massachusetts, U.S.A. August 30-September 6, 1950* (Providence, R.I.: American Mathematical Society, 1952), I, 175–181.

1955 Eine Bemerkung über die Beziehungen zwischen der Relativitätstheorie und der idealistischen Philosophie (German translation of *Gödel 1949a* by Hans Hartmann), in *Schilpp 1955*, 406–412.

1958 Über eine bisher noch nicht benützte Erweiterung des finiten Standpunktes, *Dialectica 12*, 280–287.

1962 Postscript to *Spector 1962*, 27.

1964 Revised and expanded version of *Gödel 1947*, in *Benacerraf and Putnam 1964*, 258–273.

1964a Reprint, with some alterations, of *Gödel 1944*, in *Benacerraf and Putnam 1964*, 211–232.

1965 Expanded version of *Gödel 1934*, in *Davis 1965*, 39–74.

1967 English translation of *Gödel 1931* by Jean van Heijenoort, in *van Heijenoort 1967*, 596–616.

1968 Reprint, with some alterations, of *Gödel 1946*, in *Klibansky 1968*, 250–253.

1972 On an extension of finitary mathematics which has not yet been used (to have appeared in *Dialectica*; first published in the present volume), revised and expanded English translation of *Gödel 1958*.

1972a Some remarks on the undecidability results (to have appeared in *Dialectica*; first published in the present volume).

1972b Reprint, with some alterations, of *Gödel 1944*, in *Pears 1972*, 192–226.

1974 Untitled remarks, in *Robinson 1974*, x.

1980 On a hitherto unexploited extension of the finitary standpoint, English translation of *1958* by Wilfrid Hodges and Bruce Watson, *Journal of philosophical logic 9*, 133–142.

1986 *Collected works*, volume I: *Publications 1929–1936*, edited by Solomon Feferman, John W. Dawson, Jr., Stephen C. Kleene, Gregory H. Moore, Robert M. Solovay, and Jean van Heijenoort (New York and Oxford: Oxford University Press).

Goldblatt, Robert

1978 Arithmetical necessity, provability and intuitionistic logic, *Theoria 44*, 38–46.

Goldfarb, Warren D.

1971 Review of *Skolem 1970*, *The journal of philosophy 68*, 520–530.

1979 Logic in the twenties: The nature of the quantifier, *The journal of symbolic logic 44*, 351–368.

1981 The undecidability of the second-order unification problem, *Theoretical computer science 13*, 225–230.

1984 The Gödel class with identity is unsolvable, *Bulletin of the American Mathematical Society 10*, 113–115.

1984a The unsolvability of the Gödel class with identity, *The journal of symbolic logic 49*, 1237–1252.

See also Aanderaa, Stål, and Warren D. Goldfarb.

See also Dreben, Burton, and Warren D. Goldfarb.

Goldstine, Herman H.

1972 *The computer from Pascal to von Neumann* (Princeton: Princeton University Press).

Gonseth, Ferdinand

1941 (ed.) *Les entretiens de Zurich, 6-9 décembre 1938* (Zurich: Leemann).

Goodman, Nicolas D.

1976 The theory of the Gödel functionals, *The journal of symbolic logic 41*, 574–582.

1984 Epistemic arithmetic is a conservative extension of intuitionistic arithmetic, *ibid. 49*, 192–203.

Goodstein, Reuben L.

1945 Function theory in an axiom-free equation calculus, *Proceedings of the London Mathematical Society* (2) *48*, 401–434.

1957 *Recursive number theory* (Amsterdam: North-Holland).

Grassl, Wolfgang
1982 (ed.) *Friedrich Waismann, lectures on the philosophy of mathematics*, Studien zur österreichischen Philosophie, vol. 4 (Amsterdam: Rodopi).

Grattan-Guinness, Ivor
1979 In memoriam Kurt Gödel: His 1931 correspondence with Zermelo on his incompletability theorem, *Historia mathematica 6*, 294–304.

Greenberg, Marvin J.
1974 *Euclidean and non-Euclidean geometries: Development and history* (San Francisco: Freeman).
1980 Second edition of *Greenberg 1974*.

Grzegorczyk, Andrzej
1964 Recursive objects in all finite types, *Fundamenta mathematicae 54*, 73–93.
1967 Some relational systems and the associated topological spaces, *Fundamenta mathematicae 60*, 223-231.

Gurevich, Yuri, and Saharon Shelah
1983 Random models and the Gödel case of the decision problem, *The journal of symbolic logic 48*, 1120–1124.

Hacking, Ian
1963 What is strict implication?, *The journal of symbolic logic 28*, 51–71.

Hahn, Hans
1921 *Theorie der reellen Funktionen* (Berlin: Springer).
1932 *Reelle Funktionen* (Leipzig: Akademische Verlagsgesellschaft).
1980 *Empiricism, logic and mathematics: Philosophical papers*, edited by Brian McGuinness (Dordrecht: Reidel).

Hahn, Hans, Rudolf Carnap, Kurt Gödel, Arend Heyting, Kurt Reidemeister, Arnold Scholz and John von Neumann
1931 Diskussion zur Grundlegung der Mathematik, *Erkenntnis 2*, 135–151; English translation by John W. Dawson, Jr., in *Dawson 1984*.

Hajnal, András
1956 On a consistency theorem connected with the generalized continuum problem, *Zeitschrift für mathematische Logik und Grundlagen der Mathematik 2*, 131–136.

1961 On a consistency theorem connected with the generalized continuum problem, *Acta mathematica Academiae Scientiarum Hungaricae 12*, 321–376.

See also Galvin, Fred, and András Hajnal.

Hanatani, Yoshito

1975 Calculability of the primitive recursive functionals of finite type over the natural numbers, in *Diller and Müller 1975*, 152–163.

Hanf, William P., and Dana S. Scott

1961 Classifying inaccessible cardinals, *Notices of the American Mathematical Society 8*, 445.

Harrington, Leo

See Paris, Jeff, and Leo Harrington.

Hartmanis, Juris

1978 *Feasible computations and provable complexity properties*, CBMS–NSF regional conference series in applied mathematics (Philadelphia: Society for Industrial and Applied Mathematics).

Hasse, Helmut, and Heinrich Scholz

1928 *Die Grundlagenkrisis der griechischen Mathematik* (Charlottenburg: Metzner).

Hausdorff, Felix

1908 Grundzüge einer Theorie der geordneten Mengen, *Mathematische Annalen 65*, 435–505.

1914 *Grundzüge der Mengenlehre* (Leipzig: Veit); reprinted in 1949 (New York: Chelsea).

1914a Bemerkung über den Inhalt von Punktmengen, *Mathematische Annalen 75*, 428–433.

1916 Die Mächtigkeit der Borelschen Mengen, *ibid. 77*, 430–437.

1935 Third, revised edition of *Hausdorff 1914* (Berlin and Leipzig: W. de Gruyter).

See also Paul Mongré.

Hawking, Stephen W., and George F. R. Ellis

1973 *The large scale structure of space-time* (Cambridge, U.K.: Cambridge University Press).

Heims, Steve

1980 *John von Neumann and Norbert Wiener: from mathematics to the technologies of life and death* (Cambridge, Mass.: M.I.T. Press).

Henkin, Leon

1949 The completeness of the first-order functional calculus, *The journal of symbolic logic 14*, 159–166.

See also Addison, John W., Leon Henkin and Alfred Tarski.

Henkin, Leon, John W. Addison, Chen Chung Chang, William Craig, Dana S. Scott and Robert L. Vaught

1974 (eds.) *Proceedings of the Tarski symposium*, Proceedings of symposia in pure mathematics, vol. 25 (Providence, R.I.: American Mathematical Society).

Henn, Rudolph, and Otto Moeschlin

1977 The scientific work of Oskar Morgenstern, in *Mathematical economics and game theory: Essays in honor of Oskar Morgenstern*, Springer lecture notes in economics and mathematical systems, no. 141 (New York: Springer), 1–9.

Herbrand, Jacques

1930 *Recherches sur la théorie de la démonstration* (doctoral dissertation, University of Paris); English translation by Warren D. Goldfarb in *Herbrand 1971*, 44–202.

1930a Les bases de la logique hilbertienne, *Revue de métaphysique et de morale 37*, 243–255; English translation by Warren D. Goldfarb in *Herbrand 1971*, 203–214.

1931 Sur la non-contradiction de l'arithmétique, *Journal für die reine und angewandte Mathematik 166*, 1–8; English translation by Jean van Heijenoort in *van Heijenoort 1967*, 618–628, and in *Herbrand 1971*, 282–298.

1931a Sur le problème fondamental de la logique mathématique, *Sprawozdania z posiedzeń Towarzystwa Naukowego Warszawskiego wydział III, 24*, 12–56; English translation by Warren D. Goldfarb in *Herbrand 1971*, 215–271.

1968 *Ecrits logiques*, edited by Jean van Heijenoort (Paris: Presses Universitaires de France).

1971 *Logical writings*, English translation of *Herbrand 1968* by Warren D. Goldfarb (Dordrecht: Reidel).

Hewitt, Edwin

1948 Rings of real-valued continuous functions. I, *Transactions of the American Mathematical Society 64*, 45–99, 596.

Heyting, Arend

1930 Die formalen Regeln der intuitionistischen Logik, *Sitzungsberichte der Preussischen Akademie der Wissenschaften, physikalisch-mathematische Klasse*, 42–56.

1930a Die formalen Regeln der intuitionistischen Mathematik, *ibid.*, 57–71, 158–169.

1931 Die intuitionistische Grundlegung der Mathematik, *Erkenntnis 2*, 106-115; English translation by Erna Putnam and Gerald J. Massey in *Benacerraf and Putnam 1964*, 42–49.

1956 *Intuitionism: An introduction* (Amsterdam: North-Holland).

1959 (ed.) *Constructivity in mathematics. Proceedings of the colloquium held at Amsterdam, 1957* (Amsterdam: North-Holland).

See also Hahn et alii.

Hilbert, David

1899 *Grundlagen der Geometrie. Festschrift zur Feier der Enthüllung des Gauss–Weber Denkmals in Göttingen* (Leipzig: Teubner).

1900 Mathematische Probleme. Vortrag, gehalten auf dem internationalen Mathematiker-Kongress zu Paris 1900, *Nachrichten von der Königlichen Gesellschaft der Wissenschaften zu Göttingen*, 253–297; English translation by Mary W. Newson in *Bulletin of the American Mathematical Society 8* (1902), 437–479, reprinted in *Browder 1976*, 1–34.

1902 French translation, with revisions, of *Hilbert 1899.*

1918 Axiomatisches Denken, *Mathematische Annalen 78*, 405–415; reprinted in *Hilbert 1935*, 146–156.

1922 Neubegründung der Mathematik (Erste Mitteilung), *Abhandlungen aus dem mathematischen Seminar der Hamburgischen Universität 1*, 157–177; reprinted in *Hilbert 1935*, 157–177.

1923 Die logischen Grundlagen der Mathematik, *Mathematische Annalen 88*, 151–165; reprinted in *Hilbert 1935*, 178–191.

1926 Über das Unendliche, *Mathematische Annalen 95*, 161–190; English translation by Stefan Bauer-Mengelberg in *van Heijenoort 1967*, 367–392.

1928 Die Grundlagen der Mathematik, *Abhandlungen aus dem mathematischen Seminar der Hamburgischen Universität 6*, 65–85; English translation by Stefan Bauer-Mengelberg and Dagfinn Føllesdal in *van Heijenoort 1967*, 464–479.

1929 Probleme der Grundlegung der Mathematik, *Atti del Congresso internazionale dei matematici, Bologna 3–10 settembre 1928* (Bologna: Zanichelli), I, 135–141; see also *1929a.*

1929a Reprint, with emendations and additions, of *Hilbert 1929*, in *Mathematische Annalen 102*, 1–9.

1930 Naturerkennen und Logik, *Naturwissenschaften 18*, 959–963.

1930a Seventh, revised edition of *Hilbert 1899.*

1930b Reprint of *Hilbert 1929a*, in *Hilbert 1930a*, 313–323.

1931 Die Grundlegung der elementaren Zahlenlehre, *Mathematische Annalen 104*, 485–494; reprinted in part in *Hilbert 1935*, 192–195.

1931a Beweis des tertium non datur, *Nachrichten von der Gesellschaft der Wissenschaften zu Göttingen, mathematisch-physikalische Klasse*, 120–125.

1935 *Gesammelte Abhandlungen* (Berlin: Springer), vol. 3.

Hilbert, David, and Wilhelm Ackermann

1928 *Grundzüge der theoretischen Logik* (Berlin: Springer).

1938 Second, revised edition of *Hilbert and Ackermann 1928*; English translation of *1938* by Lewis M. Hammond, George G. Leckie and F. Steinhardt (New York: Chelsea, 1950).

Hilbert, David, and Paul Bernays

1934 *Grundlagen der Mathematik*, vol. I (Berlin: Springer).

1939 *Grundlagen der Mathematik*, vol. II (Berlin: Springer).

1968 Second edition of *Hilbert and Bernays 1934*.

1970 Second edition of *Hilbert and Bernays 1939*.

Hinata, Shigeru

1967 Calculability of primitive recursive functionals of finite type, *Science reports of the Tokyo Kyoiku Daigaku, section A, 9*, 218–235.

Hodges, Andrew

1983 *Alan Turing: The enigma* (London: Burnett Books; New York: Simon and Schuster).

Hoegh-Krøhn, Raphael

See Albeverio et alii.

Hoensbroech, Franz G.

1931 Beziehungen zwischen Inhalt und Umfang von Begriffen, *Erkenntnis 2*, 291–300.

Holton, Gerald, and Yehuda Elkana

1982 (eds.) *Albert Einstein: Historical and cultural perspectives. The centennial symposium in Jerusalem* (Princeton: Princeton University Press).

Hosoi, Tsutomu, and Hiroakira Ono

1973 Intermediate propositional logics (a survey), *Journal of Tsuda College 5*, 67–82.

Howard, William A.

1968 Functional interpretation of bar induction by bar recursion, *Compositio mathematica 20*, 107–124.

1970 Assignment of ordinals to terms for primitive recursive functionals of finite type, in *Myhill et alii 1970*, 443–458.

1972 A system of abstract constructive ordinals, *The journal of symbolic logic 37*, 355–374.

1980 Ordinal analysis of terms of finite type, *ibid. 45*, 493–504.

1981 Ordinal analysis of bar recursion of type zero, *Compositio mathematica 42*, 105–119.

1981a Ordinal analysis of simple cases of bar recursion, *The journal of symbolic logic 46*, 17–30.

Hubble, Edwin

1934 The distribution of extra-galactic nebulae, *Astrophysical journal 79*, 8–76.

Huntington, Edward V.

1932 A new set of independent postulates for the algebra of logic with special reference to Whitehead and Russell's *Principia mathematica, Proceedings of the National Academy of Sciences, U.S.A. 18*, 179–180.

1934 Independent postulates related to C. I. Lewis's theory of strict implication, *Mind* (n.s.) *43*, 181–198.

Hurewicz, Witold

1932 Une remarque sur l'hypothèse du continu, *Fundamenta mathematicae 19*, 8–9.

Jeans, James

1936 Man and the universe, *Scientific progress* (Sir Halley Stewart lecture, 1935), edited by James Jeans et alii, 11–38.

Jech, Thomas

1967 Non-provability of Souslin's hypothesis, *Commentationes mathematicae Universitatis Carolinae 8*, 291–305.

1974 (ed.) *Axiomatic set theory*, Proceedings of symposia in pure mathematics, vol. 13, part 2 (Providence, R.I.: American Mathematical Society).

1978 *Set theory* (New York: Academic Press).

Jensen, Ronald B.

1972 The fine structure of the constructible hierarchy, *Annals of mathematical logic 4*, 229–308.

See also Devlin, Keith J., and Ronald B. Jensen.

See also Dodd, Anthony J., and Ronald B. Jensen.

Jeroslow, Robert G.

1973 Redundancies in the Hilbert–Bernays derivability conditions for Gödel's second incompleteness theorem, *The journal of symbolic logic 38*, 359–367.

1975 Experimental logics and Δ_2^0-theories, *Journal of philosophical logic 4*, 253–267.

Johansson, Ingebrigt

1936 Der Minimalkalkül, ein reduzierter intuitionistischer Formalismus, *Compositio mathematica 4*, 119–136.

Kaczmarz, Stefan

1932 Axioms for arithmetic, *The journal of the London Mathematical Society 7*, 179–182.

Kahr, Andrew S., Edward F. Moore and Hao Wang

1962 Entscheidungsproblem reduced to the $\forall\exists\forall$ case, *Proceedings of the National Academy of Sciences, U.S.A. 48*, 365–377.

Kalmár, László

1929 Eine Bemerkung zur Entscheidungstheorie, *Acta litterarum ac scientiarum Regiae Universitatis Hungaricae Francisco-Josephinae, sectio scientiarum mathematicarum 4*, 248–252.

1932 Ein Beitrag zum Entscheidungsproblem, *ibid., 5*, 222–236.

1933 Über die Erfüllbarkeit derjenigen Zählausdrücke, welche in der Normalform zwei benachbarte Allzeichen enthalten, *Mathematische Annalen 108*, 466–484.

1934 Über einen Löwenheimschen Satz, *Acta litterarum ac scientiarum Regiae Universitatis Hungaricae Francisco-Josephinae, sectio scientiarum mathematicarum 7*, 112–121.

1955 Über ein Problem, betreffend die Definition des Begriffes der allgemein-rekursiven Funktion, *Zeitschrift für mathematische Logik und Grundlagen der Mathematik 1*, 93–96.

1967 On the role of second-order theories, in *Lakatos 1967*, 104–105.

Kanamori, Akihiro, and Menachem Magidor

1978 The evolution of large cardinal axioms in set theory, in *Müller and Scott 1978*, 99–275.

Kanamori, Akihiro, William Reinhardt and Robert M. Solovay

1978 Strong axioms of infinity and elementary embeddings, *Annals of mathematical logic 13*, 73–116.

Kant, Immanuel

1787 *Critik der reinen Vernunft*, second revised edition (Riga: Hartknoch).

Keisler, H. Jerome, and Alfred Tarski

1964 From accessible to inaccessible cardinals: Results holding for all accessible cardinal numbers and the problem of their extension to inaccessible ones, *Fundamenta mathematicae 53*, 225–308.

Ketonen, Jussi, and Robert M. Solovay

1981 Rapidly growing Ramsey functions, *Annals of mathematics 113*, 267–314.

Klanfer, Laura

1933 Über *d*-zyklische Quadrupel, *Ergebnisse eines mathematischen Kolloquiums 4*, 10.

Kleene, Stephen C.

1934 Proof by cases in formal logic, *Annals of mathematics (2) 35*, 529–544.

1935 A theory of positive integers in formal logic, *American journal of mathematics 57*, 153–173, 219–244.

1936 General recursive functions of natural numbers, *Mathematische Annalen 112*, 727–742; reprinted in *Davis 1965*, 236–252; for an erratum, a simplification, and an addendum, see *Davis 1965*, 253.

1936a λ-definability and recursiveness, *Duke mathematical journal 2*, 340–353.

1943 Recursive predicates and quantifiers, *Transactions of the American Mathematical Society 53*, 41–73; reprinted in *Davis 1965*, 254–287; for a correction and an addendum, see *Davis 1965*, 254 and 287.

1950 A symmetric form of Gödel's theorem, *Indagationes mathematicae 12*, 244–246.

1952 *Introduction to metamathematics* (Amsterdam: North-Holland; New York: Van Nostrand).

1960 Realizability and Shanin's algorithm for the constructive deciphering of mathematical sentences, *Logique et analyse* (n.s.) *3*, 154–165.

1973 Realizability: A retrospective survey, in *Mathias and Rogers 1973*, 95–112.

1976 The work of Kurt Gödel, *The journal of symbolic logic 41*, 761–778; addendum, *ibid. 43* (1978), 613.

1981 Origins of recursive function theory, *Annals of the history of computing 3*, 52–67; corrections, *Davis 1982*, footnotes 10 and 12.

1987 Gödel's impression on students of logic in the 1930s, in *Weingartner and Schmetterer 1987*, 49–64.

1987a Kurt Gödel. April 28, 1906–January 14, 1978, *Biographical memoirs, National Academy of Sciences, U.S.A., 56*, 134–178.

Kleene, Stephen C., and J. Barkley Rosser

1935 The inconsistency of certain formal logics, *Annals of mathematics (2) 36*, 630–636.

Klein, Fritz

1931 Zur Theorie der abstrakten Verknüpfungen, *Mathematische Annalen 105*, 308–323.

1932 Über einen Zerlegungssatz in der Theorie der abstrakten Verknüpfungen, *ibid. 106*, 114–130.

Klibansky, Raymond

1968 (ed.) *Contemporary philosophy, a survey. I, Logic and foundations of mathematics* (Florence: La Nuova Italia Editrice).

Kochen, Simon

See Ax, James, and Simon Kochen.

Köhler, Eckehart

198? Gödel und der Wiener Kreis: Platonismus gegen Formalismus, to appear.

Koletsos, George

1985 Functional interpretation of the β-rule, *The journal of symbolic logic 50*, 791–805.

Kolmogorov, Andrei Nikolayevich (Kolmogoroff; Колмогоров, Андрей Николаевич)

1925 On the principle of the excluded middle (Russian), *Matematicheskii sbornik 32*, 646–667; English translation by Jean van Heijenoort in *van Heijenoort 1967*, 414–437.

1932 Zur Deutung der intuitionistischen Logik, *Mathematische Zeitschrift 35*, 58–65.

König, Dénes

1926 Sur les correspondances multivoques des ensembles, *Fundamenta mathematicae 8*, 114–134.

1927 Über eine Schlussweise aus dem Endlichen ins Unendliche: Punktmengen. Kartenfärben. Verwandtschaftsbeziehungen. Schachspiel, *Acta litterarum ac scientiarum Regiae Universitatis Hungaricae Francisco-Josephinae, sectio scientiarum mathematicarum 3*, 121–130.

König, Julius

1905 Zum Kontinuum-Problem, *Mathematische Annalen 60*, 177–180, 462.

Kreisel, Georg

1951 On the interpretation of non-finitist proofs—Part I, *The journal of symbolic logic 16*, 241–267.

1952 On the interpretation of non-finitist proofs—Part II. Interpretation of number theory. Applications, *ibid. 17*, 43–58.

1953 Note on arithmetic models for consistent formulae of the predicate calculus, II, *Actes du XIème Congrès international de philosophie, Bruxelles, 20–26 août 1953* (Amsterdam: North-Holland), vol. 14, 39–49.

1958 Hilbert's programme, *Dialectica 12*, 346–372; revised version in *Benacerraf and Putnam 1964*, 157–180.

1959 Interpretation of analysis by means of constructive functionals of finite types, in *Heyting 1959*, 101–128.

1959a Inessential extensions of Heyting's arithmetic by means of functionals of finite type (abstract), *The journal of symbolic logic 24*, 284.

1959b Inessential extensions of intuitionistic analysis by functionals of finite type (abstract), *ibid.*, 284–285.

1960 Ordinal logics and the characterization of informal concepts of proof, *Proceedings of the International Congress of Mathematicians, 14–21 August 1958* (Cambridge, U.K.: Cambridge University Press), 289–299.

1960a La prédicativité, *Bulletin de la Société mathématique de France 88*, 371–391.

1962 The axiom of choice and the class of hyperarithmetic functions, *Indagationes mathematicae 24*, 307–319.

1965 Mathematical logic, *Lectures on modern mathematics*, edited by Thomas L. Saaty (New York: Wiley), vol. 3, 95–195.

1967 Mathematical logic: What has it done for the philosophy of mathematics?, in *Schoenman 1967*, 201–272.

1967a Comments on *Mostowski 1967*, in *Lakatos 1967*, 97–103.

1968 A survey of proof theory, *The journal of symbolic logic 33*, 321–388.

1968a Functions, ordinals, species, in *van Rootselaar and Staal 1968*, 143–158.

1969 Axiomatizations of nonstandard analysis that are conservative extensions of formal systems for classical standard analysis, in *Luxemburg 1969*, 93–106.

1970 Church's thesis: A kind of reducibility axiom for constructive mathematics, in *Myhill et alii 1970*, 121–150.

1972 Which number-theoretic problems can be solved in recursive progressions on Π^1_1-paths through O?, *The journal of symbolic logic 37*, 311–334.

1976 What have we learnt from Hilbert's second problem?, in *Browder 1976*, 93–130.

1980 Kurt Gödel, 28 April 1906–14 January 1978, *Biographical memoirs of Fellows of the Royal Society 26*, 148–224; corrections, *ibid. 27*, 697, and *28*, 718.

1982 Finiteness theorems in arithmetic: An application of Herbrand's theorem for Σ_2-formulas, in *Stern 1982*, 39–55.

Kreisel, Georg, and Azriel Levy

1968 Reflection principles and their use for establishing the complexity of axiomatic systems, *Zeitschrift für mathematische Logik und Grundlagen der Mathematik 14*, 97–142.

Kreisel, Georg, and Angus Macintyre

1982 Constructive logic versus algebraization I, in *Troelstra and van Dalen 1982*, 217–258.

Kreisel, Georg, and Gaisi Takeuti

1974 Formally self-referential propositions in cut-free classical analysis and related systems, *Dissertationes Mathematicae 118*, 1–50.

Kreisel, Georg, and Anne S. Troelstra

1970 Formal systems for some branches of intuitionistic analysis, *Annals of mathematical logic 1*, 229–387.

Kripke, Saul

1965 Semantical analysis of intuitionistic logic I, in *Crossley and Dummett 1965*, 92–130.

Krivine, Jean-Louis

1968 *Théorie axiomatique des ensembles* (Paris: Presses Universitaires de France); translated into English as *Krivine 1971*.

1971 *Introduction to axiomatic set theory* (Dordrecht: Reidel), English translation of *Krivine 1968* by David Miller.

Kuczyński, Jerzy

1938 O twierdzeniu Gödel (On Gödel's theorem; Polish, with a French summary), *Kwartalnik filozoficzny 15*, 74–80.

Kunen, Kenneth
1970 Some applications of iterated ultrapowers in set theory, *Annals of mathematical logic 1*, 179–227.
1971 Elementary embeddings and infinitary combinatorics, *The journal of symbolic logic 36*, 407–413.
1980 *Set theory: An introduction to independence proofs* (Amsterdam: North-Holland).
See also Barwise, Jon, H. Jerome Keisler and Kenneth Kunen.

Kuratowski, Kazimierz
1921 Sur la notion de l'ordre dans la théorie des ensembles, *Fundamenta mathematicae 2*, 161–171.
1933 *Topologie I*, Monografie Matematyczne, vol. 3 (Warsaw: Garasiński).
1948 Ensembles projectifs et ensembles singuliers, *Fundamenta mathematicae 35*, 131–140.
1951 Sur une caractérisation des alephs, *ibid. 38*, 14–17.

Kuroda, Sigekatu
1951 Intuitionistische Untersuchungen der formalistischen Logik, *Nagoya mathematical journal 2*, 35–47.

Ladrière, Jean
1957 *Les limitations internes des formalismes. Etude sur la signification du théorème de Gödel et des théorèmes apparentés dans la théorie des fondements des mathématiques* (Louvain: Nauwelaerts; Paris: Gauthier-Villars).

Lakatos, Imre
1967 (ed.) *Problems in the philosophy of mathematics. Proceedings of the International Colloquium in the Philosophy of Science, London, 1965, Volume 1* (Amsterdam: North-Holland).

Langford, Cooper H.
1927 On inductive relations, *Bulletin of the American Mathematical Society 33*, 599–607.

Laugwitz, Detlef
1978 *Infinitesimalkalkül: Kontinuum und Zahlen. Eine elementare Einführung in die Nichtstandard-Analysis* (Mannheim, Vienna, Zurich: Bibliographisches Institut).

Lawvere, William
1971 Quantifiers and sheaves, *Actes du Congrès international des mathématiciens, 1–10 septembre 1970, Nice, France* (Paris: Gauthier-Villars), vol. I, 329–334.

Lebesgue, Henri

1905 Sur les fonctions représentables analytiquement, *Journal de mathématiques pures et appliquées 60*, 139–216.

Leibniz, Gottfried W.

1890 *Die philosophischen Schriften von Gottfried Wilhelm Leibniz*, edited by C. J. Gerhardt (Berlin: Weidmann), vol. 7.

1923 *Sämtliche Schriften und Briefe*, edited by Preussischen Akademie der Wissenschaften (Darmstadt: O. Reichl), series 1, vol. 1.

Leivant, Daniel

1985 Syntactic translations and provably recursive functions, *The journal of symbolic logic, 50*, 682–688.

Lemmon, Edward J.

1977 *An introduction to modal logic*, in collaboration with Dana Scott, edited by Krister Segerberg, American Philosophical Quarterly monograph series, no. 11 (Oxford: Blackwell).

Levy, Azriel

1957 Indépendance conditionnelle de $V = L$ et d'axiomes qui se rattachent au système de M. Gödel, *Comptes rendus hebdomadaires des séances de l'Académie des Sciences, Paris 245*, 1582–1583.

1960 Axiom schemata of strong infinity in axiomatic set theory, *Pacific journal of mathematics 10*, 223–238.

1960a Principles of reflection in axiomatic set theory, *Fundamenta mathematicae 49*, 1–10.

1960b A generalization of Gödel's notion of constructibility, *The journal of symbolic logic 25*, 147–155.

1964 Measurable cardinals and the continuum hypothesis, *Notices of the American Mathematical Society 11*, 769–770.

1965 Definability in axiomatic set theory I, in *Bar-Hillel 1965*, 127–151.

See also Kreisel, Georg, and Azriel Levy.

Levy, Azriel, and Robert M. Solovay

1967 Measurable cardinals and the continuum hypothesis, *Israel journal of mathematics 5*, 234–248.

Lewis, Clarence I.

1918 *A survey of symbolic logic* (Berkeley: University of California Press); reprinted by Dover (New York).

1932 Alternative systems of logic, *The monist 42*, 481–507.

Lindenbaum, Adolf, and Andrzej Mostowski

1938 Über die Unabhängigkeit des Auswahlaxioms und einiger seiner Folgerungen, *Comptes rendus des séances de la Société des Sciences et des Lettres de Varsovie, classe III, 31*, 27–32.

Lindstrøm, Tom

See Albeverio et alii.

Löb, Martin H.

1955 Solution of a problem of Leon Henkin, *The journal of symbolic logic 20*, 115–118.

Lorenzen, Paul

1951 Algebraische und logistische Untersuchungen über freie Verbände, *The journal of symbolic logic 16*, 81–106.

1951a Die Widerspruchsfreiheit der klassischen Analysis, *Mathematische Zeitschrift 54*, 1–24.

1951b Maß und Integral in der konstruktiven Analysis, *ibid.*, 275–290.

1955 *Einführung in die operative Logik und Mathematik* (Berlin: Springer).

1969 Second edition of *Lorenzen 1955.*

Łoś, Jerzy

1955 Quelques remarques, théorèmes, et problèmes sur les classes définissables d'algèbres, *Mathematical interpretations of formal systems*, edited by Thoralf Skolem and others (Amsterdam: North-Holland), 98–113.

Löwenheim, Leopold

1915 Über Möglichkeiten im Relativkalkül, *Mathematische Annalen 76*, 447–470; English translation by Stefan Bauer-Mengelberg in *van Heijenoort 1967*, 228–251.

Luckhardt, Horst

1973 *Extensional Gödel functional interpretation. A consistency proof of classical analysis*, Springer lecture notes in mathematics, no. 306 (Berlin: Springer).

Łukasiewicz, Jan, and Alfred Tarski

1930 Untersuchungen über den Aussagenkalkül, *Sprawozdania z posiedzeń Towarzystwa Naukowego Warszawskiego*, wydział III, *23*, 30–50; English translation by J. H. Woodger in *Tarski 1956*, 38–59.

Luxemburg, W. A. J.

1969 (ed.) *Applications of model theory to algebra, analysis, and probability* (New York: Holt).

Luzin, Nikolai (Lusin, Nicolas; Лузин, Николай Николаевич)

1914 Sur un problème de M. Baire, *Comptes rendus hebdomadaires des séances de l'Académie des Sciences, Paris 158*, 1258–1261.

1917 Sur la classification de M. Baire, *ibid. 164*, 91–94.

1929 Sur les voies de la théorie des ensembles, *Atti del Congresso internazionale dei matematici, Bologna 3–10 settembre 1928* (Bologna: Zanichelli), I, 295–299.

1930 *Leçons sur les ensembles analytiques et leurs applications* (Paris: Gauthier-Villars).

1935 Sur les ensembles analytiques nuls, *Fundamenta mathematicae 25*, 109–131.

Luzin, Nikolai, and Wacław Sierpiński

1918 Sur quelques propriétés des ensembles (A), *Bulletin international de l'Académie des sciences de Cracovie, classe des sciences mathématiques et naturelles, série A*, 35–48; reprinted in *Sierpiński 1975*, 192–204.

Maaß, Wolfgang

1976 Eine Funktionalinterpretation der prädikativen Analysis, *Archiv für mathematische Logik und Grundlagenforschung 18*, 27–46.

Mac Lane, Saunders

1961 Locally small categories and the foundations of set theory, *Infinitistic methods. Proceedings of the symposium on foundations of mathematics, Warsaw, 2–9 September 1959* (Warsaw: PWN; Oxford: Pergamon), 25–43.

MacDowell, Robert, and Ernst Specker

1961 Modelle der Arithmetik, *Infinitistic methods, Proceedings of the symposium on foundations of mathematics, Warsaw, 2–9 September 1959* (Warsaw: PWN; Oxford: Pergamon), 257–263.

Macintyre, Angus

See Kreisel, Georg, and Angus Macintyre.

Maehara, Shôji

1954 Eine Darstellung der intuitionistischen Logik in der klassischen, *Nagoya mathematical journal 7*, 45–64.

Magidor, Menachem

1977 On the singular cardinals problem II, *Annals of mathematics 106*, 517–547.

See also Foreman, Matthew, Menachem Magidor and Saharon Shelah.

See also Kanamori, Akihiro, and Menachem Magidor.

Mahlo, Paul

1911 Über lineare transfinite Mengen, *Berichte über die Verhandlungen der Königlich Sächsischen Gesellschaft der Wissenschaften zu Leipzig, Mathematisch-physische Klasse 63*, 187–225.

1912 Zur Theorie und Anwendung der ρ_0-Zahlen, *ibid. 64*, 108–112.

1913 Zur Theorie und Anwendung der ρ_0-Zahlen. II, *ibid. 65*, 268–282.

Malament, David B.

1985 "Time travel" in the Gödel universe, in *Asquith and Kitcher 1985*, 91–100.

Malmnäs, P.-E.

See Prawitz, Dag, and P.-E. Malmnäs.

Maltsev, Anatolii Ivanovich (Malcev; Мальцев, Анатолий Иванович)

1936 Untersuchungen aus dem Gebiete der mathematischen Logik, *Matematicheskii sbornik 1*, 323–336; English translation by Benjamin F. Wells III in *Maltsev 1971*, 1–14.

1941 On a general method for obtaining local theorems in group theory (Russian), *Ivanovskii Gosudarstvennii Pedagogicheskii Institut im. D. A. Furmanova. Ivanovskoye matematicheskoye obshchestvo. Ucheniye zapiski 1*, 3–9; English translation by Benjamin F. Wells III in *Maltsev 1971*, 15–21.

1971 *The metamathematics of algebraic systems: Collected papers, 1936–1967*, edited and translated by Benjamin F. Wells III (Amsterdam: North-Holland).

Martin, Donald A.

1976 Hilbert's first problem: The continuum hypothesis, in *Browder 1976*, 81–92.

Martin, Donald A., and John R. Steel

198? (untitled manuscript).

Martin-Löf, Per

1971 Hauptsatz for the theory of species, in *Fenstad 1971*, 217–233.

Mathias, Adrian R. D., and Hartley Rogers

1973 (eds.) *Cambridge summer school in mathematical logic*, Springer lecture notes in mathematics, no. 337 (Berlin: Springer).

Matiyasevich, Yuri (Matijacevič; Матиясевич, Юри)

1970 Enumerable sets are Diophantine (Russian), *Doklady Akademii Nauk S.S.S.R. 191*, 279–282; English translation, with revisions, in *Soviet mathematics doklady 11* (1970), 354–358.

1977 Primes are enumerated by a polynomial in 10 variables (Russian; English summary), in *Theoretical applications of the methods of mathematical logic, II, Zapiski nauchnyk seminarov Leningradskogo otdeleniya Matematicheskogo Instituta im. V. A. Steklova, Akademii nauk S.S.S.R.* (*Leningrad*) *68*, 62–82, 144–145.

1979 Algorithmic unsolvability of exponential Diophantine equations with three unknowns (Russian), in *Issledovaniya po teorii algorifmov i matematicheskoi logike* (*Studies in the theory of algorithms and mathematical logic*) edited by A. A. Markov and V. I. Khomich (Moscow: Nauka), 69–78.

See also Davis, Martin, Yuri Matiyasevich and Julia Robinson.

McAloon, Kenneth

1966 *Some applications of Cohen's method* (doctoral dissertation, University of California at Berkeley).

1971 Consistency results about ordinal definability, *Annals of mathematical logic 2*, 449–467.

See also Boffa, Maurice, Dirk van Dalen and Kenneth McAloon.

McKinsey, John C. C., and Alfred Tarski

1948 Some theorems about the sentential calculi of Lewis and Heyting, *The journal of symbolic logic 13*, 1–15.

McTaggart, J. Ellis

1908 The unreality of time, *Mind* (n.s.) *17*, 457–474.

Mehrtens, Herbert

1979 *Die Entstehung der Verbandstheorie* (Hildesheim: Gerstenberg).

Meltzer, Bernard, and Donald Michie

1970 (eds.) *Machine intelligence* (Edinburgh: Edinburgh University Press), vol. 5.

Menas, Telis K.
1973 *On strong compactness and supercompactness* (doctoral dissertation, University of California at Berkeley).

Menger, Karl
1928 Untersuchungen über allgemeine Metrik, *Mathematische Annalen 100*, 75–163.
1928a Bemerkungen zu Grundlagenfragen IV. Axiomatik der endlichen Mengen und der elementargeometrischen Verknüpfungsbeziehungen, *Jahresbericht der Deutschen Mathematiker-Vereinigung 37*, 309–325.
1930 Untersuchungen über allgemeine Metrik. Vierte Untersuchung. Zur Metrik der Kurven, *Mathematische Annalen 103*, 466–501.
1931 Metrische Untersuchungen. II: Die euklidische Metrik, *Ergebnisse eines mathematischen Kolloquiums 1*, 20–22.
1932 Probleme der allgemeinen metrischen Geometrie, *ibid. 2*, 20–22.
1932a Bericht über die mengentheoretischen Überdeckungssätze, *ibid.*, 23–27.
1936 (In collaboration with Franz Alt and Otto Schreiber) New foundations of projective and affine geometry. Algebra of geometry, *Annals of mathematics (2) 37*, 456–482.
1940 On algebra of geometry and recent progress in non-Euclidean geometry, *The Rice Institute pamphlet 27*, 41–79.
1952 The formative years of Abraham Wald and his work in geometry, *Annals of mathematical statistics 23*, 14–20.
See also *Gödel 1933h*.

Menger, Karl, and Leonard M. Blumenthal
1970 *Studies in Geometry* (San Francisco: Freeman).

Minari, Pierluigi
1983 Intermediate logics. A historical outline and a guided bibliography, *Rapporto matematico 79*, 1–71 (Dipartimento di Matematica, Università degli studi di Siena).

Mints, Gregory E. (Минц, Григорий Е.)
1974 On E-theorems (Russian), *Zapiski nauchnyk seminarov Leningradskogo otdeleniya Matematicheskogo Instituta im. V. A. Steklova, Akademii nauk S.S.S.R.* (*Leningrad*) *40*, 110–118, 158–159.
1975 Finite investigations of transfinite derivations (Russian), *ibid. 49*, 67–122; English translation in *Journal of soviet mathematics 10* (1978), 548–596.

1978 On Novikov's hypothesis (Russian), *Modal and intensional logics* (Moscow), 102–106; photocopied proceedings of a conference held by the Institute of Philosophy of the Soviet Academy of Sciences.

1979 Stability of E-theorems and program verification (Russian), *Semiotika i informatika 12*, 73–77.

Mirimanoff, Dmitry

1917 Les antinomies de Russell et de Burali-Forti et le problème fondamental de la théorie des ensembles, *L'enseignement mathématique 19*, 37–52.

1917a Remarques sur la théorie des ensembles et les antinomies cantoriennes. I, *ibid.*, 209–217.

1920 Remarques sur la théorie des ensembles et les antinomies cantoriennes. II, *ibid. 21*, 29–52.

Mitchell, William J.

1974 Sets constructible from sequences of ultrafilters, *The journal of symbolic logic 39*, 57–66.

1979 Hypermeasurable cardinals, in *Boffa et alii 1979*, 303–316.

198? The core model for sequences of measures, to appear.

Moeschlin, Otto

See Henn, Rudolph, and Otto Moeschlin.

Mongré, Paul (pseudonym of Felix Hausdorff)

1898 *Das Chaos in kosmischer Auslese* (Leipzig: Naumann).

Montgomery, Deane

1963 Oswald Veblen, *Bulletin of the American Mathematical Society 69*, 26–36.

Moore, Edward F.

See Kahr, Andrew S., Edward F. Moore and Hao Wang.

Moore, Gregory H.

1980 Beyond first-order logic: The historical interplay between mathematical logic and axiomatic set theory, *History and philosophy of logic 1*, 95–137.

1982 *Zermelo's axiom of choice: Its origins, development, and influence*, Studies in the history of mathematics and physical sciences, vol. 8 (New York: Springer).

Morel, Anne

See Frayne, Thomas, Anne Morel and Dana S. Scott.

Morgenstern, Oskar
See von Neumann, John, and Oskar Morgenstern.

Moschovakis, Yiannis N.
1980 *Descriptive set theory* (Amsterdam: North-Holland).

Mostowski, Andrzej
1939 Über die Unabhängigkeit des Wohlordnungssatzes vom Ordnungsprinzip, *Fundamenta mathematicae 32*, 201–252.
1947 On definable sets of positive integers, *ibid. 34*, 81–112.
1950 Some impredicative definitions in the axiomatic set-theory, *ibid. 37*, 111–124.
1951 Review of *Wang 1950*, *The journal of symbolic logic 16*, 142–143.
1952 *Sentences undecidable in formalized arithmetic: An exposition of the theory of Kurt Gödel* (Amsterdam: North-Holland).
1955 A formula with no recursively enumerable model, *Fundamenta mathematicae 42*, 125–140.
1959 On various degrees of constructivism, in *Heyting 1959*, 178–194.
1965 *Thirty years of foundational studies: Lectures on the development of mathematical logic and the study of the foundations of mathematics in 1930–1964* (= no. 17 of *Acta philosophica fennica*); reprinted in 1966 (New York: Barnes and Noble; Oxford: Blackwell).
1967 Recent results in set theory, in *Lakatos 1967*, 82–96.
1967a Reply, in *Lakatos 1967*, 105–108.
See also Ehrenfeucht, Andrzej, and Andrzej Mostowski.
See also Lindenbaum, Adolf, and Andrzej Mostowski.
See also Tarski, Alfred, Andrzej Mostowski and Raphael M. Robinson.

Müller, Gert H.
See Diller, Justus, and Gert H. Müller.

Müller, Gert H., Arnold Oberschelp and Klaus Potthoff
1975 (eds.) $\models$ *ISILC Logic conference. Proceedings of the International Summer Institute and Logic Colloquium, Kiel 1974*, Springer lecture notes in mathematics, no. 499 (Berlin: Springer).

Müller, Gert H., and Dana S. Scott
1978 (eds.) *Higher set theory. Proceedings, Oberwolfach, Germany, April 13–23, 1977*, Springer lecture notes in mathematics, no. 669 (Berlin: Springer).

Mulvey, Christopher J.
See Fourman, Michael P., Christopher J. Mulvey and Dana S. Scott.

Mycielski, Jan
1964 On the axiom of determinateness, *Fundamenta mathematicae 53*, 205–224.
See also Ehrenfeucht, Andrzej, and Jan Mycielski.

Myhill, John
1970 Formal systems of intuitionistic analysis II: The theory of species, in *Myhill et alii 1970*, 151–162.
1974 The undefinability of the set of natural numbers in the ramified *Principia*, in *Nakhnikian 1974*, 19–27.
1974a "Embedding classical type theory in 'intuitionistic' type theory": A correction, in *Jech 1974*, 185–188.

Myhill, John, Akiko Kino and Richard E. Vesley
1970 (eds.) *Intuitionism and proof theory* (Amsterdam: North-Holland).

Myhill, John, and Dana S. Scott
1971 Ordinal definability, in *Scott 1971*, 271–278.

Nahm, Werner
See Diller, Justus, and Werner Nahm.

Nakhnikian, George
1974 (ed.) *Bertrand Russell's philosophy* (London: Duckworth).

Neder, Ludwig
1931 Über den Aufbau der Arithmetik, *Jahresbericht der Deutschen Mathematiker-Vereinigung 40*, 22–37.

Noether, Emmy, and Jean Cavaillès
1937 *Briefwechsel Cantor–Dedekind* (Paris: Hermann).

Notcutt, Bernard
1934 A set of axioms for the theory of deduction, *Mind* (n.s.) *43*, 63–77.

Novikov, Petr S. (Новиков, Петр С.)
1951 On the consistency of some theorems of descriptive set theory (Russian), *Trudy Matematicheskogo Instituta imeni V. A. Steklova 38*, 279–316.

Ono, Hiroakira
See Hosoi, Tsutomu, and Hiroakira Ono.

Parikh, Rohit
1971 Existence and feasibility in arithmetic, *The journal of symbolic logic 36*, 494–508.
1973 Some results on the length of proofs, *Transactions of the American Mathematical Society 177*, 29–36.

Paris, Jeff, and Leo Harrington
1977 A mathematical incompleteness in Peano arithmetic, in *Barwise 1977*, 1133–1142.

Parry, William T.
1933 Ein Axiomensystem für eine neue Art von Implikation (analytische Implikation), *Ergebnisse eines mathematischen Kolloquiums 4*, 5–6.
1933a Zum Lewisschen Aussagenkalkül, *ibid.*, 15–16.

Parsons, Charles D.
1970 On a number-theoretic choice schema and its relation to induction, in *Myhill et alii 1970*, 459–473.

Pasch, Moritz
1882 *Vorlesungen über neuere Geometrie* (Leipzig: Teubner).
1926 Second edition of *Pasch 1882*, with an appendix by Max Dehn (Berlin: Springer).

Peano, Giuseppe
1889 *Arithmetices principia, nova methodo exposita* (Turin: Bocca); partial English translation by Jean van Heijenoort in *van Heijenoort 1967*, 83–97.
1891 Sul concetto di numero, *Rivista di matematica 1*, 87–102, 256–267.

Pears, David F.
1972 (ed.) *Bertrand Russell: A collection of critical essays* (Garden City, N.Y.: Anchor).

Peirce, Charles S.
1897 The logic of relatives, *The monist 7*, 161–217; reprinted in *1933*, 288–345.
1933 *Collected papers of Charles Sanders Peirce*, volume III: *Exact Logic*, edited by Charles Hartshorne and Paul Weiss (Cambridge, Mass.: Harvard University Press).

1976 *The new elements of mathematics*, edited by Carolyn Eisele (The Hague: Mouton), vols. I–V.

Perelman, Charles
1936 L'antinomie de M. Gödel, *Académie royale de Belgique, Bulletin de la classe des sciences (5) 22*, 730–736.

Platek, Richard A.
1969 Eliminating the continuum hypothesis, *The journal of symbolic logic 34*, 219–225.

Pohlers, Wolfram
See Buchholz, Wilfried, et alii.

Post, Emil L.
1921 Introduction to a general theory of elementary propositions, *American journal of mathematics 43*, 163–185; reprinted in *van Heijenoort 1967*, 264–283.
1936 Finite combinatory processes—formulation 1, *The journal of symbolic logic 1*, 103–105; reprinted in *Davis 1965*, 288–291.
1941 Absolutely unsolvable problems and relatively undecidable propositions: Account of an anticipation, in *Davis 1965*, 338–433.
1944 Recursively enumerable sets of positive integers and their decision problems, *Bulletin of the American Mathematical Society 50*, 284–316.
1953 A necessary condition for definability for transfinite von Neumann–Gödel set theory sets, with an application to the problem of the existence of a definable well-ordering of the continuum (preliminary report), *ibid. 59*, 246.

Powell, William C.
1975 Extending Gödel's negative interpretation to *ZF*, *The journal of symbolic logic 40*, 221–229.

Prawitz, Dag
1971 Ideas and results in proof theory, in *Fenstad 1971*, 235–307.

Prawitz, Dag, and P.-E. Malmnäs
1968 A survey of some connections between classical, intuitionistic and minimal logic, in *Schmidt et alii 1968*, 215–229.

Presburger, Mojżesz

1930 Über die Vollständigkeit eines gewissen Systems der Arithmetik ganzer Zahlen, in welchem die Addition als einzige Operation hervortritt, *Sprawozdanie z I Kongresu matematyków krajów słowiańskich, Warszawa 1929* (Warsaw, 1930), 92–101, 395.

Princeton University

1947 *Problems of mathematics*, Princeton University bicentennial conferences, series 2, conference 2.

Putnam, Hilary

1957 Arithmetic models for consistent formulae of quantification theory, *The journal of symbolic logic 22*, 110–111; abstract of a paper presented at the 27 December 1956 meeting of the Association for Symbolic Logic.

1961 *Trial and error predicates and the solution to a problem of Mostowski's* (New York: Courant Institute).

1965 Trial and error predicates and the solution to a problem of Mostowski, *The journal of symbolic logic 30*, 49–57.

See also Benacerraf, Paul, and Hilary Putnam.

Quine, Willard V.

1933 A theorem in the calculus of classes, *The journal of the London Mathematical Society 8*, 89–95.

1937 New foundations for mathematical logic, *American mathematical monthly 44*, 70–80.

1941 Whitehead and the rise of modern logic, in *Schilpp 1941*, 125–163; reprinted in *Quine 1966a*, 3–36.

1943 Notes on existence and necessity, *Journal of philosophy 40*, 113–127.

1947 The problem of interpreting modal logic, *The journal of symbolic logic 12*, 43–48.

1953 *From a logical point of view. 9 logico-philosophical essays* (Cambridge, Mass.: Harvard University Press).

1953a Three grades of modal involvement, *Actes du XIème Congrès international de philosophie, Bruxelles, 20–26 août 1953* (Amsterdam: North-Holland), vol. XIV, 65–81; reprinted in *Quine 1976*, 158–176.

1955 On Frege's way out, *Mind* (n.s.) *64*, 145–159.

1960 Carnap and logical truth, *Synthèse 12*, 350–374.

1963 Reprint of *Quine 1960* in *Schilpp 1963*, 385–406.

1966 *The ways of paradox and other essays* (New York: Random House).

1966a *Selected logic papers* (New York: Random House).
1976 Second, enlarged edition of *Quine 1966* (Cambridge, Mass.: Harvard University Press).
1979 Kurt Gödel (1906–1978), *Year book 1978 of the American Philosophical Society*, 81–84.

Ramsey, Frank P.
1926 The foundations of mathematics, *Proceedings of the London Mathematical Society (2) 25*, 338–384; reprinted in *Ramsey 1931*, 1–61.
1929 On a problem of formal logic, *Proceedings of the London Mathematical Society (2) 30*, 264–286; reprinted in *Ramsey 1931*, 82–111.
1931 *The foundations of mathematics and other logical essays*, edited by Richard B. Braithwaite (London: Kegan Paul).

Rasiowa, Helena, and Roman Sikorski
1950 A proof of the completeness theorem of Gödel, *Fundamenta mathematicae 37*, 193–200.
1951 A proof of the Skolem–Löwenheim theorem, *ibid. 38*, 230–232.
1953 Algebraic treatment of the notion of satisfiability, *ibid. 40*, 62–95.
1963 *The mathematics of metamathematics* (Warsaw: PWN).

Rath, Paul
1978 *Eine verallgemeinerte Funktionalinterpretation der Heyting Arithmetik endlicher Typen* (doctoral dissertation, Münster).

Rautenberg, Wolfgang
1979 *Klassische und nichtklassische Aussagenlogik* (Braunschweig: Vieweg).

Raychaudhuri, Amal K.
1979 *Theoretical cosmology* (Oxford: Clarendon Press).

Reid, Constance
1970 *Hilbert* (New York: Springer).

Reidemeister, Kurt
See Hahn et alii.

Reinhardt, William N.
1974 Remarks on reflection principles, large cardinals, and elementary embeddings, in *Jech 1974*, 189–205.
See also Kanamori, Akihiro, William Reinhardt and Robert M. Solovay.

Robertson, Howard P.
1933 Relativistic cosmology, *Reviews of modern physics 5*, 62–90.

Robinson, Abraham
1965 Formalism 64, in *Bar-Hillel 1965*, 228–246.
1966 *Non-standard analysis* (Amsterdam: North-Holland).
1974 Second edition of *Robinson 1966.*
1975 Concerning progress in the philosophy of mathematics, in *Rose and Shepherdson 1975*, 41–52.
See also Bar-Hillel et alii.

Robinson, Julia
See Davis, Martin, Yuri Matiyasevich and Julia Robinson.

Robinson, Raphael M.
1937 The theory of classes. A modification of von Neumann's system, *The journal of symbolic logic 2*, 29–36.
See also Tarski, Alfred, Andrzej Mostowski and Raphael M. Robinson.

Rose, Harvey E., and John C. Shepherdson
1975 (eds.) *Logic colloquium '73* (Amsterdam: North-Holland).

Rosser, J. Barkley
1935 A mathematical logic without variables. I, *Annals of mathematics (2) 36*, 127–150.
1935a A mathematical logic without variables. II, *Duke mathematics journal 1*, 328–355.
1936 Extensions of some theorems of Gödel and Church, *The journal of symbolic logic 1*, 87–91; reprinted in *Davis 1965*, 230–235.
1937 Gödel theorems for non-constructive logics, *The journal of symbolic logic 2*, 129–137.
1939 An informal exposition of proofs of Gödel's theorems and Church's theorem, *ibid. 4*, 53–60; reprinted in *Davis 1965*, 223–230.
See also Church, Alonzo, and J. Barkley Rosser.
See also Kleene, Stephen C., and J. Barkley Rosser.

Rowbottom, Frederick
1971 Some strong axioms of infinity incompatible with the axiom of constructibility, *Annals of mathematical logic 3*, 1–44.

Russell, Bertrand
1903 *The principles of mathematics* (London: Allen and Unwin).
1906 On some difficulties in the theory of transfinite numbers and order types, *Proceedings of the London Mathematical Society (2) 4*, 29–53; reprinted in *Russell 1973*, 135-164.

1906a Les paradoxes de la logique, *Revue de métaphysique et de morale 14*, 627–650.

1908 Mathematical logic as based on the theory of types, *American journal of mathematics 30*, 222–262; reprinted in *van Heijenoort 1967*, 150–182.

1919 *Introduction to mathematical philosophy* (London: Allen and Unwin; New York: Macmillan).

1920 Second edition of *Russell 1919*.

1924 Reprint of *Russell 1920*.

1940 *An inquiry into meaning and truth* (London: Allen and Unwin).

1968 *The autobiography of Bertrand Russell, 1914–1944* (London: Allen and Unwin; Boston: Little, Brown and Co.)

1973 *Essays in analysis*, edited by Douglas Lackey (New York: Braziller).

See also Whitehead, Alfred North, and Bertrand Russell.

Sanchis, Luis E.

1967 Functionals defined by recursion, *Notre Dame journal of formal logic 8*, 161–174.

Scanlon, Thomas M.

1973 The consistency of number theory via Herbrand's theorem, *The journal of symbolic logic 38*, 29–58.

Schilpp, Paul A.

1941 (ed.) *The philosophy of Alfred North Whitehead*, Library of living philosophers, vol. 3 (Evanston: Northwestern University): second edition (New York: Tudor).

1944 (ed.) *The philosophy of Bertrand Russell*, Library of living philosophers, vol. 5 (Evanston: Northwestern University); third edition (New York: Tudor, 1951).

1949 (ed.) *Albert Einstein, philosopher-scientist*, Library of living philosophers, vol. 7 (Evanston: Library of living philosophers).

1955 (ed.) *Albert Einstein als Philosoph und Naturforscher*, German translation (with additions) of *Schilpp 1949* (Stuttgart: Kohlhammer).

1963 (ed.) *The philosophy of Rudolf Carnap*, Library of living philosophers, vol. 11 (La Salle, Illinois: Open Court; London: Cambridge University Press).

Schlesinger, Karl

1935 Über die Produktionsgleichungen der ökonomischen Wertlehre, *Ergebnisse eines mathematischen Kolloquiums 6*, 10–11.

Schmidt, H. Arnold, Kurt Schütte and Helmut J. Thiele
1968 (eds.) *Contributions to mathematical logic* (Amsterdam: North-Holland).

Schoenman, Ralph
1967 (ed.) *Bertrand Russell: Philosopher of the century* (London: Allen and Unwin).

Scholz, Arnold
See Hahn et alii.

Scholz, Heinrich
See Hasse, Helmut, and Heinrich Scholz.

Schönfinkel, Moses
See Bernays, Paul, and Moses Schönfinkel.

Schreiber, Otto
See *Menger 1936*.

Schütte, Kurt
1934 Untersuchungen zum Entscheidungsproblem der mathematischen Logik, *Mathematische Annalen 109*, 572–603.
1934a Über die Erfüllbarkeit einer Klasse von logischen Formeln, *ibid. 110*, 161–194.
1954 Kennzeichnung von Ordnungszahlen durch rekursiv erklärte Funktionen, *ibid. 127*, 15–32.
1965 Predicative well-orderings, in *Crossley and Dummett 1965*, 280–303.
1965a Eine Grenze für die Beweisbarkeit der transfiniten Induktion in der verzweigten Typenlogik, *Archiv für mathematische Logik und Grundlagenforschung 7*, 45–60.
1977 *Proof theory* (Berlin: Springer).
See also Diller, Justus, and Kurt Schütte.
See also Schmidt, H. Arnold, Kurt Schütte and Helmut J. Thiele.

Schützenberger, Marcel P. (Marco; Maurice)
1945 Sur certains axiomes de la théorie des structures, *Comptes rendus hebdomadaires des séances de l'Académie des Sciences, Paris 221*, 218–220.

Schwichtenberg, Helmut
1973 *Einige Anwendungen von unendlichen Termen und Wertfunktionalen* (Habilitationsschrift, Münster).
1975 Elimination of higher type levels in definitions of primitive recursive functionals by means of transfinite recursion, in *Rose and Shepherdson 1975*, 279–303.

1977 Proof theory: Some applications of cut-elimination, in *Barwise 1977*, 867–895.

1979 On bar recursion of types 0 and 1, *The journal of symbolic logic 44*, 325–329.

Scott, Dana S.

1961 Measurable cardinals and constructible sets, *Bulletin de l'Académie polonaise des sciences, série des sciences mathématiques, astronomiques, et physiques 9*, 521–524.

1971 (ed.) *Axiomatic set theory*, Proceedings of symposia in pure mathematics, vol. 13, part 1 (Providence, R.I.: American Mathematical Society).

See also Fourman, Michael P., Christopher J. Mulvey and Dana S. Scott.

See also Fourman, Michael P., and Dana S. Scott.

See also Frayne, Thomas, Anne Morel and Dana S. Scott.

See also Hanf, William P., and Dana S. Scott.

See also Henkin et alii.

See also *Lemmon 1977*.

See also Müller, Gert H., and Dana S. Scott.

See also Myhill, John, and Dana S. Scott.

Scott, Philip J.

1978 The "Dialectica" interpretation and categories, *Zeitschrift für mathematische Logik und Grundlagen der Mathematik 24*, 553–575.

Sheffer, Henry M.

1926 Review of *Whitehead and Russell 1925*, *Isis 8*, 226–231.

Shelah, Saharon

1974 Infinite abelian groups, Whitehead problem and some constructions, *Israel journal of mathematics 18*, 243–256.

1982 *Proper forcing*, Springer lecture notes in mathematics, no. 940 (Berlin: Springer).

See also Foreman, Matthew, Menachem Magidor and Saharon Shelah.

See also Gurevich, Yuri, and Saharon Shelah.

Shepherdson, John C.

1951 Inner models for set theory, part I, *The journal of symbolic logic 16*, 161–190.

1952 Inner models for set theory, part II, *ibid. 17*, 225–237.

1953 Inner models for set theory, part III, *ibid. 18*, 145–167.

See also Rose, Harvey E., and John C. Shepherdson.

Shoenfield, Joseph R.
1959 On the independence of the axiom of constructibility, *American journal of mathematics 81*, 537–540.
1961 The problem of predicativity, in *Bar-Hillel et alii 1961*, 132–139.
1967 *Mathematical logic* (Reading, Mass.: Addison-Wesley).

Sieg, Wilfried
See Buchholz, Wilfried, et alii.

Sierpiński, Wacław
1919 Sur un théorème équivalent à l'hypothèse du continu ($2^{\aleph_0} = \aleph_1$), *Biuletyn Polskiej Akademii Umiejętności, Kraków* (= *Bulletin international de l'Académie des sciences et des lettres, Cracovie*), 1–3; reprinted in *Sierpiński 1975*, 272–274.
1924 Sur l'hypothèse du continu ($2^{\aleph_0} = \aleph_1$), *Fundamenta mathematicae 5*, 177–187; reprinted in *Sierpiński 1975*, 527–536.
1934 *Hypothèse du continu*, Monografie Matematyczne, vol. 4 (Warsaw: Garasiński).
1934a Sur une extension de la notion de l'homéomorphie, *Fundamenta mathematicae 22*, 270–275; reprinted in *Sierpiński 1976*, 201–206.
1935 Sur une hypothèse de M. Lusin, *Fundamenta mathematicae 25*, 132–135; reprinted in *Sierpiński 1976*, 269–272.
1935a Sur deux ensembles linéaires singuliers, *Annali della Scuola Normale Superiore di Pisa (2) 4*, 43–46.
1951 Sur quelques propositions concernant la puissance du continu, *Fundamenta mathematicae 38*, 1–13; reprinted in *Sierpiński 1976*, 654–664.
1956 Second, expanded edition of *Sierpiński 1934* (New York: Chelsea).
1975 *Oeuvres choisies*, Tome II: *Théorie des ensembles et ses applications. Travaux des années 1908–1929* (Warsaw: PWN).
1976 *Oeuvres choisies*, Tome III: *Théorie des ensembles et ses applications. Travaux des années 1930–1966* (Warsaw: PWN).
See also Braun, Stefania, and Wacław Sierpiński.
See also Luzin, Nikolai, and Wacław Sierpiński.

Sierpiński, Wacław, and Alfred Tarski
1930 Sur une propriété caractéristique des nombres inaccessibles, *Fundamenta mathematicae 15*, 292–300; reprinted in *Sierpiński 1976*, 29–35.

Sikorski, Roman

1951 A characterization of alephs, *Fundamenta mathematicae 38*, 18–22.

See also Rasiowa, Helena, and Roman Sikorski.

Silver, Jack H.

1971 Some applications of model theory in set theory, *Annals of mathematical logic 3*, 45–110.

1971a The consistency of the GCH with the existence of a measurable cardinal, in *Scott 1971*, 391–395.

1971b Measurable cardinals and Δ^1_3 well-orderings, *Annals of mathematics (2) 94*, 414–446.

1975 On the singular cardinals problem, *Proceedings of the International Congress of Mathematicians, Vancouver 1974*, vol. I, 265–268.

Skolem, Thoralf

1920 Logisch-kombinatorische Untersuchungen über die Erfüllbarkeit oder Beweisbarkeit mathematischer Sätze nebst einem Theoreme über dichte Mengen, *Skrifter utgit av Videnskapsselskapet i Kristiania*, I. *Matematisk-naturvidenskabelig klasse*, no. 4, 1–36; reprinted in *Skolem 1970*, 103–136; partial English translation by Stefan Bauer-Mengelberg in *van Heijenoort 1967*, 252–263.

1923 Begründung der elementaren Arithmetik durch die rekurrierende Denkweise ohne Anwendung scheinbarer Veränderlichen mit unendlichem Ausdehnungsbereich, *Skrifter utgit av Videnskapsselskapet i Kristiania*, I. *Matematisk-naturvidenskabelig klasse*, no. 6, 1–38; reprinted in *Skolem 1970*, 153–188; English translation by Stefan Bauer-Mengelberg in *van Heijenoort 1967*, 302–333.

1923a Einige Bemerkungen zur axiomatischen Begründung der Mengenlehre, *Matematikerkongressen i Helsingfors 4–7 Juli 1922, Den femte skandinaviska matematikerkongressen, Redogörelse* (Helsinki: Akademiska Bokhandeln), 217–232; reprinted in *Skolem 1970*, 137–152; English translation by Stefan Bauer-Mengelberg in *van Heijenoort 1967*, 290–301.

1928 Über die mathematische Logik, *Norsk matematisk tidsskrift 10*, 125–142; reprinted in *Skolem 1970*, 189–206; English translation by Stefan Bauer-Mengelberg and Dagfinn Føllesdal in *van Heijenoort 1967*, 508–524.

1929 Über einige Grundlagenfragen der Mathematik, *Skrifter utgitt av Det Norske Videnskaps-Akademi i Oslo, I. Matematisk-naturvidenskapelig klasse*, no. 4, 1–49; reprinted in *Skolem 1970*, 227–273.

1930 Einige Bemerkungen zu der Abhandlung von E. Zermelo: "Über die Definitheit in der Axiomatik", *Fundamenta mathematicae 15*, 337–341; reprinted in *Skolem 1970*, 275–279.

1931 Über einige Satzfunktionen in der Arithmetik, *Skrifter utgitt av Det Norske Videnskaps-Akademi i Oslo, I. Matematisk-naturvidenskapelig klasse*, no. 7, 1–28: reprinted in *Skolem 1970*, 281–306.

1932 Über die symmetrisch allgemeinen Lösungen im identischen Kalkul, *Skrifter utgitt av Det Norske Videnskaps-Akademi i Oslo, I. Matematisk-naturvidenskapelig klasse*, no. 6, 1–32; also appeared in *Fundamenta mathematicae 18*, 61–76; reprinted in *Skolem 1970*, 307–336.

1933 Ein kombinatorischer Satz mit Anwendung auf ein logisches Entscheidungsproblem, *Fundamenta mathematicae 20*, 254–261; reprinted in *Skolem 1970*, 337–344.

1933a Über die Unmöglichkeit einer vollständigen Charakterisierung der Zahlenreihe mittels eines endlichen Axiomensystems, *Norsk matematisk forenings skrifter*, series 2, no. 10, 73–82; reprinted in *Skolem 1970*, 345-354.

1934 Über die Nicht-charakterisierbarkeit der Zahlenreihe mittels endlich oder abzählbar unendlich vieler Aussagen mit ausschließlich Zahlenvariablen, *Fundamenta mathematicae 23*, 150–161; reprinted in *Skolem 1970*, 355–366.

1938 Review of *Hilbert and Ackermann 1938*, *Norsk matematisk tidsskrift 20*, 67–69.

1970 *Selected works in logic*, edited by Jens E. Fenstad (Oslo: Universitetsforlaget).

Slisenko, Anatol O. (Слисенко, Анатоль О.)

1970 (ed.) *Studies in constructive mathematics and mathematical logic, part II*, Seminars in mathematics, V. A. Steklov Mathematical Institute, vol. 8 (New York: Consultants Bureau).

Smoryński, Craig A.

1977 The incompleteness theorems, in *Barwise 1977*, 821–865.

Smullyan, Arthur F.

1948 Modality and description, *The journal of symbolic logic 13*, 31–37.

Smullyan, Raymond M.

1958 Undecidability and recursive inseparability, *Zeitschrift für mathematische Logik und Grundlagen der Mathematik 4*, 143–147.

Solomon, Martin K.

1981 A connection between Blum speedable sets and Gödel's speed-up theorem (unpublished typescript).

Solovay, Robert M.

1963 Independence results in the theory of cardinals. I, II, *Notices of the American Mathematical Society 10*, 595.

1965 $2^{\aleph_0}$ can be anything it ought to be, in *Addison et alii 1965*, 435.

1965a Measurable cardinals and the continuum hypothesis, *Notices of the American Mathematical Society 12*, 132.

1967 A nonconstructible Δ^1_3 set of integers, *Transactions of the American Mathematical Society 127*, 50–75.

1969 On the cardinality of Σ^1_2 sets of reals, in *Bulloff et alii 1969*, 58–73.

1970 A model of set theory in which every set of reals is Lebesgue measurable, *Annals of Mathematics (2) 92*, 1–56.

1974 Strongly compact cardinals and the GCH, in *Henkin et alii 1974*, 365–372.

1976 Provability interpretations of modal logic, *Israel journal of mathematics 25*, 287–304.

See also Kanamori, Akihiro, William Reinhardt and Robert M. Solovay.
See also Ketonen, Jussi, and Robert M. Solovay.
See also Levy, Azriel, and Robert M. Solovay.

Solovay, Robert M., and Stanley Tennenbaum

1971 Iterated Cohen extensions and Souslin's problem, *Annals of mathematics (2) 94*, 201–245.

Specker, Ernst

See MacDowell, Robert, and Ernst Specker.

Spector, Clifford

1957 Recursive ordinals and predicative set theory, in *Summaries of talks presented at the Summer Institute for Symbolic Logic. Cornell University* (Institute for Defense Analysis), *1957*, 377–382.

1962 Provably recursive functionals of analysis: A consistency proof of analysis by an extension of principles formulated in current intuitionistic mathematics, in *Dekker 1962*, 1–27.

See also Feferman, Solomon, and Clifford Spector.

Statman, Richard

1978 Bounds for proof-search and speed-up in the predicate calculus, *Annals of mathematical logic 15*, 225–287.

1981 Speed-up by theories with infinite models, *Proceedings of the American Mathematical Society 81*, 465–469.

Steel, John R.
See Martin, Donald A., and John R. Steel.

Stein, Martin
1976 *Interpretationen der Heyting-Arithmetik endlicher Typen* (doctoral dissertation, Münster).
1978 Interpretationen der Heyting-Arithmetik endlicher Typen, *Archiv für mathematische Logik und Grundlagenforschung 19*, 175–189.
1980 Interpretations of Heyting's arithmetic—an analysis by means of a language with set symbols, *Annals of mathematical logic 19*, 1–31.
1981 A general theorem on existence theorems, *Zeitschrift für mathematische Logik und Grundlagen der Mathematik 27*, 435–452.

Stern, Jacques
1982 (ed.) *Proceedings of the Herbrand Symposium. Logic colloquium '81* (Amsterdam: North-Holland).

Straus, Ernst G.
1982 Reminiscences, in *Holton and Elkana 1982*, 417–423.

Surányi, János
1950 Contributions to the reduction theory of the decision problem. Second paper. Three universal, one existential quantifiers, *Acta Mathematica Academiae Scientiarum Hungaricae 1*, 261–271.

Suslin, Mikhail (Souslin; Суслин, Михаил)
1920 Problème 3, *Fundamenta mathematicae 1*, 223.

Tait, William W.
1965 Infinitely long terms of transfinite type, in *Crossley and Dummett 1965*, 176–185.
1965a Functionals defined by transfinite recursion, *The journal of symbolic logic 30*, 155–174.
1967 Intensional interpretations of functionals of finite type. I, *ibid. 32*, 198–212.
1971 Normal form theorem for bar recursive functions of finite type, in *Fenstad 1971*, 353–367.

Takeuti, Gaisi
1955 On the fundamental conjecture of GLC I, *Journal of the Mathematical Society of Japan 7*, 249–275.

1957 Ordinal diagrams, *Journal of the Mathematical Society of Japan 9*, 386–394.
1960 Ordinal diagrams II, *ibid. 12*, 385–391.
1961 Remarks on Cantor's absolute, *Journal of the Mathematical Society of Japan 13*, 197–206.
1967 Consistency proofs of subsystems of classical analysis, *Annals of mathematics (2) 86*, 299–348.
1975 *Proof theory* (Amsterdam: North-Holland).
1978 Gödel numbers of product spaces, in *Müller and Scott 1978*, 461–471.

See also Kreisel, Georg, and Gaisi Takeuti.

Tarski, Alfred

1924 Sur les principes de l'arithmétique des nombres ordinaux (transfinis), *Polskie Towarzystwo Matematyczne (Cracow), Rocznik (=Annales de la Société Polonaise de Mathématique) 3*, 148–149.
1925 Quelques théorèmes sur les alephs, *Fundamenta mathematicae 7*, 1–14.
1930 Über einige fundamentale Begriffe der Metamathematik, *Sprawozdania z posiedzeń Towarzystwa Naukowego Warszawskiego, wydział III, 23*, 22–29; English translation by Joseph H. Woodger, with revisions, in *Tarski 1956*, 30–37.
1932 Der Wahrheitsbegriff in den Sprachen der deduktiven Disziplinen, *Anzeiger der Akademie der Wissenschaften in Wien 69*, 23–25.
1933 Einige Betrachtungen über die Begriffe der ω-Widerspruchsfreiheit und der ω-Vollständigkeit, *Monatshefte für Mathematik und Physik 40*, 97–112; English translation by Joseph H. Woodger in *Tarski 1956*, 279–295.
1933a Pojecie prawdy w jezykach nauk dedukcyjnych (The concept of truth in the languages of deductive sciences), *Prace Towarzystwa Naukowego Warszawskiego, wydział III*, no. 34; English translation by Joseph H. Woodger in *Tarski 1956*, 152–278.
1935 Der Wahrheitsbegriff in den formalisierten Sprachen, *Studia philosophica* (Lemberg), *1*, 261–405; German translation by L. Blaustein of *Tarski 1933a*.
1935a Grundzüge des Systemenkalküls, Erster Teil, *Fundamenta mathematicae 25*, 503–526.
1936 Grundzüge des Systemenkalküls, Zweiter Teil, *ibid. 26*, 283–301.
1938 Über unerreichbare Kardinalzahlen, *ibid. 30*, 68–89.

1944 The semantic conception of truth and the foundations of semantics, *Philosophy and phenomenological research 4*, 341–376.
1949 On essential undecidability, *The journal of symbolic logic 14*, 75–76.
1952 Some notions and methods on the borderline of algebra and metamathematics, *Proceedings of the International Congress of Mathematicians, Cambridge, Massachusetts, August 30–September 6, 1950* (Providence, R.I.: American Mathematical Society), vol. 1, 705–720.
1956 *Logic, semantics, metamathematics: Papers from 1923 to 1938*, translated into English and edited by Joseph H. Woodger (Oxford: Clarendon Press).
1962 Some problems and results relevant to the foundations of set theory, *Logic, methodology, and philosophy of science. Proceedings of the 1960 International Congress*, edited by Ernest Nagel, Patrick Suppes, and Alfred Tarski (Stanford: Stanford University Press), 125–135.
1983 Second edition of *Tarski 1956*, edited by John Corcoran.
See also Addison, John W., Leon Henkin and Alfred Tarski.
See also Feferman, Solomon, and Alfred Tarski.
See also Keisler, H. Jerome, and Alfred Tarski.
See also Łukasiewicz, Jan, and Alfred Tarski.
See also McKinsey, John C. C., and Alfred Tarski.
See also Sierpiński, Wacław, and Alfred Tarski.

Tarski, Alfred, Andrzej Mostowski and Raphael M. Robinson
1953 *Undecidable theories* (Amsterdam: North-Holland).

Taussky-Todd, Olga
1987 Remembrances of Kurt Gödel, in *Weingartner and Schmetterer 1987*, 29–41.

Tennenbaum, Stanley
1968 Souslin's problem, *Proceedings of the National Academy of Sciences, U.S.A. 59*, 60–63.
See also Solovay, Robert M., and Stanley Tennenbaum.

Thomas, Ivo
1962 Finite limitations on Dummett's LC, *Notre Dame journal of formal logic 3*, 170–174.

Troelstra, Anne S.
1973 *Metamathematical investigation of intuitionistic arithmetic and analysis*, Springer lecture notes in mathematics, no. 344 (Berlin: Springer).

1977 *Choice sequences* (Oxford: Clarendon Press).
See also Kreisel, Georg, and Anne S. Troelstra.

Troelstra, Anne S., and Dirk van Dalen
1982 (eds.) *The L. E. J. Brouwer centenary symposium* (Amsterdam: North-Holland).

Turing, Alan M.
1937 On computable numbers, with an application to the Entscheidungsproblem, *Proceedings of the London Mathematical Society* (*2*) *42*, 230–265; correction, *ibid. 43*, 544–546; reprinted as *Turing 1965*.
1939 Systems of logic based on ordinals, *Proceedings of the London Mathematical Society* (*2*) *45*, 161–228; reprinted in *Davis 1965*, 155–222.
1965 Reprint of *Turing 1937*, in *Davis 1965*, 116–154.
1970 Intelligent machinery, in *Meltzer and Michie 1970*, 1–24.

Ulam, Stanislaw
1958 John von Neumann, 1903–1957, *Bulletin of the American Mathematical Society 64*, no. 3, part 2 (May supplement), 1–49.
1976 *Adventures of a mathematician* (New York: Scribner's).

Vacca, Giovanni
1903 La logica di Leibniz, *Rivista di matematica 8*, 64–74.

van Dalen, Dirk, Daniel Lascar and Timothy J. Smiley
1982 (eds.) *Logic Colloquium '80* (Amsterdam: North-Holland).

van Heijenoort, Jean
1967 (ed.) *From Frege to Gödel: A source book in mathematical logic, 1879–1931* (Cambridge, Mass.: Harvard University Press).
1967a Logic as calculus and logic as language, *Boston studies in the philosophy of science 3*, 440–446; reprinted in *1985*, 11–16.
1982 L'oeuvre logique de Jacques Herbrand et son contexte historique, in *Stern 1982*, 57–85; English translation in *1985*, 99–121.
1985 *Selected essays* (Naples: Bibliopolis).

van Rootselaar, Bob, and J. F. Staal
1968 (eds.) *Logic, methodology and philosophy of science III. Proceedings of the Third International Congress for Logic, Methodology and Philosophy of Science, Amsterdam 1967* (Amsterdam: North-Holland).

Vaught, Robert L.
1974 Model theory before 1945, in *Henkin et alii 1974*, 153–172.
See also Henkin et alii.

Vesley, Richard E.
1972 Choice sequences and Markov's principle, *Compositio mathematica 24*, 33–53.
See also Myhill, John, Akiko Kino and Richard E. Vesley.

Vogel, Helmut
1977 Ein starker Normalisationssatz für die bar-rekursiven Funktionale, *Archiv für mathematische Logik und Grundlagenforschung 18*, 81–84.
See also Diller, Justus, and Helmut Vogel.

von Juhos, Béla
1930 *Das Problem der mathematischen Wahrscheinlichkeit* (Munich: Reinhardt).

von Neumann, John
1925 Eine Axiomatisierung der Mengenlehre, *Journal für die reine und angewandte Mathematik 154*, 219–240; correction, *ibid. 155*, 128; reprinted in *von Neumann 1961*, 34–56; English translation by Stefan Bauer-Mengelberg and Dagfinn Føllesdal in *van Heijenoort 1967*, 393–413.
1927 Zur Hilbertschen Beweistheorie, *Mathematische Zeitschrift 26*, 1–46; reprinted in *von Neumann 1961*, 256–300.
1928 Über die Definition durch transfinite Induktion und verwandte Fragen der allgemeinen Mengenlehre, *Mathematische Annalen 99*, 373–391; reprinted in *von Neumann 1961*, 320–338.
1928a Die Axiomatisierung der Mengenlehre, *Mathematische Zeitschrift 27*, 669–752; reprinted in *von Neumann 1961*, 339–422.
1929 Über eine Widerspruchfreiheitsfrage in der axiomatischen Mengenlehre, *Journal für die reine und angewandte Mathematik 160*, 227–241; reprinted in *von Neumann 1961*, 494–508.
1931 Die formalistische Grundlegung der Mathematik, *Erkenntnis 2*, 116–121; English translation by Erna Putnam and Gerald J. Massey in *Benacerraf and Putnam 1964*, 50–54.
1961 *Collected works*, volume I: *Logic, theory of sets, and quantum mechanics*, edited by A. H. Taub (Oxford: Pergamon).
1966 *Theory of self-reproducing automata*, edited by Arthur W. Burks (Urbana: University of Illinois Press).
See also Hahn et alii.

von Neumann, John, and Oskar Morgenstern
1944 *Theory of games and economic behavior* (Princeton: Princeton University Press).
1947 Second edition of *von Neumann and Morgenstern 1944*.
1953 Third edition of *von Neumann and Morgenstern 1944*.

Waismann, Friedrich
1967 *Wittgenstein und der Wiener Kreis*, edited by Brian McGuinness (Oxford: Blackwell).

Wajsberg, Mordechaj
1933 Ein erweiterter Klassenkalkül, *Monatshefte für Mathematik und Physik 40*, 113–126.

Wald, Abraham
1931 Axiomatik des Zwischenbegriffes in metrischen Räumen, *Mathematische Annalen 104*, 476–484.
1932 Axiomatik des metrischen Zwischenbegriffes, *Ergebnisse eines mathematischen Kolloquiums 2*, 17–18.
1935 Über die eindeutige positive Lösbarkeit der neuen Produktionsgleichungen, *ibid. 6*, 12–18.
1936 Über die Produktionsgleichungen der ökonomischen Wertlehre (II. Mitteilung), *ibid. 7*, 1–6.
See also *Gödel 1933h*.

Wang, Hao
1950 Remarks on the comparison of axiom systems, *Proceedings of the National Academy of Sciences, U.S.A. 36*, 448–453.
1951 Arithmetic models for formal systems, *Methodos 3*, 217–232.
1954 The formalization of mathematics, *The journal of symbolic logic 19*, 241–266: reprinted in *Wang 1962*, 559–584.
1959 Ordinal numbers and predicative set theory, *Zeitschrift für mathematische Logik und Grundlagen der Mathematik 5*, 216–239; reprinted in *Wang 1962*, 624–651.
1962 *A survey of mathematical logic* (Peking: Science Press; also Amsterdam: North-Holland, 1963); reprinted as *Logic, computers and sets* (New York: Chelsea, 1970).
1970 A survey of Skolem's work in logic, in *Skolem 1970*, 17–52.
1974 *From mathematics to philosophy* (New York: Humanities Press).
1978 Kurt Gödel's intellectual development, *The mathematical intelligencer 1*, 182–184.

1981 Some facts about Kurt Gödel, *The journal of symbolic logic 46*, 653–659.

See also Kahr, Andrew S., Edward F. Moore and Hao Wang.

Weingartner, Paul, and Leopold Schmetterer

1987 (eds.) *Gödel remembered. Salzburg 10–12 July 1983* (Naples: Bibliopolis).

Weintraub, E. Roy

1983 On the existence of a competitive equilibrium: 1930–1954, *The journal of economic literature 21*, 1–39.

Wernick, Georg

1929 Die Unabhängigkeit des zweiten distributiven Gesetzes von den übrigen Axiomen der Logistik, *Journal für die reine und angewandte Mathematik 161*, 123–134.

Weyl, Hermann

1918 *Das Kontinuum. Kritische Untersuchungen über die Grundlagen der Analysis* (Leipzig: Veit).

1932 Second edition of *Weyl 1918*.

1946 Review of *Schilpp 1944*, *American mathematical monthly 53*, 208–214; reprinted in *Weyl 1968*, 599–605.

1968 *Gesammelte Abhandlungen*, edited by K. Chandrasekharan (Berlin: Springer), vol. 4.

Whitehead, Alfred North, and Bertrand Russell

1910 *Principia mathematica* (Cambridge, U.K.: Cambridge University Press), vol. 1.

1912 *Principia mathematica*, vol. 2.

1913 *Principia mathematica*, vol. 3.

1925 Second edition of *Whitehead and Russell 1910*.

Wittgenstein, Ludwig

1921 Logisch-philosophische Abhandlung, *Annalen der Naturphilosophie 14*, 185–262.

1922 *Tractatus logico-philosophicus*, English translation of *Wittgenstein 1921* (London: Routledge and Kegan Paul).

Woodin, W. Hugh

See Foreman, Matthew, and W. Hugh Woodin.

Yasugi, Mariko

1963 Intuitionistic analysis and Gödel's interpretation, *Journal of the Mathematical Society of Japan 15*, 101–112.

Young, William Henry

1903 Zur Lehre der nicht abgeschlossenen Punktmengen, *Berichte über die Verhandlungen der Königlich Sächsischen Gesellschaft der Wissenschaften zu Leipzig, mathematisch-physische Klasse 55*, 287–293.

Zemanek, Heinz

1978 Oskar Morgenstern (1902–1977)—Kurt Gödel (1906–1978), *Elektronische Rechenanlagen 20*, 209–211.

Zermelo, Ernst

1904 Beweis, daß jede Menge wohlgeordnet werden kann. (Aus einem an Herrn Hilbert gerichteten Briefe), *Mathematische Annalen 59*, 514–516; English translation by Stefan Bauer-Mengelberg in *van Heijenoort 1967*, 139–141.

1908 Untersuchungen über die Grundlagen der Mengenlehre. I, *Mathematische Annalen 65*, 261–281; English translation by Stefan Bauer-Mengelberg in *van Heijenoort 1967*, 199–215.

1908a Neuer Beweis für die Möglichkeit einer Wohlordnung, *Mathematische Annalen 65*, 107–128; English translation by Stefan Bauer-Mengelberg in *van Heijenoort 1967*, 183–198.

1929 Über den Begriff der Definitheit in der Axiomatik, *Fundamenta mathematicae 14*, 339–344.

1930 Über Grenzzahlen und Mengenbereiche: Neue Untersuchungen über die Grundlagen der Mengenlehre, *ibid. 16*, 29–47.

Addenda and corrigenda to Volume I of these *Collected works*

The following list includes updatings of references that have appeared since the publication of Volume I of these *Collected works.* With regard to the References, note also that "Schur" (page 471) is listed without a first name because the editors have been unable to determine to which of the two mathematicians with that name Bernays was referring.

	Text as printed	*Addition/Correction*
vii, 20	article *1972*	article *1958*
4, 5	*198?*	*1987*
7, 39	*198?*	*1987*
12, 32	passsage	passage
15, 30	incapacited	incapacitated
30, 10	the present volume	Volumes I and II of these *Collected works*
30, 44	even positive integer	even integer $n > 2$
35, note a	*1985*	*1987a*
35, note a	*Taussky-Todd 198?*	*Taussky-Todd 1987*
38, 36	Spring (?)	7 April
45, 24–25	propositional formula	formula of the propositional calculus
119, 34	*neccessary*	*necessary*
135, 14	extention	extension
140, 21	", and	, and
171, 15	folllowing	following
183, 13	χ_1	χ_i
210, 43	Hilbertischen	Hilbertschen
211, 29	125).	125.)
213, 21	tranlation	translation
227, 7	adaption	adaptation
257, 10	conjuction	conjunction
282, 17	we have	$(\neg F)' := \neg F'$, we have
284, 32	*1969* (page 313,	*1936* (footnote 17; see also *1969*, page 313,
285, 7	*198?*	*1985*
307, 18	322	323
344, 45	β_1)	β_1
354, 12	$<$, *then*	, *then*
359, 32	ocurrences	occurrences
369, 10	k_i	k_1

386, 6	Lewis'	Lewis's
387, 8	Lewis'	Lewis's
395, 38	less	fewer
407, 29	1985 (eds.)	1986
407, 30	New York	Orlando
414, 19	*problems,*	*problems*
414, 36	, to appear	253–271
415, 11	Grundlagen ... historischen	Grundlegung ... historischer
415, 25	*ISLIC*	*ISILC*
417, 10	the	The
418, 17	*Synthèse*	*Synthese*
419, 37–38	um ... Stetigkeit	Delete from title
419, 39	, to appear	*31*, 3–29
422, 40	, *1947*	in *1947*
430, 6	Lewis'	Lewis's
432, 6	1985	1987a
432, 6	*Proceedings of the*	*Biographical memoirs,*
432, 7	to appear	*56*, 134–178
432, 8	198?	1987
432, 8–9	*Gödels ... Weltbild*	*Gödel remembered*
432, 11	to appear	edited by Paul Weingartner and Leopold Schmetterer (Naples: Bibliopolis), 49–64
432, 26	Kolestos	Koletsos
432, 27	198?	1985
432, 28	, to appear	*50*, 791–805
434, 3	*Staal and van Rootselaar 1968*	*van Rootselaar and Staal 1968*
434, 14	MacIntyre	Macintyre
434, 24	*Introduction à la théorie*	*Théorie*
435, 26	198?	1985
435, 27	, to appear	*50*, 682–688
437, 27–28	MacIntyre	Macintyre
442, 27	17	16
442, 32	*neuer*	*neuere*
444, 11	welchen	welchem
453, 10	Staal, J.F., and B. van Rootselaar	van Rootselaar, B. and J.F. Staal
453, 10–14		Move reference to page 456
455, 24	Ernst	Ernest
455, 36	198?	1987

455, 36–37	*Gödels . . . Weltbild*	*Gödel remembered*
455, 38	to appear	edited by Paul Weingartner and Leopold Schmetterer (Naples: Bibliopolis), 29–41
467, 30	*Kolestos 198?*	*Koletsos 1985*
468, 15	*MacIntyre*	*Macintyre*
468, 23	*Lauckhardt*	*Luckhardt*
468, 39	MacIntyre	Macintyre
468, 43	*Leivant 198?*	*Leivant 1985*
469, 35	Ernst	Ernest
470, 10	Parry . . . 267	Parry . . . 266, 267
474, 16,21	*Rusell*	*Russell*

Index